T0320349

Mathematical Methods in Dynamical Systems

The art of applying mathematics to real-world dynamical problems such as structural dynamics, fluid dynamics, wave dynamics, robot dynamics, etc. can be extremely challenging. Various aspects of mathematical modelling that may include deterministic or uncertain (fuzzy, interval, or stochastic) scenarios, along with integer or fractional order, are vital to understanding these dynamical systems. *Mathematical Methods in Dynamical Systems* offers problem-solving techniques and includes different analytical, semi-analytical, numerical, and machine intelligence methods for finding exact and/or approximate solutions of governing equations arising in dynamical systems. It provides a singular source of computationally efficient methods to investigate these systems and includes coverage of various industrial applications in a simple yet comprehensive way.

Mathematical Methods in Dynamical Systems

Edited by
S. Chakraverty and Subrat Kumar Jena

CRC Press
Taylor & Francis Group
Boca Raton London New York

CRC Press is an imprint of the
Taylor & Francis Group, an **informa** business

Cover image: Shutterstock

First edition published 2023
by CRC Press
6000 Broken Sound Parkway NW, Suite 300, Boca Raton, FL 33487-2742

and by CRC Press
4 Park Square, Milton Park, Abingdon, Oxon, OX14 4RN

CRC Press is an imprint of Taylor & Francis Group, LLC

© 2023 Taylor & Francis Group, LLC

Library of Congress Cataloging-in-Publication Data
Names: Chakraverty, Snehashish, editor. | Jena, Subrat Kumar, editor.
Title: Mathematical methods in dynamical systems / edited by S. Chakraverty and Subrat Kumar Jena.
Description: First edition. | Boca Raton : CRC Press, 2023. | Includes bibliographical references and
index. | Identifiers: LCCN 2022045309 (print) | LCCN 2022045310 (ebook) |
ISBN 9781032356860 (hbk) | ISBN 9781032356877 (pbk) | ISBN 9781003328032 (ebk)
Subjects: LCSH: System analysis--Mathematics. | Differentiable dynamical systems--Mathematical
models. | Dynamics--Mathematics.
Classification: LCC T57.62 .M3795 2023 (print) | LCC T57.62 (ebook) |
DDC 620.001/13--dc23/eng/20221129
LC record available at https://lccn.loc.gov/2022045309
LC ebook record available at https://lccn.loc.gov/2022045310

ISBN: 978-1-032-35686-0 (hbk)
ISBN: 978-1-032-35687-7 (pbk)
ISBN: 978-1-003-32803-2 (ebk)

DOI: 10.1201/9781003328032

Typeset in Times
by KnowledgeWorks Global Ltd.

Contents

Preface

The art of applying mathematics to real-world dynamical problems such as structural dynamics, fluid dynamics, wave dynamics, robot dynamics, data science, actuarial science, etc. is a challenging area. Various aspects of mathematical modelling that may include deterministic or uncertain (viz., fuzzy or interval or stochastic) scenarios along with integer or fractional order are important to understand the dynamical systems. In this book, problem-solving techniques has been discussed with the help of different analytical, semi-analytical, numerical, and machine intelligence methods for finding exact and/or approximate solutions of governing equations arising in dynamical systems. In this context, this book may be an honest attempt to address dynamical systems considering various aspects of mathematical modelling. The aim is, therefore, to address the titled subject including recent investigation by different researchers in a simple and comprehensive way.

This book is organized into 14 chapters that cover various types of dynamical models. Chapters 1–4 emphasize on dynamical problems that arise in the field of structural dynamics. The dynamic nature of wave propagation, fluid interaction, and viscoelastic behavior are discussed in Chapters 5–8. Chapter 9 addresses the dynamical systems based on fractional modelling, whereas Chapter 10 analyzes dynamical problems involving robotics. Chapters 11–12 illustrate uncertainty modelling in structural dynamics, while Chapter 13 demonstrates a dynamical problem relating to finance. Finally, Chapter 14 covers the modelling of dynamical systems using machine intelligence.

In Chapter 1, an equivalent single layer theory has been presented for the dynamic analysis of doubly curved shells characterized by an arbitrary geometry and laminated with generally anisotropic layers having arbitrary orientations and variable thicknesses. The physical domain has been described by employing curvilinear principal coordinates, whereas a distortion of the geometry has been performed by means of generalized blending functions based on a non-uniform rational B-splines description of the shell edges. Chapter 2 deals with the dynamic analysis of functionally graded thin nanoplates with variable thickness and subjected to the thermal environment using Eringen's nonlocal elasticity theory. By assuming the temperature-dependent mechanical properties of the plate material, rule of mixture is used to compute the effective properties of functionally graded material. Three different types of thickness variations, i.e., linear, parabolic, and quadratic are considered for the analysis.

A porous functionally graded magneto-electro-elastic energy harvesting system has been discussed in Chapter 3 via an analytical approach. The material properties are assumed to follow modified power-law model. The effects of main parameters such as the type of substrate material, the type of gradation, and the gradient index are studied, with varying values of the resistances. The working and design principles of a mass resonator sensor and its related inverse problems are reviewed comprehensively in Chapter 4. In this chapter, development of the mass resonator sensor is reviewed and the unsolved challenging problems/issues are also discussed.

Chapter 5 is dedicated to axial-wave propagation of fluid-filled carbon nanorods with elastic support. Eringen's nonlocal continuum theory has been used in this chapter to capture nanoscale effects. The dispersion curve, phase speed, and group speed study reveal the influence of elastic stiffness and nonlocal parameters.

In Chapter 6, the transport phenomenon of Marangoni boundary layer flow for the carbon nanotubes has been revealed. The enhancement in the heat transport properties is due to the inclusion of Hamilton Crosser conductivity model. Furthermore, the carbon nanotube nanoparticle also enhances the thermophysical properties and boost up the study to a great extent. The governing equations are solved by employing the approximate analytical technique such as differential transform method and Adomian decomposition method, while the validation is presented using traditional Runge-Kutta fourth-order based on shooting technique. The min-max game theory dilemma of a sequential fluid-structure model is discussed in Chapter 7. Chapter 8 is aimed to investigate the behavior of numerical solutions for integro PDEs, which include fractional time derivatives which are important in viscoelastic dynamical systems. The Caputo derivative is discretized using the L1 scheme, whereas the spatial derivative is approximated using a second-order finite difference scheme. Further, the composite trapezoidal rule is used in lieu of the Volterra operator.

In Chapter 9, a new neural network framework for solving distributed order fractional dynamical systems has been proposed by introducing the developed Lagrange functions. Additionally, the efficiency and accuracy of the presented method are considered by proposing seven numerical examples of both ordinary and partial fractional differential equations. Since the examined model in this research is a distributed order fractional derivative model and it is the general form of many other differential equations, it can be used for solving other differential equations too. It is well known that irrespective of the joint signature, namely, serial or articulated, single-link manipulator has a great application in the field of robotics. As such, for the study of dynamics of single-link manipulators, the conjugate problem is studied in Chapter 10 based on Timoshenko beam theory. This chapter investigates the said problem in particular for a single-link semi-compliant flexible robotic manipulator. Chapter 11 presents to handle epistemic uncertainties taken in the form of intervals where the uncertain width possesses percentage error of the crisp value. Further, finite element method is used to study various structural problems involving beam and truss elements.

Two strategies for solving fuzzy generalized and fuzzy standard eigenvalue problems using the double parametric form along with the idea of expectation are investigated in Chapter 12. Triangular and trapezoidal fuzzy numbers are employed to handle fuzzy uncertainty in this case. To support the suggested method, two numerical examples and an application problem of spring-mass systems have been provided. The purpose of Chapter 13 is to predict the dynamical exchange price of Bitcoin with respect to Indian currency using various machine learning techniques. The results show that the support vector regression model performs best to predict the dynamicity of exchange rate of Bitcoin with respect to Indian national rupees (BTC-INR), with an accuracy of 88.15%, in comparison to other machine learning techniques, such as linear regression and polynomial regression models. Finally, Chapter 14 is dedicated to develop machine intelligence-based method, viz. neural networks for finding the solution of dynamical problems, in particular the third-order Emden–Fowler model. Here, a nature-inspired machine learning technique named "curriculum learning" has been used during the training of the neural network.

This book is primarily aimed at undergraduates, graduates, postgraduates, researchers as well as industry personals of the related fields throughout the globe. It explores real-world dynamical problems arising in structural dynamics, fluid dynamics, wave dynamics, robot dynamics, data science, and actuarial science by examining various aspects of mathematical modelling, including deterministic or uncertain scenarios, as well as integer or fractional-order models. Additionally, several analytical, semi-analytical, numerical, and machine intelligence strategies for obtaining precise and approximate solutions to governing equations in dynamical systems have been presented.

The editors would like to express their gratitude to all the contributors who delivered their chapters on time. Further, the editors would like to express their appreciation to the whole team of Taylor & Francis Group/CRC Press for their effort, assistance, and cooperation in order to ensure the success of this challenging project and also for timely publication of this significant book.

Editors
S. Chakraverty and S. K. Jena

Acknowledgement

The first editor, **Prof. S. Chakraverty**, extends his gratitude to his beloved parents, the late Sh. Birendra K Chakraborty and the late Parul Chakraborty, for their blessings. He also thanks his wife, Shewli, and daughters, Shreyati and Susprihaa, for their support and inspiration during the project. The support of NIT Rourkela administration is also acknowledged with appreciation.

Dr. Subrat Kumar Jena, the second editor, is immensely grateful to his family members, specifically Sh. Ullash Chandra Jena, Sh. Durga Prasad Jena, Sh. Laxmidhara Jena, Smt. Urbasi Jena, Smt. Renu Bala Jena, Smt. Arati Jena, and his sisters Jyotrimayee, Truptimayee, and Nirupama, for their unwavering love, constant motivation, and unrelenting support. Dr. Subrat Kumar Jena expresses his indebtedness to his younger brother, Dr. Rajarama Mohan Jena, and sister-in-law, Dr. Sujata Swain, for their encouragement, love, and support. He extends his gratitude to Miss Bhubaneswari Mishra from the Department of Mathematics, National Institute of Technology Rourkela, for her unconditional love, moral support, faith, and encouragement, which reinvigorated his research vigor and motivated him to achieve beyond his expectations. He also expresses his deep appreciation to his Ph.D. supervisor, Prof. S. Charaverty, for his invaluable guidance and support during the early stages of his research. Finally, the second editor acknowledges the administration of the Indian Institute of Technology Delhi and his Postdoc mentor, Prof. S. Pradyumn, for their assistance and support.

The two editors extend their sincere gratitude and acknowledgment to the reviewers for their invaluable feedback and appreciation provided during the development of the book proposal. They express their utmost appreciation to all the chapter contributors for their timely submissions and excellent efforts. They are also immensely grateful to the entire team at CRC Press for their unwavering support, cooperation, and assistance, which has enabled them to publish this book according to schedule. Lastly, they are deeply indebted to the authors and researchers cited in the bibliography/reference sections at the end of each chapter.

About the Authors

Prof. S. Chakraverty has 30 years of experience as a researcher and teacher. Presently, he is working in the Department of Mathematics (Applied Mathematics Group), National Institute of Technology, Rourkela, Odisha, as a senior (higher administrative grade) professor. Prior to this, he was with CSIR-Central Building Research Institute, Roorkee, India. After completing graduation from St. Columba's College (Ranchi University), his career started from the University of Roorkee (now Indian Institute of Technology, Roorkee) and did M.Sc. (Mathematics) and M.Phil. (Computer Applications) from the said institute securing the first positions in the university. Dr. Chakraverty received his Ph.D. from IIT-Roorkee in 1993. Thereafter, he did his post-doctoral research at the Institute of Sound and Vibration Research (ISVR), University of Southampton, U.K., and at the Faculty of Engineering and Computer Science, Concordia University, Canada. He was also a visiting professor at Concordia and McGill universities, Canada, during 1997–1999 and visiting professor at the University of Johannesburg, Johannesburg, South Africa, during 2011–2014. He has authored/co-authored/edited 31 books, published 430 research papers (till date) in journals and conferences, and two books are ongoing. He is in the editorial boards of various international journals, book series, and conferences. Prof. Chakraverty is the chief editor of *International Journal of Fuzzy Computation and Modelling* (IJFCM), Inderscience Publisher, Switzerland (http://www.inderscience. com/ijfcm), associate editor of *Computational Methods in Structural Engineering, Frontiers in Built Environment* and *Curved and Layered Structures* (De Gruyter) and happens to be the editorial board member of *Springer Nature Applied Sciences, IGI Research Insights Books, Springer Book Series of Modeling and Optimization in Science and Technologies, Coupled Systems Mechanics* (Techno Press), *Journal of Composites Science* (MDPI), *Engineering Research Express* (IOP), and *Applications and Applied Mathematics: An International Journal.* He is also the reviewer of around 50 national and international journals of repute and he was the president of the section of mathematical sciences (including Statistics) of Indian Science Congress (2015–2016) and was the vice president – Orissa Mathematical Society (2011–2013). Prof. Chakraverty is a recipient of prestigious awards, viz. Indian National Science Academy (INSA) nomination under International Collaboration/Bilateral Exchange Program (with the Czech Republic), Platinum Jubilee ISCA Lecture Award (2014), CSIR Young Scientist Award (1997), BOYSCAST Fellow. (DST), UCOST Young Scientist Award (2007, 2008), Golden Jubilee Director's (CBRI) Award (2001), INSA International Bilateral Exchange Award ([2010–2011] selected but could not undertake [2015] selected), Roorkee University Gold Medals (1987, 1988) for first positions in M.Sc. and M.Phil. (Computer Application).

He is in the list of 2% world scientists recently (2020 and 2021) in Artificial Intelligence and Image Processing category based on an independent study done by Stanford University scientists.

Professor Chakraverty received IOP Publishing Top-Cited Paper Award for one of the most-cited articles from India, published across the entire IOP Publishing journal portfolio in the past three years (2018–2020). It also features in the top 1% of most-cited papers in the materials subject category. This data is from the citations recorded in Web of Science.

He has already guided 22 Ph.D. students and 12 are ongoing. Prof. Chakraverty has undertaken around 16 research projects as a principle investigator funded by international and national agencies totaling about Rs. 1.5 crores. He has hoisted around eight international students with different international/national fellowships to work in his group as PDF, Ph.D., visiting researchers for different periods. A good number of international and national conferences, workshops, and training programs have also been organized by him. His present research area includes differential equations (ordinary, partial, and fractional), numerical analysis and computational methods, structural dynamics (FGM, nano) and fluid dynamics, mathematical and uncertainty modelling, soft computing and machine intelligence (artificial neural network, fuzzy, interval, and affine computations).

Dr. Subrat Kumar Jena is currently working as a Postdoctoral Fellow in the department of Applied Mechanics at the Indian Institute of Technology Delhi. He has recently completed his Honorary Post-doctoral Fellowship in the field of "Computational Solid Mechanics via Numerical Solution," at the Nonlinear Multifunctional Composites – Analysis & Design (NMCAD) Lab in the Department of Aerospace Engineering, Indian Institute of Science (IISc), Bengaluru, India. He completed his Ph.D. in the Department of Mathematics, National Institute of Technology Rourkela, Rourkela, India. Subrat does research in Structural Dynamics, Computational Solid Mechanics, Applied Mathematics, Mathematical Modelling, Uncertainty Modelling, etc. He attained M.Sc. in Mathematics and Scientific Computing from Motilal Nehru National Institute of Technology Allahabad (NIT Allahabad), Prayagraj, India. Till now, he has published 32 research papers in peer-reviewed international journal, 2 international conference papers, 9 book chapters, and 2 books (2 books are in press, 2 books are ongoing).

He has been awarded an IOP Publishing Top-Cited Paper Award 2021 and 2022 from India, published across the entire IOP Publishing journal portfolio in the past three years i.e., 2018–2020 and 2019–2021, respectively, and the paper was among the top 1% of most-cited papers in the materials subject category. One of his paper published in ZAMM – Journal of Applied Mathematics and Mechanic (Wiley) – is among the top cited papers for the year 2020–2021. Three of his papers have been recognized as the best-cited papers published in Curved and Layered Structures Journal (De Gruyter). He was also featured in Stanford University's Top 2% Most Influential Scientists List in 2021 for the year 2020.

Additionally, he has been listed in Shell Buckling website as "Shell Buckling People" for his substantial contributions to the fields of shell buckling mainly in the area of static or dynamic analysis of general structures. He also serves as a reviewer and guest editor for many prestigious international journals and has reviewed over 60 manuscripts so far. Further, he is continuing collaborative research works with renowned researchers from India, Italy, Estonia, Turkey, Iran, Poland, UK, and Canada.

Contributors

S. Chakraverty
Department of Mathematics
National Institute of Technology
Rourkela, India

S.M. Chithra
Department of Mathematics
R.M.K. College of Engineering and
    Technology
Tamil Nadu, India

Rossana Dimitri
Department of Innovation Engineering
School of Engineering
University of Salento
Lecce, Italy

Dineshkumar Harursampath
Department of Aerospace Engineering
Indian Institute of Science (IISc)
Bangalore, India

Rohtas Kumar
Risk and Analytics
CRISIL Ltd.
Powai, Mumbai, India

Vinyas Mahesh
Department of Mechanical Engineering
National Institute of Technology
Silchar, Assam, India

Vishwas Mahesh
Department of Aerospace Engineering
Indian Institute of Science (IISc)
Bangalore, India

Arjun Siddharth Mangalasseri
Department of Aerospace Engineering
Indian Institute of Science (IISc)
Bangalore, India

Bhubaneswari Mishra
Department of Mathematics
National Institute of Technology
Rourkela, India

S.R. Mishra
Department of Mathematics
Siksha 'O' Anusandhan Deemed to be
    University
Bhubaneswar, India

Mohammad Mahdi Moayeri
Institute for Cognitive and Brain
    Sciences (ICBS)
Shahid Beheshti University
Tehran, Iran

J. Mohapatra
Department of Mathematics
National Institute of Technology
Rourkela, India

Sriram Mukunda
Department of Mechanical Engineering
Nitte Meenakshi Institute of
    Technology
Bangalore, India

Sukanta Nayak
Department of Mathematics
School of Advanced Sciences
VIT-AP University
Amaravati, Andhra Pradesh, India

Kourosh Parand
Department of Statistics and Actuarial
    Science
University of Waterloo
Waterloo, Canada

Sathiskumar A. Ponnusami
Department of Mechanical Engineering
   and Aeronautics
University of London
London, United Kingdom

Jamal Amani Rad
Institute for Cognitive and Brain
   Sciences (ICBS)
Shahid Beheshti University
Tehran, Iran

Priya Rao
Department of Mathematics
National Institute of Technology
Rourkela, India

Amir Hosein Hadian Rasanan
Institute for Cognitive and Brain
   Sciences (ICBS)
Shahid Beheshti University
Tehran, Iran

P.K. Ratha
Department of Mathematics
Siksha 'O' Anusandhan Deemed to be
   University
Bhubaneswar, India

Debanik Roy
Department of Atomic Energy
Bhaba Atomic Research Centre and
   Homi Bhabha National Institute
Mumbai, India

Arup Kumar Sahoo
Department of Mathematics
National Institute of Technology
Rourkela, India

Mrutyunjaya Sahoo
Department of Mathematics
National Institute of Technology
Rourkela, India

Rahul Saini
Department of Applied Mechanics
Indian Institute of Technology
New Delhi, India

S. Santra
Department of Mathematics
National Institute of Technology
Rourkela, India

V. Senthilkumar
CSIR Fourth Paradigm Institute
Belur Campus
Bangalore, India

Shravani V. Shetgaonkar
Department of Mathematics
School of Advanced Sciences
VIT-AP University
Amaravati, Andhra Pradesh, India

Francesco Tornabene
Department of Innovation Engineering
School of Engineering
University of Salento
Lecce, Italy

R.S. Tripathy
Department of Mathematics
Siksha 'O' Anusandhan Deemed to be
   University
Bhubaneswar, India

Matteo Viscoti
Department of Innovation Engineering
School of Engineering
University of Salento
Lecce, Italy

Yin Zhang
State Key Laboratory of Nonlinear
   Mechanics (LNM)
Institute of Mechanics
Chinese Academy of Sciences
Beijing, China

1 Dynamical Problems for Generally Anisotropic Shells with the GDQ Method

Francesco Tornabene, Matteo Viscoti, and Rossana Dimitri

CONTENTS

1.1 INTRODUCTION

Anisotropic materials are frequently employed in the manufacturing process of many structures characterized by complex geometries [1–3]. In this perspective, new theoretical and computational strategies are required to model even more complicated structural members [4, 5]. Referring to doubly curved shell structures, different two-dimensional approaches should be usually followed [6–9], such as Layer-Wise (LW) [10, 11], Equivalent Layer-Wise (ELW) [12] and Equivalent Single Layer (ESL) [13] approaches. According to the first strategy, each lamina of the stacking sequence is described using a reference middle surface, where the governing equations are derived. Eventually, they are assembled taking into account the interlaminar compatibility conditions [14–16]. The unknown field variable is described from the Degrees of Freedom (DOFs) associated with the interlaminar displacements.

DOI: 10.1201/9781003328032-1

The ELW methodology differs from LW as a single reference middle surface is provided for the entire structure. Last but not least, the ESL framework accounts for the reference surface located in the middle thickness, whereas the displacement field variable is taken alongside the surface at issue [17–20]. As a matter of fact, parametrization of the reference surface of the structure is a key element for the correct theoretical development of the formulation. Another interesting aspect consists in the axiomatic assumptions made for the field variable along the shell thickness. Some inceptive works on the ESL first-order formulations for shell structures can be found in [21], based on a first-order displacement field dispersion. In [22–25], free vibration analysis is performed on plates and curved composite structures by employing the First-Order Shear Deformation Theory (FSDT). However, in some cases, the influence of shear effects cannot be ignored for the evaluation of the structural response of structures with curvatures, especially in the presence of softcore materials [26–29]. Accordingly, refined through-the-thickness kinematic assumptions should be considered, as applied in [30, 31], based on the Third-Order Shear Deformation Theory (TSDT), accounting for a third-order in-plane displacement field power expansion. A milestone in the ESL theoretical framework of shells is the generalized formulation provided in [32–34], based on a generalized set of thickness functions, regardless of the actual choice of the displacement field assumption. Further theories can be found in [35–39], based on the use of trigonometric expressions. In the case of laminated materials, the thickness functions should be assessed to include interlaminar phenomena. To this end, in [40–44], an effective strategy is presented for a stepwise inclination of the actual in-plane displacement field distribution based on the layer sequence. Moreover, a refined zigzag theory for laminated composite structures is proposed in [45, 46], where the expression of the through-the-thickness functions is derived from the shear properties of the stacking sequence.

A proper selection of the thickness functions must be performed to get accurate results, even for structures made of unconventional materials. The use of higher-order theories can allow for the stretching and warping effects that are normally depicted only by three-dimensional models [47–49]. The present approach is demonstrated to be an effective strategy also for multifield simulations like bending analyses in thermal environment [50, 51]. Moreover, the lamination schemes embedding softcore layers with various material symmetries can be properly analysed only if a set of higher-order kinematic assumptions is assumed for each direction in which reduced stiffness is traced. In this perspective, the dynamic assessment of lattice pantographic panels characterized by complex geometries can be easily performed starting from the development of an equivalent continuum model [52–54].

Referring to doubly curved shell structures, the governing equations are usually determined following an energy approach [55, 56]. As far as the static problem is concerned, the well-known Minimum Potential Principle is adopted, whereas the dynamic equilibrium equations can be derived from the Hamiltonian Principle. As a matter of fact, both strong and weak formulations of the differential problem can be developed. In the first case, both Dirichlet and Neumann boundary conditions emerge [57, 58], whereas the external constraints are outlined for the weak form by elimination of the corresponding DOFs. An effective strategy to derive the governing equations can be found in [59], where a wide range of structural theories are reconducted to a generalized scheme.

Once the fundamental relations have been derived for the physical domain, the analytical or semi-analytical solution can be found in some cases [60, 61], whereas a numerical implementation of the problem is preferred. Different from classical-weighted residual methods, the Generalized Differential Quadrature (GDQ) is a powerful tool for this task because it accounts for the description of the derivative of a function as a weighted sum of the values assumed by the function itself within the definition domain [17, 62–64]. The accuracy of the model comes from the selection of the computational grid points from the continuum interval, together with the calculation of the weighting coefficients [65]. Accordingly, the GDQ procedure is based on the well-known Weierstrass interpolation theorem; therefore, the coefficients are calculated by employing a generalized procedure based on the properties of the selected interpolating polynomials. Nevertheless, the GDQ coefficients can be employed to compute integrals, leading to the so-called Generalized Integral Quadrature (GIQ) method [66]. Belonging to the class of spectral collocation methods, both the GDQ and GIQ have been demonstrated to be effective strategies for the numerical assessment of static and dynamic structural problems based on a reduced number of DOFs (see [67–70] for plates, singly curved and doubly curved structures). Moreover, the GDQ formulations have been provided for laminated appliances employing composite orthotropic layups, Functionally Graded Materials (FGMs) and nanomaterials [71–74]. A formulation for the dynamic analysis of doubly curved shell structures is reported in [47, 75], by employing generally anisotropic materials within the lamination scheme. Excellent accuracy has been outlined with respect to high computationally demanding simulations. In [76], the GDQ method has been successfully applied to the transient analysis of curved frames of general shape, whereas in [77], possible damage has been introduced within the structures via the GDQ implementation. Moreover, the GDQ method best fits all the requests of the so-called IsoGeometric Analysis (IGA), accounting for a NURBS description of the physical domain, as well as the interpolating functions [78–80].

In this chapter, we investigate different laminated doubly curved shell structures based on higher-order theories and a refined ESL theory. A generalized approach is considered to define the kinematic field within a large systematic investigation, where the reference surface of the shell is described in curvilinear principal coordinates, and arbitrarily shaped domains are described by a proper selection of the generalized blending functions based on a NURBS parametrization of the structural edges. An effective homogenization of a generally anisotropic lamination scheme is provided, accounting for a higher-order description of the warping acting along each principal direction. Moreover, all the interlaminar issues are described by the zigzag functions. The fundamental governing equations are derived from the Hamiltonian Principle, accounting for the natural set of boundary conditions. The numerical implementation of the structural problem is developed via the GDQ and GIQ approaches, accounting for non-uniform computational grids. A series of validating examples are proposed, where a free vibration analysis is performed for structures with different curvatures and geometric properties. The first mode frequencies and shapes of various structures provided by the present approach are compared to predictions from finite elements based on refined parabolic brick meshes, with a perfect alignment among results. The proposed formulation shows its effectiveness due to its accurate predictions even with a reduced computational cost. The higher-order ESL theory presented in this chapter has been implemented in the software DiQuMASPAB, as developed by the authors for the structural assessment of anisotropic doubly curved shells [81].

1.2 DOUBLY CURVED SHELL GEOMETRY

A doubly curved shell structure can be depicted as a three-dimensional solid within the Euclidean space. For this reason, its geometry is univocally defined with three parameters $\alpha_1, \alpha_2, \alpha_3$. If we denote with O the origin of a global Cartesian coordinate system, the unit vectors of which are $\mathbf{e}_1, \mathbf{e}_2, \mathbf{e}_3$, the position vector $\mathbf{R}$ of an arbitrary point of the three-dimensional shell reads as follows:

$$\mathbf{R}(\alpha_1,\alpha_2,\alpha_3) = f_1(\alpha_1,\alpha_2,\alpha_3)\mathbf{e}_1 + f_2(\alpha_1,\alpha_2,\alpha_3)\mathbf{e}_2 + f_3(\alpha_1,\alpha_2,\alpha_3)\mathbf{e}_3 \qquad (1.1)$$

f_1, f_2, f_3 being the three continuous functions. In order to give a physical meaning to Eqn. (1.1), the variables $\alpha_1, \alpha_2, \alpha_3$ should be defined in a three-dimensional interval, namely $(\alpha_1,\alpha_2,\alpha_3) \in \left[\alpha_1^0,\alpha_1^1\right] \times \left[\alpha_2^0,\alpha_2^1\right] \times \left[\alpha_3^0,\alpha_3^1\right]$, where α_i^0, α_i^1 with $i = 1,2,3$ are its extremes.

Employing the parametrization of Eqn. (1.1), it is possible to assess the ESL geometric representation of a doubly curved shell introducing a reference surface $\mathbf{r}(\alpha_1,\alpha_2)$ located in the middle thickness of the structure, as it has been shown in Figure 1.1. In particular, the parametrization in terms of α_1, α_2 is intended to be referred to as a curvilinear principal reference system. As a consequence, the association $\alpha_3 = \zeta$ refers in each point to the outward normal direction, whose unit vector $\mathbf{n}(\alpha_1,\alpha_2)$ is defined as follows:

$$\mathbf{n} = \frac{\mathbf{r}_{,1} \times \mathbf{r}_{,2}}{\left|\mathbf{r}_{,1} \times \mathbf{r}_{,2}\right|} \qquad (1.2)$$

where $\mathbf{r}_{,i} = \partial \mathbf{r}/\partial \alpha_i$ is the partial derivative of $\mathbf{r}(\alpha_1,\alpha_2)$ with respect to the principal coordinate $\alpha_i = \alpha_1,\alpha_2$. In this way, Eqn. (1.1) turns into the following expression [17]:

$$\mathbf{R}(\alpha_1,\alpha_2,\zeta) = \mathbf{r}(\alpha_1,\alpha_2) + \frac{h(\alpha_1,\alpha_2)}{2} z\mathbf{n}(\alpha_1,\alpha_2) \qquad (1.3)$$

setting $z = 2\zeta/h$ a dimensionless thickness parameter, with $z \in [-1, 1]$. In this way, an orthogonal curvilinear reference system $O'\alpha_1\alpha_2\zeta$ made of principal coordinates is sought from the geometric features of the shell object of analysis. Accordingly, the overall thickness of the shell structure is identified with $h(\alpha_1,\alpha_2)$.

Taking into account the ESL assessment of the shell geometry according to Eqn. (1.3), the main curvature radii of the reference surface $\mathbf{r}(\alpha_1,\alpha_2)$, denoted by

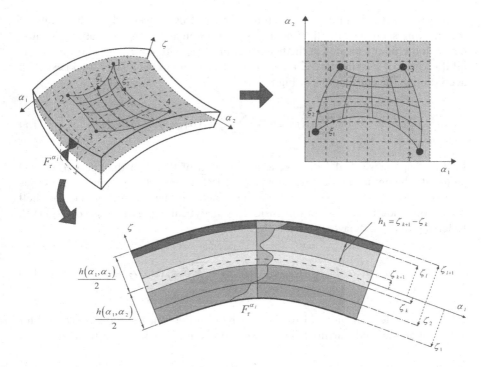

FIGURE 1.1   Representation of a laminated doubly curved shell structure employing curvilinear principal coordinates α_1, α_2. ESL assessment of the displacement field variable employing a set of generalized thickness functions $F_\tau^{\alpha_i}$ for each $\alpha_i = \alpha_1, \alpha_2, \alpha_3$ direction and for each τ-th kinematic expansion order. NURBS-based mapping of the physical domain employing generalized blending functions.

$R_i(\alpha_1, \alpha_2)$ for $i = 1, 2$, can be defined for each point of $\mathbf{r}$ according to the following expression [17]:

$$R_i(\alpha_1, \alpha_2) = -\frac{\mathbf{r}_{,i} \cdot \mathbf{r}_{,i}}{\mathbf{r}_{,ii} \cdot \mathbf{n}} \quad \text{for} \quad i = 1, 2 \tag{1.4}$$

As it is well known, since the parametrization in terms of α_1, α_2 accounts for a principal coordinate set, the quantities provided in Eqn. (1.4) lead to the maximum and minimum curvature values in each point of the shell [17]. Moreover, the Lamè parameters $A_i(\alpha_1, \alpha_2)$ with $i = 1, 2$ [17] are computed too:

$$A_i(\alpha_1, \alpha_2) = \sqrt{\mathbf{r}_{,i} \cdot \mathbf{r}_{,i}} \quad \text{for} \quad i = 1, 2 \tag{1.5}$$

Referring to the three-dimensional solid, the through-the-thickness curvature index $H_i(\alpha_1, \alpha_2)$ should be introduced for each α_1, α_2 shell principal direction [17]:

$$H_i(\alpha_1, \alpha_2) = 1 + \frac{\zeta}{R_i} \quad \text{for} \quad i = 1, 2 \tag{1.6}$$

If we consider a laminated structure composed by l layers, the k-th lamina of the stacking sequence, the thickness of which is denoted by h_k, is spatially located between the points located at the heights ζ_k and ζ_{k+1} with $k = 1,...,l$. Thus, the overall thickness of the structure can be obtained as a sum of each layer thickness [17], according to the conventions of Figure 1.1:

$$h(\alpha_1,\alpha_2) = \sum_{k=1}^{l} h_k(\alpha_1,\alpha_2) = \sum_{k=1}^{l} \left(\zeta_{k+1}(\alpha_1,\alpha_2) - \zeta_k(\alpha_1,\alpha_2) \right) \qquad (1.7)$$

In this chapter, we deal with a general variation of the shell thickness $\overline{h}(\alpha_1,\alpha_2)$ alongside the entire physical domain $\left[\alpha_1^0,\alpha_1^1\right] \times \left[\alpha_2^0,\alpha_2^1\right]$. To this purpose, a series of scaling δ_f and shift $\overline{\delta}$ parameters are introduced so that the thickness of the shell can be obtained from a set of dimensionless analytical expressions $\phi_f(\alpha_1,\alpha_2)$ [17]. Referring to the k-th layer, it gives:

$$h_k(\alpha_1,\alpha_2) = h_k^0 \overline{h}(\alpha_1,\alpha_2) = h_k^0 \left(1 + \sum_{f=1}^{4} \delta_f \phi_f(\alpha_1,\alpha_2) + \overline{\delta} \right) \qquad (1.8)$$

setting h_k^0 a reference value of the layer height. In the following, the $\phi_f(\alpha_1,\alpha_2)$ for $f = 1,..,4$ are defined starting from the following dimensionless coordinate $\overline{\alpha}_i$:

$$\overline{\alpha}_i = \frac{\alpha_i - \alpha_i^0}{\alpha_i^1 - \alpha_i^0} \quad \text{for } i = 1,2 \qquad (1.9)$$

The thickness variation laws implemented in Eqn. (1.8) account for a power and sinusoidal expression [17]:

$$\phi_j(\alpha_1,\alpha_2) = \begin{cases} \overline{\alpha}_i^{p_j} & \\ \left(\sin\left(\pi \left(n_j \overline{\alpha}_i + \alpha_{jm} \right) \right) \right)^{p_j} & \end{cases} \quad \text{for} \quad \begin{matrix} j = i = 1 \\ j = i = 2 \end{matrix}$$

$$\qquad (1.10)$$

$$\phi_j(\alpha_1,\alpha_2) = \begin{cases} \left(1 - \overline{\alpha}_i\right)^{p_j} & \\ \left(\sin\left(\pi \left(n_j \left(1 - \overline{\alpha}_i\right) + \alpha_{jm} \right) \right) \right)^{p_j} & \end{cases} \quad \text{for} \quad \begin{matrix} j = 3, i = 1 \\ j = 4, i = 2 \end{matrix}$$

where α_{jm} belongs to the closed interval $[0,1]$, whereas n_j and p_j are properly introduced as governing parameters belonging to the natural and real numbers, respectively.

Referring to an arbitrary curve $\mathbf{C}(u)$ lying on the reference surface $\mathbf{r}(\alpha_1,\alpha_2)$ expressed in principal coordinates, it is possible to define a local coordinate system for each $u \in [a,b]$ composed by the differential properties of the curve at issue. In particular, $\mathbf{n}_s$ unit vector is parallel to the tangent direction, $\mathbf{n}_s$ is the principal normal of the curve, whereas $\mathbf{n}_\zeta$ identifies the bi-normal unit vector. Such vectors are defined from their cosine directors $n_{n1}, n_{n2}, n_{n3}, n_{s1}, n_{s2}, n_{s3}, n_{\zeta 1}, n_{\zeta 2}, n_{\zeta 3}$ computed with respect to $\alpha_1, \alpha_2, \zeta$ coordinate axes:

$$\mathbf{n}_n = \begin{bmatrix} n_{n1} & n_{n2} & n_{n3} \end{bmatrix}^T$$

$$\mathbf{n}_s = \begin{bmatrix} n_{s1} & n_{s2} & n_{s3} \end{bmatrix}^T \qquad (1.11)$$

$$\mathbf{n}_\zeta = \begin{bmatrix} n_{\zeta 1} & n_{\zeta 2} & n_{\zeta 3} \end{bmatrix}^T$$

Since the curve belongs to the reference surface at issue, it gives $n_{\zeta 3} = 1$ and $n_{n3} = n_{s3} = n_{\zeta 1} = n_{\zeta 2} = 0$.

In this chapter, the curve $\mathbf{C}(u)$ is computed using NURBS [17], setting $[a,b] = [0,1]$:

$$\mathbf{C}(u) = \sum_{i=0}^{n} R_{i,p}(u) \mathbf{P}_i \qquad (1.12)$$

In the previous equation, a set of n control points $\mathbf{P}_i$ has been selected with $i = 1,...,n$, each of them associated with a rational basis function of the p-th degree denoted by $R_{i,p}(u)$. The latter can be computed as follows, setting $w_i > 0$ non-negative weighting coefficients and $N_{i,p}(u)$ a B-spline function of order $(p+1)$:

$$R_{i,p}(u) = \frac{N_{i,p}(u) w_i}{\displaystyle\sum_{i=0}^{n} N_{i,p}(u) w_i} \qquad (1.13)$$

Basis functions employed in Eqn. (1.13) are defined using a recursive procedure [17], taking into account a step function for $p = 0$:

$$N_{i,0}(u) = \begin{cases} 1 & \text{if} \quad u_i \leq u < u_{i+1} \\ 0 & \text{otherwise} \end{cases} \qquad (1.14)$$

$$N_{i,p}(u) = \frac{u - u_i}{u_{i+p} - u_i} N_{i,p-1}(u) + \frac{u_{i+p+1} - u}{u_{i+p+1} - u_{i+1}} N_{i+1,p-1}(u)$$

As can be seen, if it is assumed $w_i = 1$ for $i = 1,...,n$, Eqn. (1.14) gives $R_{i,p}(u) = N_{i,p}(u)$.

In Eqn. (1.14), the well-known knot vector $\mathbf{V}$ of multiplicity $p+1$ is employed, taking into account a discretization of $[a,b]$ interval characterized by m control points:

$$\mathbf{V} = \left[\underbrace{a, ..., a}_{p+1}, u_{p+1}, ..., u_{m-p-1}, \underbrace{b, ..., b}_{p+1} \right] \qquad (1.15)$$

1.3   GENERALIZED BLENDING FUNCTIONS

For an arbitrarily shaped structure, the reference surface introduced in Eqn. (1.3) cannot be described employing the previously discussed orthogonal principal coordinates α_1, α_2, since the physical domain is not a rectangular interval (Figure 1.1). As a consequence, a distortion procedure should be adopted, employing the following set of generalized blending functions $\alpha_i(\xi_1, \xi_2)$ for $i = 1, 2$ [17]:

$$\alpha_1(\xi_1, \xi_2) = \frac{1}{2}\left((1-\xi_2)\bar{\alpha}_{1(1)}(\xi_1) + (1+\xi_1)\bar{\alpha}_{1(2)}(\xi_2) + (1+\xi_2)\bar{\alpha}_{1(3)}(\xi_1) + (1-\xi_1)\bar{\alpha}_{1(4)}(\xi_2)\right) +$$

$$-\frac{1}{4}\left((1-\xi_1)(1-\xi_2)\alpha_{1(1)} + (1+\xi_1)(1-\xi_2)\alpha_{1(2)} + (1+\xi_1)(1+\xi_2)\alpha_{1(3)} + (1-\xi_1)(1+\xi_2)\alpha_{1(4)}\right)$$

$$\alpha_2(\xi_1, \xi_2) = \frac{1}{2}\left((1-\xi_2)\bar{\alpha}_{2(1)}(\xi_1) + (1+\xi_1)\bar{\alpha}_{2(2)}(\xi_2) + (1+\xi_2)\bar{\alpha}_{2(3)}(\xi_1) + (1-\xi_1)\bar{\alpha}_{2(4)}(\xi_2)\right) +$$

$$-\frac{1}{4}\left((1-\xi_1)(1-\xi_2)\alpha_{2(1)} + (1+\xi_1)(1-\xi_2)\alpha_{2(2)} + (1+\xi_1)(1+\xi_2)\alpha_{2(3)} + (1-\xi_1)(1+\xi_2)\alpha_{2(4)}\right)$$

$$(1.16)$$

In this way, the distorted domain is described in terms of a set of natural coordinates ξ_1, ξ_2, coming across the definition of a dimensionless rectangular computational domain, namely $(\xi_1, \xi_2) \in [-1,1] \times [-1,1]$. The relations reported in Eqn. (1.16) account for the location $(\alpha_{1(p)}, \alpha_{2(p)})$ for $p = 1,...,4$ within the principal coordinate system of the four corners of the physical domain. Moreover, symbols $(\bar{\alpha}_{1(q)}, \bar{\alpha}_{2(q)})$ for $q = 1,...,4$ denote the shell edges computed by employing Eqn. (1.12), expressed with respect to the principal parametric lines. For the sake of completeness, it should be recalled that in absence of a physical domain distortion, the blending relations of Eqn. (1.16) lead back to a scaling operation up to the dimensionless interval. Accordingly, the following nomenclature can be introduced for a smart identification of shell edges (Figure 1.1):

$$\text{West edge (W)} \quad \rightarrow \quad \xi_2 = -1 \quad \rightarrow \quad (\alpha_1, \alpha_2) = \left(\bar{\alpha}_{1(1)}(\xi_1), \bar{\alpha}_{2(1)}(\xi_1)\right)$$

$$\text{South edge (S)} \quad \rightarrow \quad \xi_1 = 1 \quad \rightarrow \quad (\alpha_1, \alpha_2) = \left(\bar{\alpha}_{1(2)}(\xi_2), \bar{\alpha}_{2(2)}(\xi_2)\right)$$

$$\text{East edge (E)} \quad \rightarrow \quad \xi_2 = 1 \quad \rightarrow \quad (\alpha_1, \alpha_2) = \left(\bar{\alpha}_{1(3)}(\xi_1), \bar{\alpha}_{2(3)}(\xi_1)\right) \qquad (1.17)$$

$$\text{North edge (N)} \quad \rightarrow \quad \xi_1 = -1 \quad \rightarrow \quad (\alpha_1, \alpha_2) = \left(\bar{\alpha}_{1(4)}(\xi_2), \bar{\alpha}_{2(4)}(\xi_2)\right)$$

Now, it is possible to assess the first-order partial derivatives with respect to α_1, α_2 in terms of the natural coordinates ξ_1, ξ_2. Employing the well-known chain rule, one gets:

$$\begin{bmatrix} \dfrac{\partial}{\partial \alpha_1} \\[2ex] \dfrac{\partial}{\partial \alpha_2} \end{bmatrix} = \begin{bmatrix} \dfrac{\partial \xi_1}{\partial \alpha_1} & \dfrac{\partial \xi_2}{\partial \alpha_1} \\[2ex] \dfrac{\partial \xi_1}{\partial \alpha_2} & \dfrac{\partial \xi_2}{\partial \alpha_2} \end{bmatrix} \begin{bmatrix} \dfrac{\partial}{\partial \xi_1} \\[2ex] \dfrac{\partial}{\partial \xi_2} \end{bmatrix} \qquad (1.18)$$

In the following, the partial derivatives with respect to ξ_1, ξ_2 are expressed as:

$$
\begin{bmatrix} \dfrac{\partial}{\partial \xi_1} \\[2ex] \dfrac{\partial}{\partial \xi_2} \end{bmatrix} = \begin{bmatrix} \dfrac{\partial \alpha_1}{\partial \xi_1} & \dfrac{\partial \alpha_2}{\partial \xi_1} \\[2ex] \dfrac{\partial \alpha_1}{\partial \xi_2} & \dfrac{\partial \alpha_2}{\partial \xi_2} \end{bmatrix} \begin{bmatrix} \dfrac{\partial}{\partial \alpha_1} \\[2ex] \dfrac{\partial}{\partial \alpha_2} \end{bmatrix} = \mathbf{J} \begin{bmatrix} \dfrac{\partial}{\partial \alpha_1} \\[2ex] \dfrac{\partial}{\partial \alpha_2} \end{bmatrix}
\tag{1.19}
$$

with $\mathbf{J}$ being the Jacobian matrix of the coordinate transformation, the determinant of which can be computed as:

$$
\det(\mathbf{J}) = \frac{\partial \alpha_1}{\partial \xi_1} \frac{\partial \alpha_2}{\partial \xi_2} - \frac{\partial \alpha_2}{\partial \xi_1} \frac{\partial \alpha_1}{\partial \xi_2}
\tag{1.20}
$$

When the parametrization is such that $\det(\mathbf{J}) \neq 0$, the inverse form of the Jacobian matrix $\mathbf{J}$ can be pursued, leading to the following relation:

$$
\mathbf{J}^{-1} = \frac{1}{\det(\mathbf{J})} \begin{bmatrix} \dfrac{\partial \alpha_2}{\partial \xi_2} & -\dfrac{\partial \alpha_2}{\partial \xi_1} \\[2ex] -\dfrac{\partial \alpha_1}{\partial \xi_2} & \dfrac{\partial \alpha_1}{\partial \xi_1} \end{bmatrix} = \begin{bmatrix} \xi_{1,\alpha_1} & \xi_{2,\alpha_1} \\[1ex] \xi_{1,\alpha_2} & \xi_{2,\alpha_2} \end{bmatrix}
\tag{1.21}
$$

Starting from Eqn. (1.21), the inverse transformation of Eqn. (1.19) is performed, thus giving:

$$
\begin{bmatrix} \dfrac{\partial}{\partial \alpha_1} \\[2ex] \dfrac{\partial}{\partial \alpha_2} \end{bmatrix} = \mathbf{J}^{-1} \begin{bmatrix} \dfrac{\partial}{\partial \xi_1} \\[2ex] \dfrac{\partial}{\partial \xi_2} \end{bmatrix} = \begin{bmatrix} \dfrac{\partial \xi_1}{\partial \alpha_1} & \dfrac{\partial \xi_2}{\partial \alpha_1} \\[2ex] \dfrac{\partial \xi_1}{\partial \alpha_2} & \dfrac{\partial \xi_2}{\partial \alpha_2} \end{bmatrix} \begin{bmatrix} \dfrac{\partial}{\partial \xi_1} \\[2ex] \dfrac{\partial}{\partial \xi_2} \end{bmatrix}
\tag{1.22}
$$

where $\xi_{i,\alpha_j} = \partial \xi_i / \partial \alpha_j$, setting $i, j = 1, 2$. In the same way, the second-order derivatives in the mapped domain can be expressed with respect to the ξ_1, ξ_2 principal coordinates. The mixed second-order derivatives read as follows [17]:

$$
\frac{\partial^2}{\partial \alpha_1 \partial \alpha_2} = \xi_{1,\alpha_1} \xi_{1,\alpha_2} \frac{\partial^2}{\partial \xi_1^2} + \xi_{2,\alpha_1} \xi_{2,\alpha_2} \frac{\partial^2}{\partial \xi_2^2} + \left(\xi_{1,\alpha_1} \xi_{2,\alpha_2} + \xi_{1,\alpha_2} \xi_{2,\alpha_1} \right) \frac{\partial^2}{\partial \xi_1 \partial \xi_2}
$$
$$
+ \xi_{1,\alpha_1 \alpha_2} \frac{\partial}{\partial \xi_1} + \xi_{2,\alpha_1 \alpha_2} \frac{\partial}{\partial \xi_2}
\tag{1.23}
$$

where coefficients $\xi_{j,\alpha_1\alpha_2}$ with $j = 1,2$ are defined as:

$$\xi_{1,\alpha_1\alpha_2} = \frac{1}{\det(\mathbf{J})^2}\left(-\frac{\partial\alpha_2}{\partial\xi_2}\frac{\partial^2\alpha_1}{\partial\xi_1\partial\xi_2} + \frac{\partial\alpha_2}{\partial\xi_2}\frac{\partial\alpha_1}{\partial\xi_2}\frac{\det(\mathbf{J}_{\xi_1})}{\det(\mathbf{J})} + \frac{\partial\alpha_2}{\partial\xi_1}\frac{\partial^2\alpha_1}{\partial\xi_2^2} - \frac{\partial\alpha_2}{\partial\xi_1}\frac{\partial\alpha_1}{\partial\xi_2}\frac{\det(\mathbf{J}_{\xi_2})}{\det(\mathbf{J})}\right)$$

$$\xi_{2,\alpha_1\alpha_2} = \frac{1}{\det(\mathbf{J})^2}\left(-\frac{\partial\alpha_2}{\partial\xi_1}\frac{\partial^2\alpha_1}{\partial\xi_1\partial\xi_2} + \frac{\partial\alpha_1}{\partial\xi_2}\frac{\partial\alpha_1}{\partial\xi_1}\frac{\det(\mathbf{J}_{\xi_1})}{\det(\mathbf{J})} + \frac{\partial\alpha_2}{\partial\xi_2}\frac{\partial^2\alpha_1}{\partial\xi_1^2} - \frac{\partial\alpha_2}{\partial\xi_1}\frac{\partial\alpha_1}{\partial\xi_1}\frac{\det(\mathbf{J}_{\xi_2})}{\det(\mathbf{J})}\right)$$

$$(1.24)$$

with $\det(\mathbf{J}_{\xi_i})$ being the derivatives with respect to the natural coordinates $\xi_i = \xi_1, \xi_2$ of the determinant of the $\mathbf{J}$ matrix of Eqn. (1.20), which are calculated according to the following expressions:

$$\det(\mathbf{J}_{\xi_1}) = \frac{\partial\alpha_1}{\partial\xi_1}\frac{\partial^2\alpha_2}{\partial\xi_1\partial\xi_2} - \frac{\partial\alpha_2}{\partial\xi_1}\frac{\partial^2\alpha_1}{\partial\xi_1\partial\xi_2} + \frac{\partial\alpha_2}{\partial\xi_2}\frac{\partial^2\alpha_1}{\partial\xi_1^2} - \frac{\partial\alpha_1}{\partial\xi_2}\frac{\partial^2\alpha_2}{\partial\xi_1^2}$$

$$\det(\mathbf{J}_{\xi_2}) = -\frac{\partial\alpha_1}{\partial\xi_2}\frac{\partial^2\alpha_2}{\partial\xi_1\partial\xi_2} + \frac{\partial\alpha_2}{\partial\xi_2}\frac{\partial^2\alpha_1}{\partial\xi_1\partial\xi_2} - \frac{\partial\alpha_2}{\partial\xi_1}\frac{\partial^2\alpha_1}{\partial\xi_2^2} + \frac{\partial\alpha_1}{\partial\xi_1}\frac{\partial^2\alpha_2}{\partial\xi_2^2}$$

$$(1.25)$$

Similarly, the derivatives with respect to the α_i principal directions for $i = 1,2$ take the following form:

$$\frac{\partial^2}{\partial\alpha_i^2} = \xi_{1,\alpha_i}^2\frac{\partial^2}{\partial\xi_1^2} + \xi_{2,\alpha_i}^2\frac{\partial^2}{\partial\xi_2^2} + 2\xi_{1,\alpha_i}\xi_{2,\alpha_i}\frac{\partial^2}{\partial\xi_1\partial\xi_2} + \xi_{1,\alpha_i\alpha_i}\frac{\partial}{\partial\xi_1} + \xi_{2,\alpha_i\alpha_i}\frac{\partial}{\partial\xi_2} \quad \text{for } i = 1,2$$

$$(1.26)$$

The extended expressions of coefficients $\xi_{1,\alpha_i\alpha_i}$ for $i = 1,2$ occurring in Eqn. (1.26) are thus reported:

$$\xi_{1,\alpha_1\alpha_1} = \frac{1}{\det(\mathbf{J})^2}\left(\frac{\partial\alpha_2}{\partial\xi_2}\frac{\partial^2\alpha_2}{\partial\xi_1\partial\xi_2} - \left(\frac{\partial\alpha_2}{\partial\xi_2}\right)^2\frac{\det(\mathbf{J}_{\xi_1})}{\det(\mathbf{J})} - \frac{\partial\alpha_2}{\partial\xi_1}\frac{\partial^2\alpha_2}{\partial\xi_2^2} + \frac{\partial\alpha_2}{\partial\xi_1}\frac{\partial\alpha_2}{\partial\xi_2}\frac{\det(\mathbf{J}_{\xi_2})}{\det(\mathbf{J})}\right)$$

$$\xi_{1,\alpha_2\alpha_2} = \frac{1}{\det(\mathbf{J})^2}\left(\frac{\partial\alpha_1}{\partial\xi_2}\frac{\partial^2\alpha_1}{\partial\xi_1\partial\xi_2} - \left(\frac{\partial\alpha_1}{\partial\xi_2}\right)^2\frac{\det(\mathbf{J}_{\xi_1})}{\det(\mathbf{J})} - \frac{\partial\alpha_1}{\partial\xi_1}\frac{\partial^2\alpha_1}{\partial\xi_2^2} + \frac{\partial\alpha_1}{\partial\xi_1}\frac{\partial\alpha_1}{\partial\xi_2}\frac{\det(\mathbf{J}_{\xi_2})}{\det(\mathbf{J})}\right)$$

$$(1.27)$$

Accordingly, quantities $\xi_{2,\alpha_i\alpha_i}$ for $i = 1,2$ read as follows:

$$\xi_{2,\alpha_1\alpha_1} = \frac{1}{\det(\mathbf{J})^2}\left(-\frac{\partial\alpha_2}{\partial\xi_2}\frac{\partial^2\alpha_2}{\partial\xi_1^2} + \frac{\partial\alpha_2}{\partial\xi_2}\frac{\partial\alpha_2}{\partial\xi_1}\frac{\det(\mathbf{J}_{\xi_1})}{\det(\mathbf{J})} + \frac{\partial\alpha_2}{\partial\xi_1}\frac{\partial^2\alpha_2}{\partial\xi_1\partial\xi_2} - \left(\frac{\partial\alpha_2}{\partial\xi_1}\right)^2\frac{\det(\mathbf{J}_{\xi_2})}{\det(\mathbf{J})}\right)$$

$$\xi_{2,\alpha_2\alpha_2} = \frac{1}{\det(\mathbf{J})^2}\left(-\frac{\partial\alpha_1}{\partial\xi_2}\frac{\partial^2\alpha_1}{\partial\xi_1^2} + \frac{\partial\alpha_1}{\partial\xi_2}\frac{\partial\alpha_1}{\partial\xi_1}\frac{\det(\mathbf{J}_{\xi_1})}{\det(\mathbf{J})} + \frac{\partial\alpha_1}{\partial\xi_1}\frac{\partial^2\alpha_1}{\partial\xi_1\partial\xi_2} - \left(\frac{\partial\alpha_1}{\partial\xi_1}\right)^2\frac{\det(\mathbf{J}_{\xi_2})}{\det(\mathbf{J})}\right)$$

$$(1.28)$$

1.4   DISPLACEMENTS AND KINEMATIC RELATIONS

In the previous section, the three-dimensional nature of a doubly curved shell has been outlined. In particular, in Eqn. (1.1), a curvilinear set of coordinates has been defined starting from the shell geometry. In this perspective, a three-dimensional displacement field $\mathbf{U}(\alpha_1,\alpha_2,\alpha_3,t) = \begin{bmatrix} U_1 & U_2 & U_3 \end{bmatrix}^T$ is associated with each point of the structure, and its components are taken along the principal directions α_1, α_2 and $\alpha_3 = \zeta$. Following the ESL approach, each component of $\mathbf{U}(\alpha_1,\alpha_2,\alpha_3,t)$ acting along the α_i principal direction is expanded in a generalized form up to an arbitrary $(N+1)$-th order via the introduction of a set of functions denoted by $F_\tau^{\alpha_i}(\zeta)$ for $\tau = 0,\ldots,N+1$ depending from the thickness coordinate ζ (Figure 1.1), as extensively explained in [17]. In a compact notation, one gets:

$$
\begin{bmatrix} U_1 \\ U_2 \\ U_3 \end{bmatrix} = \sum_{\tau=0}^{N+1} \begin{bmatrix} F_\tau^{\alpha_1} & 0 & 0 \\ 0 & F_\tau^{\alpha_2} & 0 \\ 0 & 0 & F_\tau^{\alpha_3} \end{bmatrix} \begin{bmatrix} u_1^{(\tau)} \\ u_2^{(\tau)} \\ u_3^{(\tau)} \end{bmatrix} \quad \Leftrightarrow \quad \mathbf{U}(\alpha_1,\alpha_2,\alpha_3,t) = \sum_{\tau=0}^{N+1} \mathbf{F}_\tau(\zeta) \mathbf{u}^{(\tau)}(\alpha_1,\alpha_2,t)
$$

$$(1.29)$$

As a consequence, for all $\tau = 0,\ldots,N+1$ kinematic expansion orders, the generalized displacement field vector $\mathbf{u}^{(\tau)}(\alpha_1,\alpha_2,t)$ lying on the reference surface $\mathbf{r}(\alpha_1,\alpha_2)$ is introduced, accounting for the spatial and time dependence of the structural problem. Employing the generalized approach of Eqn. (1.29), it is possible to develop a generalized theory embedding in itself an arbitrary kinematic assumption. Moreover, all the classical ESL theories that can be found in literature are taken into account too. In this perspective, the well-known FSDT displacement field comes out if the following thickness functions are adopted [17]:

$$
F_0^{\alpha_1} = 1, \quad F_0^{\alpha_2} = 1, \quad F_0^{\alpha_3} = 1
$$

$$
F_1^{\alpha_1} = \zeta, \quad F_1^{\alpha_2} = \zeta, \quad F_1^{\alpha_3} = 0
$$

$$(1.30)$$

In the same way, the TSDT [17] is obtained if the following assumptions are considered within Eqn. (1.29):

$$
F_0^{\alpha_1} = 1, \quad F_0^{\alpha_2} = 1, \quad F_0^{\alpha_3} = 1
$$

$$
F_1^{\alpha_1} = \zeta, \quad F_1^{\alpha_2} = \zeta, \quad F_1^{\alpha_3} = 0
$$

$$
F_2^{\alpha_1} = \zeta^2, \quad F_2^{\alpha_2} = \zeta^2, \quad F_2^{\alpha_3} = 0
$$

$$
F_3^{\alpha_1} = \zeta^3, \quad F_3^{\alpha_2} = \zeta^3, \quad F_3^{\alpha_3} = 0
$$

$$(1.31)$$

In this chapter, an effective zigzag function [17] has been applied for each $\alpha_i = \alpha_1, \alpha_2, \alpha_3$, associated to the $\tau = (N+1)$-th order of Eqn. (1.29):

$$F_{N+1}^{\alpha_i}(\zeta) = (-1)^k z_k = (-1)^k \left(\frac{2}{\zeta_{k+1} - \zeta_k} \zeta - \frac{\zeta_{k+1} + \zeta_k}{\zeta_{k+1} - \zeta_k} \right) \quad \text{for} \quad i = 1, 2, 3 \quad (1.32)$$

where k is the actual layer of the stacking sequence, located between ζ_k and ζ_{k+1} for $k = 1, ..., l$, with l being the total number of laminae. In this way, a series of complicated coupling effects between two adjacent laminae alongside the thickness direction can be easily predicted even though a two-dimensional ESL formulation is considered. However, a generalized power polynomial expansion of the displacement field is adopted for the arbitrary τ-th kinematic expansion order. To sum up, for Eqn. (1.29) one gets the following expression for each three-dimensional displacement field thickness function $F_\tau^{\alpha_i}(\zeta)$:

$$F_\tau^{\alpha_i}(\zeta) = \begin{cases} \zeta^\tau & \text{for } \tau = 0, ..., N \\ (-1)^k z_k & \text{for } \tau = N+1 \end{cases} \quad (1.33)$$

In order to identify the through-the-thickness displacement field assumption, the nomenclature $\text{ED}(Z) - N$ is adopted. Accordingly, letter "E" means that an ESL approach is followed for the description of the field variable, whereas "D" tells that a higher-order assumption is taken for the three-dimensional displacement field vector. Furthermore, N denotes the maximum kinematic expansion order. When the generalized thickness functions embed the zigzag expression presented in Eqn. (1.33), the capital letter "Z" is attached. For the sake of completeness, we report the extended version of the three-dimensional displacement field components for the EDZ4 theory as follows:

$$U_1 = u_1^{(0)} + \zeta u_1^{(1)} + \zeta^2 u_1^{(2)} + \zeta^3 u_1^{(3)} + \zeta^4 u_1^{(4)} + (-1)^k z_k u_1^{(5)}$$
$$U_2 = u_2^{(0)} + \zeta u_2^{(1)} + \zeta^2 u_2^{(2)} + \zeta^3 u_2^{(3)} + \zeta^4 u_2^{(4)} + (-1)^k z_k u_2^{(5)} \quad (1.34)$$
$$U_3 = u_3^{(0)} + \zeta u_3^{(1)} + \zeta^2 u_3^{(2)} + \zeta^3 u_3^{(3)} + \zeta^4 u_3^{(4)} + (-1)^k z_k u_3^{(5)}$$

where a power series up to the fourth order is adopted, together with the piecewise inclination in the profile of the thickness functions.

Starting from the displacement field assumption of Eqn. (1.29), the kinematic relations for a doubly curved shell should be assessed within the ESL framework. To this purpose, the three-dimensional strain vector $\varepsilon(\alpha_1, \alpha_2, \alpha_3, t) = \begin{bmatrix} \varepsilon_1 & \varepsilon_2 & \gamma_{12} & \gamma_{13} & \gamma_{23} & \varepsilon_3 \end{bmatrix}^T$ written for a doubly curved solid employing principal coordinates should be kept in mind [17]:

$$\varepsilon = \mathbf{D}\mathbf{U} = \mathbf{D}_\zeta \left(\sum_{i=1}^{3} \mathbf{D}_\Omega^{\alpha_i} \right) \mathbf{U} \quad (1.35)$$

where the kinematic differential operator $\mathbf{D}$ is split into a thickness operator $\mathbf{D}_\zeta$ and three operators $\mathbf{D}_\Omega^{\alpha_i} = \mathbf{D}_\Omega^{\alpha_1}, \mathbf{D}_\Omega^{\alpha_2}, \mathbf{D}_\Omega^{\alpha_3}$. The former embeds the derivations with respect to the thickness coordinate ζ, as follows:

$$
\mathbf{D}_\zeta =
\begin{bmatrix}
\dfrac{1}{H_1} & 0 & 0 & 0 & 0 & 0 & 0 & 0 & 0 \\[2mm]
0 & \dfrac{1}{H_2} & 0 & 0 & 0 & 0 & 0 & 0 & 0 \\[2mm]
0 & 0 & \dfrac{1}{H_1} & \dfrac{1}{H_2} & 0 & 0 & 0 & 0 & 0 \\[2mm]
0 & 0 & 0 & 0 & \dfrac{1}{H_1} & 0 & \dfrac{\partial}{\partial \zeta} & 0 & 0 \\[2mm]
0 & 0 & 0 & 0 & 0 & \dfrac{1}{H_2} & 0 & \dfrac{\partial}{\partial \zeta} & 0 \\[2mm]
0 & 0 & 0 & 0 & 0 & 0 & 0 & 0 & \dfrac{\partial}{\partial \zeta}
\end{bmatrix}
\tag{1.36}
$$

However, $\mathbf{D}_\Omega^{\alpha_i}$ for $i = 1,2,3$, the kinematic operators account for the partial derivatives with respect to α_1, α_2, as follows:

$$
\mathbf{D}_\Omega^{\alpha_1} = \begin{bmatrix} \bar{\mathbf{D}}_\Omega^{\alpha_1} & \mathbf{0} & \mathbf{0} \end{bmatrix}
$$
$$
\mathbf{D}_\Omega^{\alpha_2} = \begin{bmatrix} \mathbf{0} & \bar{\mathbf{D}}_\Omega^{\alpha_2} & \mathbf{0} \end{bmatrix}
\tag{1.37}
$$
$$
\mathbf{D}_\Omega^{\alpha_3} = \begin{bmatrix} \mathbf{0} & \mathbf{0} & \bar{\mathbf{D}}_\Omega^{\alpha_3} \end{bmatrix}
$$

An extended version of the differential operators $\bar{\mathbf{D}}_\Omega^{\alpha_i}$ for $i = 1,2,3$ takes the following form:

$$
\bar{\mathbf{D}}_\Omega^{\alpha_1} = \begin{bmatrix} \dfrac{1}{A_1}\dfrac{\partial}{\partial\alpha_1} & \dfrac{1}{A_1 A_2}\dfrac{\partial A_2}{\partial\alpha_1} & -\dfrac{1}{A_1 A_2}\dfrac{\partial A_1}{\partial\alpha_2} & \dfrac{1}{A_2}\dfrac{\partial}{\partial\alpha_2} & -\dfrac{1}{R_1} & 0 & 1 & 0 & 0 \end{bmatrix}^T
$$

$$
\bar{\mathbf{D}}_\Omega^{\alpha_2} = \begin{bmatrix} \dfrac{1}{A_1 A_2}\dfrac{\partial A_1}{\partial\alpha_2} & \dfrac{1}{A_2}\dfrac{\partial}{\partial\alpha_2} & \dfrac{1}{A_1}\dfrac{\partial}{\partial\alpha_1} & -\dfrac{1}{A_1 A_2}\dfrac{\partial A_2}{\partial\alpha_1} & 0 & -\dfrac{1}{R_2} & 0 & 1 & 0 \end{bmatrix}^T
\tag{1.38}
$$

$$
\bar{\mathbf{D}}_\Omega^{\alpha_3} = \begin{bmatrix} \dfrac{1}{R_1} & \dfrac{1}{R_2} & 0 & 0 & \dfrac{1}{A_1}\dfrac{\partial}{\partial\alpha_1} & \dfrac{1}{A_2}\dfrac{\partial}{\partial\alpha_2} & 0 & 0 & 1 \end{bmatrix}^T
$$

Starting from the kinematic relations (35) for a three-dimensional solid, we can redefine them in a two-dimensional sense. Referring to any τ-th order of the unknown field variable expansion, a vector of ESL-generalized strain components is introduced, namely $\boldsymbol{\varepsilon}^{(\tau)\alpha_i}(\alpha_1,\alpha_2,t)=\left[\varepsilon_1^{(\tau)\alpha_i}\ \varepsilon_2^{(\tau)\alpha_i}\ \gamma_1^{(\tau)\alpha_i}\ \gamma_2^{(\tau)\alpha_i}\ \gamma_{13}^{(\tau)\alpha_i}\ \gamma_{23}^{(\tau)\alpha_i}\ \omega_{13}^{(\tau)\alpha_i}\ \omega_{23}^{(\tau)\alpha_i}\ \varepsilon_3^{(\tau)\alpha_i}\right]^T$, whose components are located on the reference middle surface $\mathbf{r}(\alpha_1,\alpha_2)$. Substituting the unified assessment of the displacement field of Eqn. (1.29) into the three-dimensional relations of Eqn. (1.35), the definition of $\boldsymbol{\varepsilon}^{(\tau)\alpha_i}(\alpha_1,\alpha_2,t)$ for $\tau=0,...,N+1$ vector is sought [17]:

$$\boldsymbol{\varepsilon}=\sum_{\tau=0}^{N+1}\sum_{i=1}^{3}\mathbf{D}_\zeta\mathbf{D}_\Omega^{\alpha_i}\mathbf{F}_\tau\mathbf{u}^{(\tau)}=\sum_{\tau=0}^{N+1}\sum_{i=1}^{3}\mathbf{Z}^{(\tau)\alpha_i}\mathbf{D}_\Omega^{\alpha_i}\mathbf{u}^{(\tau)}=\sum_{\tau=0}^{N+1}\sum_{i=1}^{3}\mathbf{Z}^{(\tau)\alpha_i}\boldsymbol{\varepsilon}^{(\tau)\alpha_i} \qquad (1.39)$$

In the previous equation, the three-dimensional strain vector $\boldsymbol{\varepsilon}$ has been linked to the ESL strain vector $\boldsymbol{\varepsilon}^{(\tau)\alpha_i}$ due to the introduction of a matrix $\mathbf{Z}^{(\tau)\alpha_i}$ referred to each τ-th generalized expansion and $\alpha_i=\alpha_1,\alpha_2,\alpha_3$ principal direction. In the following, an extended version of the matrix at issue has been reported, accounting for the shell curvature thickness parameters $H_i(\alpha_1,\alpha_2)$ for $i=1,2$ defined in Eqn. (1.6) and the thickness function $F_\tau^{\alpha_i}$ referred to $\alpha_i=\alpha_1,\alpha_2,\alpha_3$:

$$\mathbf{Z}^{(\tau)\alpha_i}=\begin{bmatrix} \dfrac{F_\tau^{\alpha_i}}{H_1} & 0 & 0 & 0 & 0 & 0 & 0 & 0 & 0 \\[2mm] 0 & \dfrac{F_\tau^{\alpha_i}}{H_2} & 0 & 0 & 0 & 0 & 0 & 0 & 0 \\[2mm] 0 & 0 & \dfrac{F_\tau^{\alpha_i}}{H_1} & \dfrac{F_\tau^{\alpha_i}}{H_2} & 0 & 0 & 0 & 0 & 0 \\[2mm] 0 & 0 & 0 & 0 & \dfrac{F_\tau^{\alpha_i}}{H_1} & 0 & \dfrac{\partial F_\tau^{\alpha_i}}{\partial\zeta} & 0 & 0 \\[2mm] 0 & 0 & 0 & 0 & 0 & \dfrac{F_\tau^{\alpha_i}}{H_2} & 0 & \dfrac{\partial F_\tau^{\alpha_i}}{\partial\zeta} & 0 \\[2mm] 0 & 0 & 0 & 0 & 0 & 0 & 0 & 0 & \dfrac{\partial F_\tau^{\alpha_i}}{\partial\zeta} \end{bmatrix}$$

for $i=1,2,3$

$$(1.40)$$

The generalized strain components of vector $\varepsilon^{(\tau)\alpha_i}$ read as follows [17]:

$$\varepsilon_1^{(\tau)\alpha_1} = \frac{1}{A_1}\frac{\partial u_1^{(\tau)}}{\partial \alpha_1}, \qquad \varepsilon_1^{(\tau)\alpha_2} = \frac{u_2^{(\tau)}}{A_1 A_2}\frac{\partial A_1}{\partial \alpha_2}, \qquad \varepsilon_1^{(\tau)\alpha_3} = \frac{u_3^{(\tau)}}{R_1}$$

$$\varepsilon_2^{(\tau)\alpha_1} = \frac{u_1^{(\tau)}}{A_1 A_2}\frac{\partial A_2}{\partial \alpha_1}, \qquad \varepsilon_2^{(\tau)\alpha_2} = \frac{1}{A_2}\frac{\partial u_2^{(\tau)}}{\partial \alpha_2}, \qquad \varepsilon_2^{(\tau)\alpha_3} = \frac{u_3^{(\tau)}}{R_2}$$

$$\gamma_1^{(\tau)\alpha_1} = -\frac{u_1^{(\tau)}}{A_1 A_2}\frac{\partial A_1}{\partial \alpha_2}, \qquad \gamma_1^{(\tau)\alpha_2} = \frac{1}{A_1}\frac{\partial u_2^{(\tau)}}{\partial \alpha_1}, \qquad \gamma_1^{(\tau)\alpha_3} = 0$$

$$\gamma_2^{(\tau)\alpha_1} = \frac{1}{A_2}\frac{\partial u_1^{(\tau)}}{\partial \alpha_2}, \qquad \gamma_2^{(\tau)\alpha_2} = -\frac{u_2^{(\tau)}}{A_1 A_2}\frac{\partial A_2}{\partial \alpha_1}, \qquad \gamma_2^{(\tau)\alpha_3} = 0$$

$$\gamma_{13}^{(\tau)\alpha_1} = -\frac{u_1^{(\tau)}}{R_1}, \qquad \gamma_{13}^{(\tau)\alpha_2} = 0, \qquad \gamma_{13}^{(\tau)\alpha_3} = \frac{1}{A_1}\frac{\partial u_3^{(\tau)}}{\partial \alpha_1}$$

$$\gamma_{23}^{(\tau)\alpha_1} = 0, \qquad \gamma_{23}^{(\tau)\alpha_2} = -\frac{u_2^{(\tau)}}{R_2}, \qquad \gamma_{23}^{(\tau)\alpha_3} = \frac{1}{A_2}\frac{\partial u_3^{(\tau)}}{\partial \alpha_2}$$

$$\omega_{13}^{(\tau)\alpha_1} = u_1^{(\tau)}, \qquad \omega_{13}^{(\tau)\alpha_2} = 0, \qquad \omega_{13}^{(\tau)\alpha_3} = 0$$

$$\omega_{23}^{(\tau)\alpha_1} = 0, \qquad \omega_{23}^{(\tau)\alpha_2} = u_2^{(\tau)}, \qquad \omega_{23}^{(\tau)\alpha_3} = 0$$

$$\varepsilon_3^{(\tau)\alpha_1} = 0, \qquad \varepsilon_3^{(\tau)\alpha_2} = 0, \qquad \varepsilon_3^{(\tau)\alpha_3} = u_3^{(\tau)}$$

$$(1.41)$$

For more details on the ESL assessment of the kinematic relations, the interested reader can refer to [17].

1.5   ANISOTROPIC CONSTITUTIVE RELATIONS

We now focus on the assessment within the ESL framework of the mechanical properties and constitutive law of each layer. Then, an equivalent relationship referred to the middle surface is derived accounting for the entire lamination scheme made of l layers.

Referring to the arbitrary k-th lamina of the continuum solid, the anisotropic constitutive relation reads as follows:

$$\hat{\sigma}^{(k)} = \mathbf{E}^{(k)}\hat{\varepsilon}^{(k)} \quad \text{for} \quad k = 1,...,l \tag{1.42}$$

where $\hat{\sigma}^{(k)} = \begin{bmatrix} \hat{\sigma}_1^{(k)} & \hat{\sigma}_2^{(k)} & \hat{\tau}_{12}^{(k)} & \hat{\tau}_{13}^{(k)} & \hat{\tau}_{23}^{(k)} & \hat{\sigma}_3^{(k)} \end{bmatrix}^T$ and $\hat{\varepsilon}^{(k)} = \begin{bmatrix} \hat{\varepsilon}_1^{(k)} & \hat{\varepsilon}_2^{(k)} & \hat{\gamma}_{12}^{(k)} & \hat{\gamma}_{13}^{(k)} & \hat{\gamma}_{23}^{(k)} & \hat{\varepsilon}_3^{(k)} \end{bmatrix}^T$ are the three-dimensional stress and strain vectors referred to the so-called material reference system, denoted by $O'\hat{\alpha}_1^{(k)}\hat{\alpha}_2^{(k)}\hat{\zeta}^{(k)}$. In index notation, the stiffness matrix $\mathbf{E}^{(k)}$ of the k-th layer reads as follows [17]:

$$
\mathbf{E}^{(k)} = \begin{bmatrix}
E_{11}^{(k)} & E_{12}^{(k)} & E_{16}^{(k)} & E_{14}^{(k)} & E_{15}^{(k)} & E_{13}^{(k)} \\
E_{12}^{(k)} & E_{22}^{(k)} & E_{26}^{(k)} & E_{24}^{(k)} & E_{25}^{(k)} & E_{23}^{(k)} \\
E_{16}^{(k)} & E_{26}^{(k)} & E_{66}^{(k)} & E_{46}^{(k)} & E_{56}^{(k)} & E_{36}^{(k)} \\
E_{14}^{(k)} & E_{24}^{(k)} & E_{46}^{(k)} & E_{44}^{(k)} & E_{45}^{(k)} & E_{34}^{(k)} \\
E_{15}^{(k)} & E_{25}^{(k)} & E_{56}^{(k)} & E_{45}^{(k)} & E_{55}^{(k)} & E_{35}^{(k)} \\
E_{13}^{(k)} & E_{23}^{(k)} & E_{36}^{(k)} & E_{34}^{(k)} & E_{35}^{(k)} & E_{33}^{(k)}
\end{bmatrix} \quad \text{for} \quad k = 1,...,l \quad (1.43)
$$

where $E_{ij}^{(k)}$ for $i,j = 1,...,6$ are the stiffness components referred to the i-th component of $\hat{\sigma}^{(k)}$ and the j-th component of the three-dimensional strain vector $\hat{\varepsilon}^{(k)}$. Generally speaking, they are provided for a three-dimensional continuum, thus giving $E_{ij}^{(k)} = C_{ij}^{(k)}$. In plane-stress conditions $\left(\hat{\sigma}_3^{(k)} = 0 \right)$, all the stiffness constants should be replaced with their reduced values $E_{ij}^{(k)} = Q_{ij}^{(k)}$, which are computed according to the following relation:

$$
Q_{ij}^{(k)} = C_{ij}^{(k)} - \frac{C_{j3}^{(k)} C_{i3}^{(k)}}{C_{33}^{(k)}} \quad \text{for} \quad i,j = 1,...,6, \ k = 1,...,l \quad (1.44)
$$

Referring to the material reference system $O'\hat{\alpha}_1^{(k)}\hat{\alpha}_2^{(k)}\hat{\zeta}^{(k)}$, it should be stated that the $\hat{\zeta}^{(k)}$ axis is intended to be oriented along the outward normal direction of the shell, thus giving $\hat{\zeta}^{(k)} = \zeta$. When a through-the-thickness homogenization of the shell mechanical properties is performed, the constitutive relationship of Eqn. (1.42), for each k-th layer, should be referred to the geometric reference system that has been previously introduced from the differential geometry properties of the shell. For this reason, an arbitrary angle ϑ_k is assessed between $\hat{\alpha}_1^{(k)}$ and α_1 axis, associated with a rotation orthogonal matrix $\mathbf{T}^{(k)}(\vartheta_k)$. The anisotropic stiffness matrix $\mathbf{E}^{(k)}$ for $k = 1,...,l$ in Eqn. (1.42) can thus be rotated from the material to the geometric reference system leading to the following definition of matrix $\bar{\mathbf{E}}^{(k)}$ [17]:

$$
\bar{\mathbf{E}}^{(k)} = \mathbf{T}^{(k)}\mathbf{E}^{(k)}\left(\mathbf{T}^{(k)} \right)^T \quad (1.45)
$$

Eqn. (1.42) can be written with respect to the shell reference system $O'\alpha_1\alpha_2\zeta$ as follows:

$$
\sigma^{(k)} = \bar{\mathbf{E}}^{(k)}\varepsilon^{(k)} \iff
\begin{bmatrix}
\sigma_1^{(k)} \\
\sigma_2^{(k)} \\
\tau_{12}^{(k)} \\
\tau_{13}^{(k)} \\
\tau_{23}^{(k)} \\
\sigma_3^{(k)}
\end{bmatrix}
=
\begin{bmatrix}
\bar{E}_{11}^{(k)} & \bar{E}_{12}^{(k)} & \bar{E}_{16}^{(k)} & \bar{E}_{14}^{(k)} & \bar{E}_{15}^{(k)} & \bar{E}_{13}^{(k)} \\
\bar{E}_{12}^{(k)} & \bar{E}_{22}^{(k)} & \bar{E}_{26}^{(k)} & \bar{E}_{24}^{(k)} & \bar{E}_{25}^{(k)} & \bar{E}_{23}^{(k)} \\
\bar{E}_{16}^{(k)} & \bar{E}_{25}^{(k)} & \bar{E}_{66}^{(k)} & \bar{E}_{46}^{(k)} & \bar{E}_{56}^{(k)} & \bar{E}_{36}^{(k)} \\
\bar{E}_{14}^{(k)} & \bar{E}_{24}^{(k)} & \bar{E}_{46}^{(k)} & \bar{E}_{44}^{(k)} & \bar{E}_{45}^{(k)} & \bar{E}_{34}^{(k)} \\
\bar{E}_{15}^{(k)} & \bar{E}_{25}^{(k)} & \bar{E}_{56}^{(k)} & \bar{E}_{45}^{(k)} & \bar{E}_{55}^{(k)} & \bar{E}_{35}^{(k)} \\
\bar{E}_{13}^{(k)} & \bar{E}_{23}^{(k)} & \bar{E}_{36}^{(k)} & \bar{E}_{34}^{(k)} & \bar{E}_{35}^{(k)} & \bar{E}_{33}^{(k)}
\end{bmatrix}
\begin{bmatrix}
\varepsilon_1^{(k)} \\
\varepsilon_2^{(k)} \\
\gamma_{12}^{(k)} \\
\gamma_{13}^{(k)} \\
\gamma_{23}^{(k)} \\
\varepsilon_3^{(k)}
\end{bmatrix}
\tag{1.46}
$$

$$\text{for} \quad k = 1, \dots, l$$

Accordingly, $\bar{E}_{ij}^{(k)}$ provides a linear elastic relationship between the j-th and i-th components of the three-dimensional strain and stress vector, respectively, defined with respect to $O'\alpha_1\alpha_2\zeta$.

Now, a generalized constitutive relationship is provided for the entire lamination scheme, accounting for the ESL-unified assessment of the displacement field variable of Eqn. (1.29). Taking into account the constitutive behaviour defined for each lamina according to Eqn. (1.46), from the computation of the virtual variation $\delta\Phi$ of the elastic strain energy Φ provided by the three-dimensional vectors $\sigma^{(k)}$ and $\varepsilon^{(k)}$, through-the-thickness integrations can be collected in the generalized stress resultant vector $\mathbf{S}^{(\tau)\alpha_i}(\alpha_1,\alpha_2,t) = \begin{bmatrix} N_1^{(\tau)\alpha_i} & N_2^{(\tau)\alpha_i} & N_{12}^{(\tau)\alpha_i} & N_{21}^{(\tau)\alpha_i} & T_1^{(\tau)\alpha_i} & T_2^{(\tau)\alpha_i} & P_1^{(\tau)\alpha_i} & P_2^{(\tau)\alpha_i} & S_3^{(\tau)\alpha_i} \end{bmatrix}^T$, defined for each $\tau = 0, \dots, N+1$ and $\alpha_i = \alpha_1, \alpha_2, \alpha_3$ [17], the components of which account as:

$$
N_1^{(\tau)\alpha_i} = \sum_{k=1}^{l} \int_{\zeta_k}^{\zeta_{k+1}} \sigma_1^{(k)} F_\tau^{\alpha_i} H_2 \, d\zeta, \quad
N_{21}^{(\tau)\alpha_i} = \sum_{k=1}^{l} \int_{\zeta_k}^{\zeta_{k+1}} \tau_{12}^{(k)} F_\tau^{\alpha_i} H_1 \, d\zeta,
$$

$$
P_1^{(\tau)\alpha_i} = \sum_{k=1}^{l} \int_{\zeta_k}^{\zeta_{k+1}} \tau_{13}^{(k)} \frac{\partial F_\tau^{\alpha_i}}{\partial \zeta} H_1 H_2 \, d\zeta
$$

$$
N_2^{(\tau)\alpha_i} = \sum_{k=1}^{l} \int_{\zeta_k}^{\zeta_{k+1}} \sigma_2^{(k)} F_\tau^{\alpha_i} H_1 \, d\zeta, \quad
T_1^{(\tau)\alpha_i} = \sum_{k=1}^{l} \int_{\zeta_k}^{\zeta_{k+1}} \tau_{13}^{(k)} F_\tau^{\alpha_i} H_2 \, d\zeta,
\tag{1.47}
$$

$$
P_2^{(\tau)\alpha_i} = \sum_{k=1}^{l} \int_{\zeta_k}^{\zeta_{k+1}} \tau_{23}^{(k)} \frac{\partial F_\tau^{\alpha_i}}{\partial \zeta} H_1 H_2 \, d\zeta
$$

$$
N_{12}^{(\tau)\alpha_i} = \sum_{k=1}^{l} \int_{\zeta_k}^{\zeta_{k+1}} \tau_{12}^{(k)} F_\tau^{\alpha_i} H_2 \, d\zeta, \quad
T_2^{(\tau)\alpha_i} = \sum_{k=1}^{l} \int_{\zeta_k}^{\zeta_{k+1}} \tau_{23}^{(k)} F_\tau^{\alpha_i} H_1 \, d\zeta,
$$

$$
S_3^{(\tau)\alpha_i} = \sum_{k=1}^{l} \int_{\zeta_k}^{\zeta_{k+1}} \sigma_3^{(k)} \frac{\partial F_\tau^{\alpha_i}}{\partial \zeta} H_1 H_2 \, d\zeta
$$

Based on the unified expression of Eqn. (1.29), the ESL version of the linear elastic behaviour for an anisotropic-laminated shell provides, for each τ-th kinematic expansion order, the expression of the generalized stress resultant vector $\mathbf{S}^{(\tau)\alpha_i}$ in terms of the generalized strain vector $\boldsymbol{\varepsilon}^{(\eta)\alpha_i}$ defined in Eqn. (1.39):

$$\mathbf{S}^{(\tau)\alpha_i} = \sum_{\eta=0}^{N+1}\sum_{j=1}^{3} \mathbf{A}^{(\tau\eta)\alpha_i\alpha_j}\boldsymbol{\varepsilon}^{(\eta)\alpha_j} \quad \text{for} \quad \begin{array}{l} \tau = 0,...,N+1, \\ \alpha_i = \alpha_1,\alpha_2,\alpha_3 \end{array} \tag{1.48}$$

For $\tau,\eta = 0,...,N+1$ and $\alpha_i,\alpha_j = \alpha_1,\alpha_2,\alpha_3$, the generalized stiffness matrix $\mathbf{A}^{(\tau\eta)\alpha_i\alpha_j}$ is defined as follows [17]:

$$\mathbf{A}^{(\tau\eta)\alpha_i\alpha_j} = \sum_{k=1}^{l}\left(\mathbf{Z}^{(\tau)\alpha_i}\right)^T \overline{\mathbf{E}}^{(k)}\mathbf{Z}^{(\eta)\alpha_j} H_1 H_2 d\zeta \quad \text{for} \quad \begin{array}{l} \tau,\eta = 0,...,N+1, \\ \alpha_i,\alpha_j = \alpha_1,\alpha_2,\alpha_3 \end{array} \tag{1.49}$$

As can be seen, the dimension of the matrix $\mathbf{A}^{(\tau\eta)\alpha_i\alpha_j}$ is 9×9 and assumes the following extended version:

$\mathbf{A}^{(\tau\eta)\alpha_i\alpha_j} =$

$$\begin{bmatrix}
A_{11(20)11}^{(\tau\eta)[00]\alpha_i\alpha_j} & A_{12(11)12}^{(\tau\eta)[00]\alpha_i\alpha_j} & A_{16(20)13}^{(\tau\eta)[00]\alpha_i\alpha_j} & A_{16(11)14}^{(\tau\eta)[00]\alpha_i\alpha_j} & A_{14(20)}^{(\tau\eta)[00]\alpha_i\alpha_j} & A_{15(11)}^{(\tau\eta)[00]\alpha_i\alpha_j} & A_{14(10)}^{(\tau\eta)[01]\alpha_i\alpha_j} & A_{15(10)}^{(\tau\eta)[01]\alpha_i\alpha_j} & A_{13(10)}^{(\tau\eta)[01]\alpha_i\alpha_j} \\
A_{12(11)}^{(\tau\eta)[00]\alpha_i\alpha_j} & A_{22(02)}^{(\tau\eta)[00]\alpha_i\alpha_j} & A_{26(11)}^{(\tau\eta)[00]\alpha_i\alpha_j} & A_{26(02)}^{(\tau\eta)[00]\alpha_i\alpha_j} & A_{24(11)}^{(\tau\eta)[00]\alpha_i\alpha_j} & A_{25(02)}^{(\tau\eta)[00]\alpha_i\alpha_j} & A_{24(01)}^{(\tau\eta)[01]\alpha_i\alpha_j} & A_{25(01)}^{(\tau\eta)[01]\alpha_i\alpha_j} & A_{23(01)}^{(\tau\eta)[01]\alpha_i\alpha_j} \\
A_{16(20)}^{(\tau\eta)[00]\alpha_i\alpha_j} & A_{26(11)}^{(\tau\eta)[00]\alpha_i\alpha_j} & A_{66(20)}^{(\tau\eta)[00]\alpha_i\alpha_j} & A_{66(11)}^{(\tau\eta)[00]\alpha_i\alpha_j} & A_{46(20)}^{(\tau\eta)[00]\alpha_i\alpha_j} & A_{56(11)}^{(\tau\eta)[00]\alpha_i\alpha_j} & A_{46(10)}^{(\tau\eta)[01]\alpha_i\alpha_j} & A_{56(10)}^{(\tau\eta)[01]\alpha_i\alpha_j} & A_{36(10)}^{(\tau\eta)[01]\alpha_i\alpha_j} \\
A_{16(11)}^{(\tau\eta)[00]\alpha_i\alpha_j} & A_{26(02)}^{(\tau\eta)[00]\alpha_i\alpha_j} & A_{66(11)}^{(\tau\eta)[00]\alpha_i\alpha_j} & A_{66(02)}^{(\tau\eta)[00]\alpha_i\alpha_j} & A_{46(11)}^{(\tau\eta)[00]\alpha_i\alpha_j} & A_{56(02)}^{(\tau\eta)[00]\alpha_i\alpha_j} & A_{46(01)}^{(\tau\eta)[01]\alpha_i\alpha_j} & A_{56(01)}^{(\tau\eta)[01]\alpha_i\alpha_j} & A_{36(01)}^{(\tau\eta)[01]\alpha_i\alpha_j} \\
A_{14(20)}^{(\tau\eta)[00]\alpha_i\alpha_j} & A_{24(11)}^{(\tau\eta)[00]\alpha_i\alpha_j} & A_{46(20)}^{(\tau\eta)[00]\alpha_i\alpha_j} & A_{46(11)}^{(\tau\eta)[00]\alpha_i\alpha_j} & A_{44(20)}^{(\tau\eta)[00]\alpha_i\alpha_j} & A_{45(11)}^{(\tau\eta)[00]\alpha_i\alpha_j} & A_{44(10)}^{(\tau\eta)[01]\alpha_i\alpha_j} & A_{45(10)}^{(\tau\eta)[01]\alpha_i\alpha_j} & A_{34(10)}^{(\tau\eta)[01]\alpha_i\alpha_j} \\
A_{15(11)}^{(\tau\eta)[00]\alpha_i\alpha_j} & A_{25(02)}^{(\tau\eta)[00]\alpha_i\alpha_j} & A_{56(11)}^{(\tau\eta)[00]\alpha_i\alpha_j} & A_{56(02)}^{(\tau\eta)[00]\alpha_i\alpha_j} & A_{45(11)}^{(\tau\eta)[00]\alpha_i\alpha_j} & A_{55(02)}^{(\tau\eta)[00]\alpha_i\alpha_j} & A_{45(10)}^{(\tau\eta)[01]\alpha_i\alpha_j} & A_{55(01)}^{(\tau\eta)[01]\alpha_i\alpha_j} & A_{35(01)}^{(\tau\eta)[01]\alpha_i\alpha_j} \\
A_{14(10)}^{(\tau\eta)[10]\alpha_i\alpha_j} & A_{24(01)}^{(\tau\eta)[10]\alpha_i\alpha_j} & A_{46(10)}^{(\tau\eta)[10]\alpha_i\alpha_j} & A_{46(01)}^{(\tau\eta)[10]\alpha_i\alpha_j} & A_{44(10)}^{(\tau\eta)[10]\alpha_i\alpha_j} & A_{45(10)}^{(\tau\eta)[10]\alpha_i\alpha_j} & A_{44(00)}^{(\tau\eta)[11]\alpha_i\alpha_j} & A_{45(00)}^{(\tau\eta)[11]\alpha_i\alpha_j} & A_{34(00)}^{(\tau\eta)[11]\alpha_i\alpha_j} \\
A_{15(10)}^{(\tau\eta)[10]\alpha_i\alpha_j} & A_{25(01)}^{(\tau\eta)[10]\alpha_i\alpha_j} & A_{56(10)}^{(\tau\eta)[10]\alpha_i\alpha_j} & A_{56(01)}^{(\tau\eta)[10]\alpha_i\alpha_j} & A_{45(10)}^{(\tau\eta)[10]\alpha_i\alpha_j} & A_{55(10)}^{(\tau\eta)[10]\alpha_i\alpha_j} & A_{45(00)}^{(\tau\eta)[11]\alpha_i\alpha_j} & A_{55(00)}^{(\tau\eta)[11]\alpha_i\alpha_j} & A_{35(00)}^{(\tau\eta)[11]\alpha_i\alpha_j} \\
A_{13(10)}^{(\tau\eta)[10]\alpha_i\alpha_j} & A_{23(01)}^{(\tau\eta)[10]\alpha_i\alpha_j} & A_{36(10)}^{(\tau\eta)[10]\alpha_i\alpha_j} & A_{36(01)}^{(\tau\eta)[10]\alpha_i\alpha_j} & A_{34(10)}^{(\tau\eta)[10]\alpha_i\alpha_j} & A_{35(01)}^{(\tau\eta)[10]\alpha_i\alpha_j} & A_{34(00)}^{(\tau\eta)[11]\alpha_i\alpha_j} & A_{35(00)}^{(\tau\eta)[11]\alpha_i\alpha_j} & A_{33(00)}^{(\tau\eta)[11]\alpha_i\alpha_j}
\end{bmatrix}$$

$$\tag{1.50}$$

The arbitrary element of $\mathbf{A}^{(\tau\eta)\alpha_i\alpha_j}$, denoted by $A_{nm(pq)}^{(\tau\eta)[fg]\alpha_i\alpha_j}$, can be computed as [17]:

$$A_{nm(pq)}^{(\tau\eta)[fg]\alpha_i\alpha_j} = \sum_{k=1}^{l}\int_{\zeta_k}^{\zeta_{k+1}} \overline{B}_{nm}^{(k)}\frac{\partial^f F_\eta^{\alpha_j}}{\partial\zeta^f}\frac{\partial^g F_\tau^{\alpha_i}}{\partial\zeta^g}\frac{H_1 H_2}{H_1^p H_2^q}d\zeta \quad \text{for} \quad \begin{array}{l} \tau,\eta = 0,...,N+1 \\[4pt] n,m = 1,...,6 \\[4pt] p,q = 0,1,2 \\[4pt] \alpha_i,\alpha_j = \alpha_1,\alpha_2,\alpha_3 \\[4pt] f,g = 0,1 \end{array} \tag{1.51}$$

In Eqn. (1.51), coefficient $\overline{B}_{nm}^{(k)}$ for $k = 1,...,l$ and $n,m = 1,..,6$ accounts for the rotated three-dimensional stiffness constants $\overline{E}_{nm}^{(k)}$ of the matrix introduced in Eqn. (1.45). In the case of constant through-the-thickness out-of-plane displacement field assumptions within Eqn. (1.29), namely $F_0^{\alpha_3} = 1$ and $F_\tau^{\alpha_3} = 0$ for $\tau = 1,...,N+1$, the shear correction factor $\kappa(\zeta)$ should be adopted:

$$\overline{B}_{nm}^{(k)} = \begin{cases} \overline{E}_{nm}^{(k)} & \text{for} \quad n,m = 1,2,3,6 \\ \kappa(\zeta)\overline{E}_{nm}^{(k)} & \text{for} \quad n,m = 4,5 \end{cases} \tag{1.52}$$

In the present contribution, a constant value has been assumed so that $\kappa(\zeta) = 5/6$.

In the case of orthotropic materials, the mechanical properties are provided with respect to the material reference system in terms of the well-known engineering constants of the k-th lamina, namely the Young Moduli $E_1^{(k)}, E_2^{(k)}, E_3^{(k)}$, shear Moduli $G_{12}^{(k)}, G_{13}^{(k)}, G_{23}^{(k)}$ and Poisson's coefficients $v_{12}^{(k)}, v_{13}^{(k)}, v_{23}^{(k)}$. For this reason, the corresponding elastic coefficients $C_{ij}^{(k)}$ with $i,j = 1,...,6$ and $k = 1,...,l$ can be computed starting from the well-known relation [17], together with the quantity $\Delta^{(k)}$, defined as:

$$\Delta^{(k)} = \frac{1 - v_{12}^{(k)}v_{21}^{(k)} - v_{23}^{(k)}v_{32}^{(k)} - v_{13}^{(k)}v_{31}^{(k)} - 2v_{21}^{(k)}v_{13}^{(k)}v_{32}^{(k)}}{E_1^{(k)}E_2^{(k)}E_3^{(k)}} \tag{1.53}$$

where

$$C_{11}^{(k)} = \frac{1 - v_{23}^{(k)}v_{32}^{(k)}}{\Delta^{(k)}E_2^{(k)}E_3^{(k)}}, \quad C_{22}^{(k)} = \frac{1 - v_{13}^{(k)}v_{31}^{(k)}}{\Delta^{(k)}E_1^{(k)}E_3^{(k)}}, \quad C_{33}^{(k)} = \frac{1 - v_{12}^{(k)}v_{21}^{(k)}}{\Delta^{(k)}E_1^{(k)}E_2^{(k)}}$$

$$C_{12}^{(k)} = \frac{v_{21}^{(k)} + v_{23}^{(k)}v_{31}^{(k)}}{\Delta^{(k)}E_2^{(k)}E_3^{(k)}} = \frac{v_{12}^{(k)} + v_{32}^{(k)}v_{13}^{(k)}}{\Delta^{(k)}E_1^{(k)}E_3^{(k)}}$$

$$C_{13}^{(k)} = \frac{v_{31}^{(k)} + v_{21}^{(k)}v_{32}^{(k)}}{\Delta^{(k)}E_2^{(k)}E_3^{(k)}} = \frac{v_{13}^{(k)} + v_{12}^{(k)}v_{23}^{(k)}}{\Delta^{(k)}E_1^{(k)}E_2^{(k)}}$$

$$C_{23}^{(k)} = \frac{v_{32}^{(k)} + v_{12}^{(k)}v_{31}^{(k)}}{\Delta^{(k)}E_1^{(k)}E_3^{(k)}} = \frac{v_{23}^{(k)} + v_{21}^{(k)}v_{13}^{(k)}}{\Delta^{(k)}E_1^{(k)}E_2^{(k)}}$$

$$C_{44}^{(k)} = G_{13}^{(k)}, \quad C_{55}^{(k)} = G_{23}^{(k)}, \quad C_{66}^{(k)} = G_{12}^{(k)}$$

$$C_{16}^{(k)} = C_{26}^{(k)} = C_{14}^{(k)} = C_{24}^{(k)} = C_{46}^{(k)} = C_{15}^{(k)} = C_{25}^{(k)} = C_{56}^{(k)} = C_{45}^{(k)} = C_{36}^{(k)} = C_{34}^{(k)} = C_{35}^{(k)} = 0 \tag{1.54}$$

Taking into account the generalized constitutive law of Eqn. (1.48), the generalized stress components can be expressed, for each $\tau = 0,...,N+1$, in terms of the generalized displacement field $\mathbf{u}^{(\tau)} = \begin{bmatrix} u_1^{(\tau)} & u_2^{(\tau)} & u_3^{(\tau)} \end{bmatrix}^T$ according to the following relation:

$$\mathbf{S}^{(\tau)\alpha_i} = \sum_{\eta=0}^{N+1}\sum_{j=1}^{3} \mathbf{A}^{(\tau\eta)\alpha_i\alpha_j} \mathbf{D}_\Omega^{\alpha_j} \mathbf{u}^{(\eta)} = \sum_{\eta=0}^{N+1}\sum_{j=1}^{3} \mathbf{O}^{(\tau\eta)\alpha_i\alpha_j} \mathbf{u}^{(\eta)} \quad \text{for} \quad \begin{matrix} \tau = 0,...,N+1, \\ \alpha_i = \alpha_1,\alpha_2,\alpha_3 \end{matrix} \quad (1.55)$$

The interested reader can refer to [17] for the extended expression of coefficients in matrix $\mathbf{O}^{(\tau\eta)\alpha_i\alpha_j}$.

1.6 GOVERNING EQUATIONS

In the present section, the fundamental differential relations are derived for the dynamic analysis of curved structures made of generally anisotropic materials.

In order to seek the complete set of governing equations, a procedure based on the computation of the Hamiltonian Principle [17] is followed, accounting for the total energy of the structure, defined by the internal elastic deformation energy Φ and kinetic energy T. Referring to a time interval $[t_1,t_2]$, one gets:

$$\delta \int_{t_1}^{t_2} (\Phi - T)dt = 0 \rightarrow \int_{t_1}^{t_2} \delta\Phi\, dt - \int_{t_1}^{t_2} \delta T\, dt = 0 \qquad (1.56)$$

Hereafter, the two integrals in Eqn. (1.56) are derived following the ESL framework, according to Eqn. (1.29).

1.6.1 ELASTIC STRAIN ENERGY

The first energetic contribution that should be included within the Hamiltonian Principle of Eqn. (1.56) accounts for the deformability of the doubly curved shell object of analysis. In particular, the total virtual variation of the elastic strain energy $\delta\Phi$ can be computed within the ESL framework in terms of the previously introduced generalized strain vector $\boldsymbol{\varepsilon}^{(\tau)\alpha_i}(\alpha_1,\alpha_2,t)$ and the generalized stress resultants $\mathbf{S}^{(\tau)\alpha_i}(\alpha_1,\alpha_2,t)$ lying on the shell reference surface $\mathbf{r}(\alpha_1,\alpha_2)$:

$$\delta\Phi = \sum_{\tau=0}^{N+1}\sum_{i=1}^{3} \iint_{\alpha_1\alpha_2} \left(\delta\boldsymbol{\varepsilon}^{(\tau)\alpha_i}\right)^T \mathbf{S}^{(\tau)\alpha_i} A_1 A_2\, d\alpha_1 d\alpha_2 \qquad (1.57)$$

Substituting the unified assessment of the displacement field (29) in Eqn. (1.57) and applying the integration by parts, $\delta\Phi$ is directly written in terms of $u_1^{(\tau)}, u_2^{(\tau)}, u_3^{(\tau)}$. One gets [17]:

$$\delta\Phi = \sum_{\tau=0}^{N+1}\int_{\alpha_1}\int_{\alpha_2}\left(-\left(\frac{\partial\left(N_1^{(\tau)\alpha_1}A_2\right)}{\partial\alpha_1} + \frac{\partial\left(N_{21}^{(\tau)\alpha_1}A_1\right)}{\partial\alpha_2} + N_{12}^{(\tau)\alpha_1}\frac{\partial A_1}{\partial\alpha_2} - N_2^{(\tau)\alpha_1}\frac{\partial A_2}{\partial\alpha_1} + \left(\frac{T_1^{(\tau)\alpha_1}}{R_1} - P_1^{(\tau)\alpha_1}\right)A_1A_2\right)\delta u_1^{(\tau)} + \right.$$

$$-\left(\frac{\partial\left(N_2^{(\tau)\alpha_2}A_1\right)}{\partial\alpha_2} + \frac{\partial\left(N_{12}^{(\tau)\alpha_2}A_2\right)}{\partial\alpha_1} + N_{21}^{(\tau)\alpha_2}\frac{\partial A_2}{\partial\alpha_1} - N_1^{(\tau)\alpha_2}\frac{\partial A_1}{\partial\alpha_2} + \left(\frac{T_2^{(\tau)\alpha_2}}{R_2} - P_2^{(\tau)\alpha_2}\right)A_1A_2\right)\delta u_2^{(\tau)} +$$

$$\left.-\left(\frac{\partial\left(T_1^{(\tau)\alpha_3}A_2\right)}{\partial\alpha_1} + \frac{\partial\left(T_2^{(\tau)\alpha_3}A_1\right)}{\partial\alpha_2} - \left(\frac{N_1^{(\tau)\alpha_3}}{R_1} + \frac{N_2^{(\tau)\alpha_3}}{R_2}\right)A_1A_2 - S_3^{(\tau)\alpha_3}A_1A_2\right)\delta u_3^{(\tau)}\right)d\alpha_1 d\alpha_2 + \tag{1.58}$$

$$+\sum_{\tau=0}^{N+1}\oint_{\alpha_2}\left(N_1^{(\tau)\alpha_1}\delta u_1^{(\tau)} + N_{12}^{(\tau)\alpha_2}\delta u_2^{(\tau)} + T_1^{(\tau)\alpha_3}\delta u_3^{(\tau)}\right)A_2\,d\alpha_2$$

$$+\sum_{\tau=0}^{N+1}\oint_{\alpha_1}\left(N_{21}^{(\tau)\alpha_1}\delta u_1^{(\tau)} + N_2^{(\tau)\alpha_2}\delta u_2^{(\tau)} + T_2^{(\tau)\alpha_3}\delta u_3^{(\tau)}\right)A_1\,d\alpha_1$$

1.6.2 KINETIC ENERGY

We now focus on the computation of the virtual variation of the kinetic energy T of a doubly curved shell within the higher-order ESL formulation. For a laminated structure composed by l laminae, if we denote with $\rho^{(k)}$ the density of the k-th layer of the lamination scheme, the following relation can be stated for a 3D solid, setting $\dot{\mathbf{U}}$ for the time derivative of the displacement vector with the general components denoted by $\dot{U}_i = \partial U_i/\partial t$ for $i = 1,2,3$:

$$\delta T = \sum_{k=1}^{l}\int_{\zeta_k}^{\zeta_{k+1}}\int_{\alpha_1\alpha_2}\rho^{(k)}\left(\delta\dot{\mathbf{U}}\right)^T\dot{\mathbf{U}}\,H_1 H_2 A_1 A_2 d\alpha_1 d\alpha_2 d\zeta \tag{1.59}$$

Referring to Eqn. (1.56) and applying the integration by parts, Eqn. (1.59) leads to the following relation:

$$\int_{t_1}^{t_2}\delta T\,dt = -\sum_{k=1}^{l}\int_{t_1}^{t_2}\int_{\zeta_k}^{\zeta_{k+1}}\int_{\alpha_1\alpha_2}\rho^{(k)}\left(\delta\mathbf{U}\right)^T\ddot{\mathbf{U}}\,H_1 H_2 A_1 A_2 d\alpha_1 d\alpha_2 d\zeta\,dt \tag{1.60}$$

where $\ddot{\mathbf{U}}$ refers to the acceleration vector with components $\ddot{U}_i = \partial^2 U_i/\partial t^2$, for $i = 1,2,3$. Based on the kinematic definition (29), Eqn. (1.59) can be defined in a unified setting [17]:

$$\int_{t_1}^{t_2}\delta T\,dt = -\sum_{i=1}^{3}\sum_{\tau=0}^{N+1}\int_{t_1}^{t_2}\int_{\alpha_1\alpha_2}\left(\left(\sum_{\eta=0}^{N+1}I_{ii}^{(\tau\eta)\alpha_i\alpha_i}\ddot{u}_i^{(\eta)}\right)\delta u_1^{(\tau)}\right)A_1 A_2 d\alpha_1 d\alpha_2 dt \tag{1.61}$$

with $\ddot{u}_i^{(\eta)} = \partial^2 u_i^{(\eta)}/\partial t^2$ being the second-order time derivative of the generalized displacement field components $u_i^{(\eta)} = u_1^{(\eta)}, u_2^{(\eta)}, u_3^{(\eta)}$ for $\eta = 0,...,N+1$. The generalized inertial terms $I_{ii}^{(\tau\eta)\alpha_i\alpha_i}$ for each $\tau, \eta = 0, ..., N+1$ and $i = 1, 2, 3$ can be computed according to the following expression:

$$I_{ij}^{(\tau\eta)\alpha_i\alpha_j} = \sum_{k=1}^{l} \int_{\zeta_k}^{\zeta_{k+1}} \rho^{(k)} F_\tau^{\alpha_i} F_\eta^{\alpha_j} H_1 H_2 \, d\zeta \quad \text{for} \quad i,j = 1,2,3, \quad \tau,\eta = 0, ..., N+1 \quad (1.62)$$

1.6.3 FUNDAMENTAL SET OF EQUATIONS

By combination of Eqns. (1.56), (1.57) and (1.61), we obtain the equilibrium equations for a doubly curved shell according to higher-order theories:

$$\sum_{i=1}^{3} \mathbf{D}_\Omega^{*\alpha_i} \mathbf{S}^{(\tau)\alpha_i} = \sum_{\eta=0}^{N+1} \mathbf{M}^{(\tau\eta)} \ddot{\mathbf{u}}^{(\eta)} \quad \text{for} \quad \tau = 0, ..., N+1 \quad (1.63)$$

$\mathbf{M}^{(\tau\eta)}$ being the generalized inertial matrix for each $\tau, \eta = 0, ..., N+1$, which reads as follows:

$$\mathbf{M}^{(\tau\eta)} = \begin{bmatrix} I_{11}^{(\tau\eta)\alpha_1\alpha_1} & 0 & 0 \\ 0 & I_{22}^{(\tau\eta)\alpha_2\alpha_2} & 0 \\ 0 & 0 & I_{33}^{(\tau\eta)\alpha_3\alpha_3} \end{bmatrix} \quad \text{for} \quad \tau, \eta = 0, ..., N+1 \quad (1.64)$$

where the coefficients $I_{ii}^{(\tau\eta)\alpha_i\alpha_i} = I_{11}^{(\tau\eta)\alpha_1\alpha_1}, I_{22}^{(\tau\eta)\alpha_2\alpha_2}, I_{33}^{(\tau\eta)\alpha_3\alpha_3}$ are computed according to Eqn. (1.62). On the other hand, the equilibrium operators $\mathbf{D}_\Omega^{*\alpha_i}$ for $i = 1, 2, 3$ read as [17]:

$$\mathbf{D}_\Omega^{*\alpha_1} = \begin{bmatrix} \bar{\mathbf{D}}_\Omega^{*\alpha_1} \\ 0 \\ 0 \end{bmatrix}, \quad \mathbf{D}_\Omega^{*\alpha_2} = \begin{bmatrix} 0 \\ \bar{\mathbf{D}}_\Omega^{*\alpha_2} \\ 0 \end{bmatrix}, \quad \mathbf{D}_\Omega^{*\alpha_3} = \begin{bmatrix} 0 \\ 0 \\ \bar{\mathbf{D}}_\Omega^{*\alpha_3} \end{bmatrix} \quad (1.65)$$

where the differential operators $\bar{\mathbf{D}}_\Omega^{*\alpha_i}$ can be expressed as:

$$\bar{\mathbf{D}}_\Omega^{*\alpha_1} = \begin{bmatrix} \dfrac{1}{A_1}\dfrac{\partial}{\partial\alpha_1} + \dfrac{1}{A_1 A_2}\dfrac{\partial A_2}{\partial\alpha_1} \\ -\dfrac{1}{A_1 A_2}\dfrac{\partial A_2}{\partial\alpha_1} \\ \dfrac{1}{A_1 A_2}\dfrac{\partial A_1}{\partial\alpha_2} \\ \dfrac{1}{A_2}\dfrac{\partial}{\partial\alpha_2} + \dfrac{1}{A_1 A_2}\dfrac{\partial A_1}{\partial\alpha_2} \\ \dfrac{1}{R_1} \\ 0 \\ -1 \\ 0 \\ 0 \end{bmatrix}^T, \quad \bar{\mathbf{D}}_\Omega^{*\alpha_2} = \begin{bmatrix} -\dfrac{1}{A_1 A_2}\dfrac{\partial A_1}{\partial\alpha_2} \\ \dfrac{1}{A_2}\dfrac{\partial}{\partial\alpha_2} + \dfrac{1}{A_1 A_2}\dfrac{\partial A_1}{\partial\alpha_2} \\ \dfrac{1}{A_1}\dfrac{\partial}{\partial\alpha_1} + \dfrac{1}{A_1 A_2}\dfrac{\partial A_2}{\partial\alpha_1} \\ \dfrac{1}{A_1 A_2}\dfrac{\partial A_2}{\partial\alpha_1} \\ 0 \\ \dfrac{1}{R_2} \\ 0 \\ -1 \\ 0 \end{bmatrix}^T, \quad \bar{\mathbf{D}}_\Omega^{*\alpha_3} = \begin{bmatrix} -\dfrac{1}{R_1} \\ -\dfrac{1}{R_2} \\ 0 \\ 0 \\ \dfrac{1}{A_1}\dfrac{\partial}{\partial\alpha_1} + \dfrac{1}{A_1 A_2}\dfrac{\partial A_2}{\partial\alpha_1} \\ \dfrac{1}{A_2}\dfrac{\partial}{\partial\alpha_2} + \dfrac{1}{A_1 A_2}\dfrac{\partial A_1}{\partial\alpha_2} \\ 0 \\ 0 \\ -1 \end{bmatrix}^T$$

$$(1.66)$$

Rearranging Eqn. (1.63) so that the generalized displacement field vector $\mathbf{u}^{(\eta)}$ for $\eta = 0,...,N+1$ is outlined, one gets the final form of the fundamental governing equations:

$$\sum_{\eta=0}^{N+1} \mathbf{L}^{(\tau\eta)} \mathbf{u}^{(\eta)} = \sum_{\eta=0}^{N+1} \mathbf{M}^{(\tau\eta)} \ddot{\mathbf{u}}^{(\eta)} \qquad \text{for } \tau = 0,...,N+1 \tag{1.67}$$

In a more expanded form, Eqn. (1.67) reads as:

$$\begin{bmatrix} \mathbf{L}^{(00)} & \mathbf{L}^{(01)} & \cdots & \mathbf{L}^{(0(N))} & \mathbf{L}^{(0(N+1))} \\ \mathbf{L}^{(10)} & \mathbf{L}^{(11)} & \cdots & \mathbf{L}^{(1(N))} & \mathbf{L}^{(1(N+1))} \\ \vdots & \vdots & \ddots & \vdots & \vdots \\ \mathbf{L}^{((N)0)} & \mathbf{L}^{((N)1)} & \cdots & \mathbf{L}^{((N)(N))} & \mathbf{L}^{((N)(N+1))} \\ \mathbf{L}^{((N+1)0)} & \mathbf{L}^{((N+1)1)} & \cdots & \mathbf{L}^{((N+1)(N))} & \mathbf{L}^{((N+1)(N+1))} \end{bmatrix} \begin{bmatrix} \mathbf{u}^{(0)} \\ \mathbf{u}^{(1)} \\ \vdots \\ \mathbf{u}^{(N)} \\ \mathbf{u}^{(N+1)} \end{bmatrix} =$$

$$= \begin{bmatrix} \mathbf{M}^{(00)} & \mathbf{M}^{(01)} & \cdots & \mathbf{M}^{(0(N))} & \mathbf{M}^{(0(N+1))} \\ \mathbf{M}^{(10)} & \mathbf{M}^{(11)} & \cdots & \mathbf{M}^{(1(N))} & \mathbf{M}^{(1(N+1))} \\ \vdots & \vdots & \ddots & \vdots & \vdots \\ \mathbf{M}^{((N)0)} & \mathbf{M}^{((N)1)} & \cdots & \mathbf{M}^{((N)(N))} & \mathbf{M}^{((N)(N+1))} \\ \mathbf{M}^{((N+1)0)} & \mathbf{M}^{((N+1)1)} & \cdots & \mathbf{M}^{((N+1)(N))} & \mathbf{M}^{((N+1)(N+1))} \end{bmatrix} \begin{bmatrix} \ddot{\mathbf{u}}^{(0)} \\ \ddot{\mathbf{u}}^{(1)} \\ \vdots \\ \ddot{\mathbf{u}}^{(N)} \\ \ddot{\mathbf{u}}^{(N+1)} \end{bmatrix} \tag{1.68}$$

For an arbitrary order of the kinematic expansion $\tau, \eta = 0,...,N+1$, the fundamental matrix $\mathbf{L}^{(\tau\eta)}$ is a 3×3 matrix of the following type, see [17]:

$$\mathbf{L}^{(\tau\eta)} = \sum_{i=1}^{3} \sum_{j=1}^{3} \mathbf{D}_{\Omega}^{*\alpha_i} \mathbf{A}^{(\tau\eta)\alpha_i\alpha_j} \mathbf{D}_{\Omega}^{\alpha_j} = \begin{bmatrix} L_{11}^{(\tau\eta)\alpha_1\alpha_1} & L_{12}^{(\tau\eta)\alpha_1\alpha_2} & L_{13}^{(\tau\eta)\alpha_1\alpha_3} \\ L_{21}^{(\tau\eta)\alpha_2\alpha_1} & L_{22}^{(\tau\eta)\alpha_2\alpha_2} & L_{23}^{(\tau\eta)\alpha_2\alpha_3} \\ L_{31}^{(\tau\eta)\alpha_3\alpha_1} & L_{32}^{(\tau\eta)\alpha_3\alpha_2} & L_{33}^{(\tau\eta)\alpha_3\alpha_3} \end{bmatrix} \tag{1.69}$$

$$\text{for } \tau, \eta = 0,...,N+1$$

For a free vibration analysis problem, for each $\tau, \eta = 0,...,N+1$, Eqn. (1.67) should be solved using harmonic solutions. From the application of the variables separation method, the generalized displacement field vector $\mathbf{u}^{(\eta)}(\alpha_1, \alpha_2, t)$ takes the following form:

$$\mathbf{u}^{(\eta)}(\alpha_1, \alpha_2, t) = \bar{\mathbf{U}}^{(\eta)}(\alpha_1, \alpha_2) e^{i\omega t} \tag{1.70}$$

where ω is the circular frequency, and $\bar{\mathbf{U}}^{(n)}(\alpha_1, \alpha_2)$ is the corresponding mode shape vector. By substitution of Eqn. (1.70) into Eqn. (1.67), the following time-independent condensed relation is provided for the modal analysis of anisotropic shells:

$$\sum_{\eta=0}^{N+1} \mathbf{L}^{(\tau\eta)} \bar{\mathbf{U}}^{(\eta)} + \omega^2 \sum_{\eta=0}^{N+1} \mathbf{M}^{(\tau\eta)} \bar{\mathbf{U}}^{(\eta)} = \mathbf{0} \qquad \text{for } \tau = 0, ..., N+1 \qquad (1.71)$$

Accordingly, if we denote with f the natural frequency, it is $\omega = 2\pi f$.

1.7   NUMERICAL IMPLEMENTATION

In this section, the fundamental relations of Eqn. (1.71) are numerically tackled via the GDQ method [17]. Referring to a generic univariate function $f(x)$ defined in a closed interval $[a, b]$ it should be recalled that the GDQ procedure accounts for the derivative of an arbitrary n-th order as a weighted sum of the values assumed by the function itself in a set of points alongside the definition interval. For a discrete grid made of I_N points, the value of the n-th order derivative of f in $x = x_i$ with $i = 1, ..., I_N$ can be computed as:

$$\left. \frac{\partial^n f(x)}{\partial x^n} \right|_{x=x_i} \cong \sum_{j=1}^{I_N} \varsigma_{ij}^{(n)} f(x_j) \qquad i = 1, 2, \ldots, I_N \qquad (1.72)$$

where $f(x_j)$ is the value assumed by the unknown function in the discrete points of the computational grid, setting $j = 1, ..., I_N$. The quadrature weighting coefficients $\varsigma_{ij}^{(n)}$ in Eqn. (1.72) for $i, j = 1, ..., I_N$ are computed starting from the well-known recursive formula based on the Weierstrass interpolation theorem, applied to the polynomials $\mathcal{L}(r)$ defined in an interval $[c, d]$. Based on a generalized approach, one gets the following relations [17]:

$$\tilde{\varsigma}_{ij}^{(0)} = \delta_{ij} = \begin{cases} 0, & i \neq j \\ 1, & i = j \end{cases}$$

$$\tilde{\varsigma}_{ij}^{(1)} = \frac{\mathcal{L}^{(1)}(r_i)}{(r_i - r_j)\mathcal{L}^{(1)}(r_j)}, \quad i \neq j, n = 1$$

$$\tilde{\varsigma}_{ij}^{(n)} = n\left(\tilde{\varsigma}_{ij}^{(1)}\tilde{\varsigma}_{ii}^{(n-1)} - \frac{\tilde{\varsigma}_{ij}^{(n-1)}}{r_i - r_j} \right), \quad i \neq j, n \geq 2$$

$$\tilde{\varsigma}_{ii}^{(n)} = -\sum_{j=1 \ j \neq i}^{I_N} \tilde{\varsigma}_{ij}^{(n)}, \quad i = j, n \geq 1$$

(1.73)

where $r_i, r_j \in [c,d]$ for $i, j = 1,...,I_N$ refer to the discrete set of collocation points. For a function $f(x)$ defined in the domain $[a,b]$, it is:

$$x_k(r) = \frac{x_{I_N} - x_1}{r_{I_N} - r_1}(r_k - r_1) + x_1, \quad k = 1,...,I_N \tag{1.74}$$

with $r_k \in [c,d]$ for $k = 1,...,I_N$, whereas $\varsigma_{ij}^{(n)}$ in Eqn. (1.72) takes the following form:

$$\varsigma_{ij}^{(n)} = \left(\frac{r_{I_N} - r_1}{x_{I_N} - x_1}\right)^n \tilde{\varsigma}_{ij}^{(n)} \quad \text{for} \quad i, j = 1,...,I_N \tag{1.75}$$

Starting from the GDQ generalized expression (72), the GIQ method [17] is introduced for the numerical assessment of integrals. Thus, the integral of a univariate function $f(x)$ defined in a closed interval $x \in [a,b]$ is computed as follows:

$$\int_a^b f(x)dx = \sum_{j=1}^{I_N} w_j^{1I_N} f(x_j) \tag{1.76}$$

where $f(x_j)$ for $j = 1,...,I_N$ accounts for the values assumed by the function in a discrete set of points belonging to the definition interval, namely $x_j \in [a,b]$, whereas the weighting coefficients $w_j^{1I_N}$ are defined as:

$$w_j^{1I_N} = w_{I_N j} - w_{1j} \tag{1.77}$$

To integrate the same function f within another interval $[c,d]$, the GIQ weighting coefficients $w_j^{1I_N}$ provided by Eqn. (1.77) can be turned into $\tilde{w}_j^{1I_N}$, so that they are referred to the actual integration limits, such that:

$$w_j^{1I_N} = \frac{b-a}{c-d} \tilde{w}_j^{1I_N} \tag{1.78}$$

The weighting coefficients $w_{1j}, w_{I_N j}$ in Eqn. (1.77) are defined from the GDQ ones $\tilde{\varsigma}_{ij}^{(1)}$ from Eqn. (1.73) referred to the $[c,d]$ interval, i.e.

$$\bar{\varsigma}_{ij}^{(1)} = \begin{cases} \dfrac{r_i - \varepsilon}{r_j - \varepsilon} \tilde{\varsigma}_{ij}^{(1)} & \text{for} \quad i \neq j \\[3mm] \dfrac{1}{r_i - \varepsilon} + \tilde{\varsigma}_{ij}^{(1)} & \text{for} \quad i = j \end{cases} \tag{1.79}$$

for each $i, j = 1,...I_N$, setting $r_i, r_j \in [c,d]$ and $\varepsilon = 1 \times 10^{-10}$. Such coefficients from Eqn. (1.79) are, thus, collected in a $I_N \times I_N$ matrix denoted by $\bar{\varsigma}^{(1)}$. At the same time, the GIQ coefficients matrix $\mathbf{W}$ can be computed in a unified manner as:

$$\mathbf{W} = \left(\bar{\varsigma}^{(1)}\right)^{-1} \tag{1.80}$$

whose arbitrary element denoted by w_{ij} for $i, j = 1, ..., I_N$. In this chapter, we apply a non-uniform Chebyshev-Gauss-Lobatto (CGL) distribution [17] to define the computational grid. Referring to a dimensionless interval $[-1, 1]$, its generic element $r_k^{(Q)}$ reads as follows, setting $k = 1, ..., I_Q$ and $Q = N, M, T$:

$$r_k^{(Q)} = -\cos\left(\frac{k-1}{I_Q-1}\pi\right) \quad \text{for} \quad k = 1, ..., I_Q, \quad r_k^{(Q)} \in [-1, 1], \quad Q = N, M, T \quad (1.81)$$

In particular, the discrete computational grid for the ESL problem at issue should be defined in the computational domain $[-1, 1] \times [-1, 1]$ described by the natural coordinates ξ_1, ξ_2. Moreover, a one-dimensional set of discrete points for each layer is defined by the thickness coordinate $\zeta_j^{(k)} \in \left[\zeta_1^{(k)}, \zeta_{I_T}^{(k)}\right] = [\zeta_k, \zeta_{k+1}]$, where $\xi_{1j}, \xi_{2j}, \zeta_j^{(k)}$ are defined in the physical domain as:

$$\xi_{1j} = r_j^{(N)} \qquad \text{for} \quad j = 1, ..., I_N$$

$$\xi_{2j} = r_j^{(M)} \qquad \text{for} \quad j = 1, ..., I_M \qquad (1.82)$$

$$\zeta_j^{(k)} = \frac{\zeta_{I_T}^{(k)} - \zeta_1^{(k)}}{r_{I_T}^{(T)} - r_1^{(T)}}\left(r_j^{(T)} - r_1^{(T)}\right) + \zeta_1^{(k)} \qquad \text{for} \quad j = 1, ..., I_T$$

Once the discrete version of the fundamental set of equations [Eqn. (1.71)] is derived, it is possible to assess a DOFs condensation so that the dimension of the emerging linear system is conveniently reduced, accounting for an effective numerical implementation of boundary conditions [17]. To this purpose, all DOFs related to the boundary points, denoted by "b", should be separated from those referred to the domain inner points, identified with subscript "d":

$$\bar{\mathbf{L}}\delta = \omega^2 \bar{\mathbf{M}}\delta \quad \Leftrightarrow \quad \begin{bmatrix} \bar{\mathbf{L}}_{bb} & \bar{\mathbf{L}}_{bd} \\ \bar{\mathbf{L}}_{db} & \bar{\mathbf{L}}_{dd} \end{bmatrix} \begin{bmatrix} \delta_b \\ \delta_d \end{bmatrix} = \omega^2 \begin{bmatrix} \mathbf{0} & \mathbf{0} \\ \mathbf{0} & \bar{\mathbf{M}}_{dd} \end{bmatrix} \begin{bmatrix} \delta_b \\ \delta_d \end{bmatrix} \qquad (1.83)$$

Accordingly, the global size of the fundamental and mass matrixes $\bar{\mathbf{L}}$ and $\bar{\mathbf{M}}$, respectively, is equal to $3I_N I_M (N+2) \times 3I_N I_M (N+2)$, whereas the modal amplitude vector δ accounts for a column vector of dimensions $3I_N I_M (N+2) \times 1$. Eliminating the sub-vectors δ_b from Eqn. (1.83), it gives:

$$\left(\bar{\mathbf{M}}_{dd}^{-1}\left(\bar{\mathbf{L}}_{dd} - \bar{\mathbf{L}}_{db}\bar{\mathbf{L}}_{bb}^{-1}\bar{\mathbf{L}}_{bd}\right) - \omega^2 \mathbf{I}\right)\delta_d = \mathbf{0} \qquad (1.84)$$

where $\mathbf{I}$ is the identity matrix of proper dimensions.

1.8 APPLICATIONS AND RESULTS

In this section, a series of examples are presented to study the free vibration response of different shell structures with different lamination schemes and curvatures, as provided by an ESL formulation accounting for various kinematic assumptions in a generalized setting. Accordingly, the NURBS-based mapping has been adopted for the physical domain distortion, accounting for both straight- and curve-mapped edges. For each case study, the dynamic behaviour of the structure has been investigated under two different external constraints configurations. Accordingly, Clamped (C) edges have been provided with all the DOFs fixed [17], namely:

$$u_n^{(\tau)} = u_s^{(\tau)} = u_\zeta^{(\tau)} = 0 \quad \text{for } \tau = 0,...,N+1, \quad \text{at } \xi_1 = -1 \text{ or } \xi_1 = 1, \ -1 \leq \xi_2 \leq 1$$

$$u_n^{(\tau)} = u_s^{(\tau)} = u_\zeta^{(\tau)} = 0 \quad \text{for } \tau = 0,...,N+1, \quad \text{at } \xi_2 = -1 \text{ or } \xi_2 = 1, \ -1 \leq \xi_1 \leq 1$$

$$(1.85)$$

On the other hand, for a Free (F) boundary condition, the generalized stress resultants acting at the shell edge have been enforced to be null, i.e. [17]:

$$N_n^{(\tau)} = 0, \ N_{ns}^{(\tau)} = 0, \ T_\zeta^{(\tau)} = 0 \quad \text{for } \tau = 0,...,N+1, \quad \text{at } \xi_1 = -1 \text{ or } \xi_1 = 1, \ -1 \leq \xi_2 \leq 1$$

$$N_{ns}^{(\tau)} = 0, \ N_n^{(\tau)} = 0, \ T_\zeta^{(\tau)} = 0 \quad \text{for } \tau = 0,...,N+1, \quad \text{at } \xi_2 = -1 \text{ or } \xi_2 = 1, \ -1 \leq \xi_1 \leq 1$$

$$(1.86)$$

Since the proposed formulation accounts for a general assessment of the lamination schemes with arbitrary layer orientation and mechanical symmetries, different materials have been employed to check for the accuracy of the formulation with respect to various layer couplings. Moreover, generally anisotropic continua as well as orthotropic and isotropic materials have been considered, the mechanical properties of which have been directly set within the DiQuMASPAB database [81]. The first class of materials refers to a triclinic material with density $\rho^{(k)} = 7750 \, \text{kg/m}^3$ and a three-dimensional stiffness matrix defined as:

$$\mathbf{C}^{(k)} = \begin{bmatrix} C_{11}^{(k)} & C_{12}^{(k)} & C_{16}^{(k)} & C_{14}^{(k)} & C_{15}^{(k)} & C_{13}^{(k)} \\ C_{12}^{(k)} & C_{22}^{(k)} & C_{26}^{(k)} & C_{24}^{(k)} & C_{25}^{(k)} & C_{23}^{(k)} \\ C_{16}^{(k)} & C_{26}^{(k)} & C_{66}^{(k)} & C_{46}^{(k)} & C_{56}^{(k)} & C_{36}^{(k)} \\ C_{14}^{(k)} & C_{24}^{(k)} & C_{46}^{(k)} & C_{44}^{(k)} & C_{45}^{(k)} & C_{34}^{(k)} \\ C_{15}^{(k)} & C_{25}^{(k)} & C_{56}^{(k)} & C_{45}^{(k)} & C_{55}^{(k)} & C_{35}^{(k)} \\ C_{13}^{(k)} & C_{23}^{(k)} & C_{36}^{(k)} & C_{34}^{(k)} & C_{35}^{(k)} & C_{33}^{(k)} \end{bmatrix} \quad (1.87)$$

$$= \begin{bmatrix} 98.84 & 53.92 & 0.03 & 1.05 & -0.1 & 50.78 \\ 53.92 & 99.19 & 0.03 & 0.55 & -0.18 & 50.87 \\ 0.03 & 0.03 & 22.55 & -0.04 & 0.25 & 0.02 \\ 1.05 & 0.55 & -0.04 & 21.1 & 0.07 & 1.03 \\ -0.1 & -0.18 & 0.25 & 0.07 & 21.14 & -0.18 \\ 50.78 & 50.87 & 0.02 & 1.03 & -0.18 & 87.23 \end{bmatrix} \text{GPa}$$

where $C_{ij}^{(k)}$ coefficients, for $i, j = 1,...,6$, have been referred to the material reference system according to the notation introduced in Eqn. (1.42), setting $\mathbf{E}^{(k)} = \mathbf{C}^{(k)}$. In the same way, a trigonal material $\left(\rho^{(k)} = 2649 \, \text{kg/m}^3\right)$ has been introduced:

$$\mathbf{C}^{(k)} = \begin{bmatrix} C_{11}^{(k)} & C_{12}^{(k)} & C_{16}^{(k)} & C_{14}^{(k)} & C_{15}^{(k)} & C_{13}^{(k)} \\ C_{12}^{(k)} & C_{22}^{(k)} & C_{26}^{(k)} & C_{24}^{(k)} & C_{25}^{(k)} & C_{23}^{(k)} \\ C_{16}^{(k)} & C_{26}^{(k)} & C_{66}^{(k)} & C_{46}^{(k)} & C_{56}^{(k)} & C_{36}^{(k)} \\ C_{14}^{(k)} & C_{24}^{(k)} & C_{46}^{(k)} & C_{44}^{(k)} & C_{45}^{(k)} & C_{34}^{(k)} \\ C_{15}^{(k)} & C_{25}^{(k)} & C_{56}^{(k)} & C_{45}^{(k)} & C_{55}^{(k)} & C_{35}^{(k)} \\ C_{13}^{(k)} & C_{23}^{(k)} & C_{36}^{(k)} & C_{34}^{(k)} & C_{35}^{(k)} & C_{33}^{(k)} \end{bmatrix} \tag{1.88}$$

$$= \begin{bmatrix} 86.74 & 6.99 & 0 & 0 & -17.91 & 11.91 \\ 6.99 & 86.74 & 0 & 0 & 17.91 & 11.91 \\ 0 & 0 & 39.88 & -17.91 & 0 & 0 \\ 0 & 0 & -17.91 & 57.94 & 0 & 0 \\ -17.91 & 17.91 & 0 & 0 & 57.94 & 0 \\ 11.91 & 11.91 & 0 & 0 & 0 & 107.20 \end{bmatrix} \text{GPa}$$

The Graphite–Epoxy orthotropic material mechanical properties have been expressed in terms of the well-known engineering constants, referring to the material reference system:

$$E_1^{(k)} = 137.90 \, \text{GPa} \qquad G_{12}^{(k)} = 7.10 \, \text{GPa} \qquad v_{12}^{(k)} = 0.30$$
$$E_2^{(k)} = 8.96 \, \text{GPa} \qquad G_{13}^{(k)} = 7.10 \, \text{GPa} \qquad v_{13}^{(k)} = 0.30 \tag{1.89}$$
$$E_3^{(k)} = 8.96 \, \text{GPa} \qquad G_{23}^{(k)} = 6.21 \, \text{GPa} \qquad v_{23}^{(k)} = 0.49$$

the density of which is equal to $\rho^{(k)} = 1450 \, \text{kg/m}^3$. A Glass–Epoxy material $\left(\rho^{(k)} = 1900 \, \text{kg/m}^3\right)$ has been also considered in the analyses:

$$E_1^{(k)} = 53.78 \, \text{GPa} \qquad G_{12}^{(k)} = 8.96 \, \text{GPa} \qquad v_{12}^{(k)} = 0.25$$
$$E_2^{(k)} = 17.93 \, \text{GPa} \qquad G_{13}^{(k)} = 8.96 \, \text{GPa} \qquad v_{13}^{(k)} = 0.25 \tag{1.90}$$
$$E_3^{(k)} = 17.93 \, \text{GPa} \qquad G_{23}^{(k)} = 3.45 \, \text{GPa} \qquad v_{23}^{(k)} = 0.34$$

The last cases consider the following isotropic materials, expressed in terms of the elastic modulus $E^{(k)}$ and Poisson's ratio $v^{(k)}$:

$$\text{Concrete} \quad \rightarrow \quad E^{(k)} = 25 \, \text{GPa} \qquad v^{(k)} = 0.20 \qquad \rho^{(k)} = 1800 \, \text{kg/m}^3$$
$$\text{Aluminium} \quad \rightarrow \quad E^{(k)} = 70 \, \text{GPa} \qquad v^{(k)} = 0.30 \qquad \rho^{(k)} = 2707 \, \text{kg/m}^3 \tag{1.91}$$
$$\text{Zirconia} \quad \rightarrow \quad E^{(k)} = 168 \, \text{GPa} \qquad v^{(k)} = 0.30 \qquad \rho^{(k)} = 5700 \, \text{kg/m}^3$$

Both straight and curved shells considered in this section are characterized by a variable thickness along the two principal directions of the physical domain, following the approach outlined in Eqn. (1.8). For each selected structure, all the geometry and mechanical information are reported in Figures 1.2–1.6.

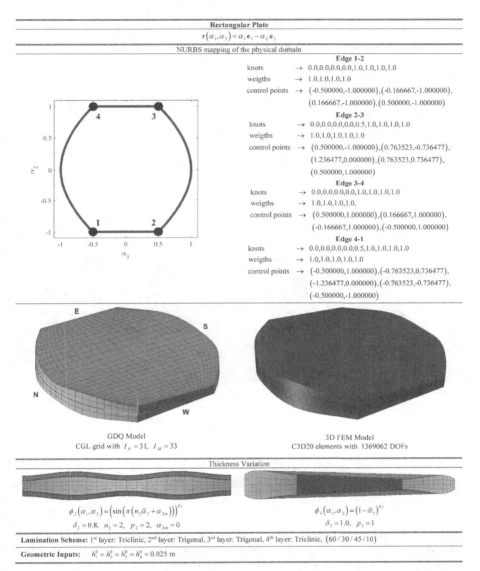

Rectangular Plate

$$r(\alpha_1,\alpha_2) = \alpha_1\,\mathbf{e}_1 - \alpha_2\,\mathbf{e}_2$$

NURBS mapping of the physical domain

Edge 1-2

knots → 0.0,0.0,0.0,0.0,0.0,1.0,1.0,1.0,1.0

weigths → 1.0,1.0,1.0,1.0

control points → $(-0.500000,-1.000000),(-0.166667,-1.000000),$

$(0.166667,-1.000000),(0.500000,-1.000000)$

Edge 2-3

knots → 0.0,0.0,0.0,0.0,0.0,0.5,1.0,1.0,1.0,1.0

weigths → 1.0,1.0,1.0,1.0,1.0

control points → $(0.500000,-1.000000),(0.763523,-0.736477),$

$(1.236477,0.000000),(0.763523,0.736477),$

$(0.500000,1.000000)$

Edge 3-4

knots → 0.0,0.0,0.0,0.0,0.0,1.0,1.0,1.0,1.0

weigths → 1.0,1.0,1.0,1.0,

control points → $(0.500000,1.000000),(0.166667,1.000000),$

$(-0.166667,1.000000),(-0.500000,1.000000)$

Edge 4-1

knots → 0.0,0.0,0.0,0.0,0.0,0.5,1.0,1.0,1.0,1.0

weigths → 1.0,1.0,1.0,1.0,1.0

control points → $(-0.500000,1.000000),(-0.763523,0.736477),$

$(-1.236477,0.000000),(-0.763523,-0.736477),$

$(-0.500000,-1.000000)$

GDQ Model
CGL grid with $I_N = 31$, $I_M = 33$

3D FEM Model
C3D20 elements with 1369062 DOFs

Thickness Variation

$$\phi_2(\alpha_1,\alpha_2) = \left(\sin\left(\pi\left(n_2\bar{\alpha}_2 + \alpha_{2m}\right)\right)\right)^{p_2}$$

$$\delta_2 = 0.8,\ n_2 = 2,\ p_2 = 2,\ \alpha_{2m} = 0$$

$$\phi_3(\alpha_1,\alpha_2) = \left(1-\bar{\alpha}_1\right)^{p_3}$$

$$\delta_3 = 1.0,\ p_3 = 1$$

Lamination Scheme: 1st layer: Triclinic, 2nd layer: Trigonal, 3rd layer: Trigonal, 4th layer: Triclinic, $(60/30/45/10)$

Geometric Inputs: $h_1^0 = h_2^0 = h_3^0 = h_4^0 = 0.025$ m

FIGURE 1.2 Mechanical and geometric properties of a rectangular plate of variable thickness and arbitrary shape laminated with generally anisotropic lamination scheme. NURBS description of the edges of the structure. Representation of the GDQ model developed with the present formulation and 3D FEM parabolic brick mesh employed for the calculation of the reference solution in terms of mode frequencies and shapes.

Skew Cone

$$\mathbf{r}(\alpha_1,\alpha_2) = R_0(\alpha_1)\cos\alpha_2\,\mathbf{e}_1 - R_0(\alpha_1)\sin\alpha_2\,\mathbf{e}_2 + \alpha_1\cos\varphi\,\mathbf{e}_3$$

$$R_0(\alpha_1) = R_b + \alpha_1\sin\varphi$$

NURBS mapping of the physical domain

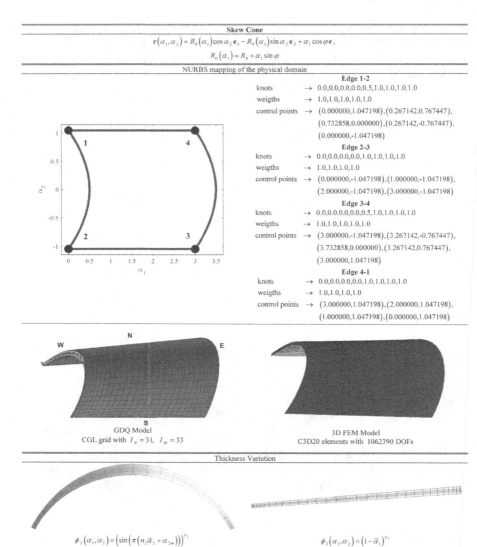

Edge 1-2

knots → 0.0,0.0,0.0,0.0,0.0,0.5,1.0,1.0,1.0,1.0,1.0

weigths → 1.0,1.0,1.0,1.0,1.0,1.0

control points → (0.000000,1.047198),(0.267142,0.767447),

(0.732858,0.000000),(0.267142,-0.767447),

(0.000000,-1.047198)

Edge 2-3

knots → 0.0,0.0,0.0,0.0,1.0,1.0,1.0,1.0

weigths → 1.0,1.0,1.0,1.0

control points → (0.000000,-1.047198),(1.000000,-1.047198),

(2.000000,-1.047198),(3.000000,-1.047198)

Edge 3-4

knots → 0.0,0.0,0.0,0.0,0.0,0.5,1.0,1.0,1.0,1.0,1.0

weigths → 1.0,1.0,1.0,1.0,1.0,1.0

control points → (3.000000,-1.047198),(3.267142,-0.767447),

(3.732858,0.000000),(3.267142,0.767447),

(3.000000,1.047198)

Edge 4-1

knots → 0.0,0.0,0.0,0.0,1.0,1.0,1.0,1.0

weigths → 1.0,1.0,1.0,1.0

control points → (3.000000,1.047198),(2.000000,1.047198),

(1.000000,1.047198),(0.000000,1.047198)

GDQ Model
CGL grid with $I_N = 31$, $I_M = 33$

3D FEM Model
C3D20 elements with 1062390 DOFs

Thickness Variation

$$\phi_2(\alpha_1,\alpha_2) = \left(\sin\left(\pi\left(n_2\bar{\alpha}_2 + \alpha_{2m}\right)\right)\right)^{p_2}$$

$$\delta_2 = 1.5, \quad n_2 = 1, \quad p_2 = 2, \quad \alpha_{2m} = 0$$

$$\phi_3(\alpha_1,\alpha_2) = \left(1 - \bar{\alpha}_1\right)^{p_3}$$

$$\delta_3 = 1.0, \quad p_3 = 1$$

Lamination Scheme: 1st layer: Graphite - Epoxy, 2nd layer: Triclinic, 3rd layer: Triclinic, 4th layer: Glass - Epoxy, $(0/45/45/30)$

Geometric Inputs: $R_b = 1.5$ m, $\varphi = 5$ deg , $h_1^0 = h_4^0 = 0.02$ m, $h_2^0 = h_5^0 = 0.01$ m

FIGURE 1.3 Mechanical and geometric properties of a skew cone of variable thickness and arbitrary shape laminated with generally anisotropic lamination scheme. NURBS description of the edges of the structure. Representation of the GDQ model developed with the present formulation and 3D FEM parabolic brick mesh employed for the calculation of the reference solution in terms of mode frequencies and shapes.

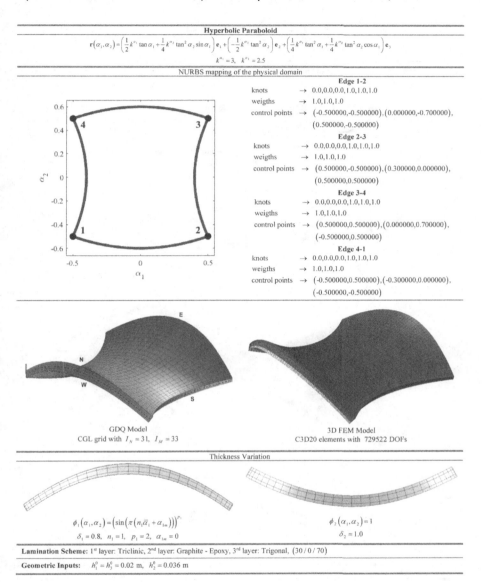

Hyperbolic Paraboloid

$$\mathbf{r}\left(\alpha_1,\alpha_2\right) = \left(\frac{1}{2}k^{\alpha_1}\tan\alpha_1 + \frac{1}{4}k^{\alpha_1}\tan^2\alpha_2\sin\alpha_1\right)\mathbf{e}_1 + \left(-\frac{1}{2}k^{\alpha_1}\tan^2\alpha_2\right)\mathbf{e}_2 + \left(\frac{1}{4}k^{\alpha_1}\tan^2\alpha_1 + \frac{1}{4}k^{\alpha_2}\tan^2\alpha_2\cos\alpha_1\right)\mathbf{e}_3$$

$$k^{\alpha_1} = 3,\quad k^{\alpha_2} = 2.5$$

NURBS mapping of the physical domain

Edge 1-2

knots → 0.0,0.0,0.0,1.0,1.0,1.0

weigths → 1.0,1.0,1.0

control points → (-0.500000,-0.500000),(0.000000,-0.700000),(0.500000,-0.500000)

Edge 2-3

knots → 0.0,0.0,0.0,1.0,1.0,1.0

weigths → 1.0,1.0,1.0

control points → (0.500000,-0.500000),(0.300000,0.000000),(0.500000,0.500000)

Edge 3-4

knots → 0.0,0.0,0.0,1.0,1.0,1.0

weigths → 1.0,1.0,1.0

control points → (0.500000,0.500000),(0.000000,0.700000),(-0.500000,0.500000)

Edge 4-1

knots → 0.0,0.0,0.0,1.0,1.0,1.0

weigths → 1.0,1.0,1.0

control points → (-0.500000,0.500000),(-0.300000,0.000000),(-0.500000,-0.500000)

GDQ Model
CGL grid with $I_N = 31$, $I_M = 33$

3D FEM Model
C3D20 elements with 729522 DOFs

Thickness Variation

$$\phi_1\left(\alpha_1,\alpha_2\right) = \left(\sin\left(\pi\left(n_1\bar{\alpha}_1 + \alpha_{1m}\right)\right)\right)^{p_1}$$

$$\delta_1 = 0.8,\quad n_1 = 1,\quad p_1 = 2,\quad \alpha_{1m} = 0$$

$$\phi_2\left(\alpha_1,\alpha_2\right) = 1$$

$$\delta_2 = 1.0$$

Lamination Scheme: 1st layer: Triclinic, 2nd layer: Graphite - Epoxy, 3rd layer: Trigonal, $(30/0/70)$

Geometric Inputs: $h_1^0 = h_3^0 = 0.02$ m, $h_2^0 = 0.036$ m

FIGURE 1.4 Mechanical and geometric properties of a hyperbolic hyperboloid of variable thickness and arbitrary shape laminated with generally anisotropic lamination scheme. NURBS description of the edges of the structure. Representation of the GDQ model developed with the present formulation and 3D FEM parabolic brick mesh employed for the calculation of the reference solution in terms of mode frequencies and shapes.

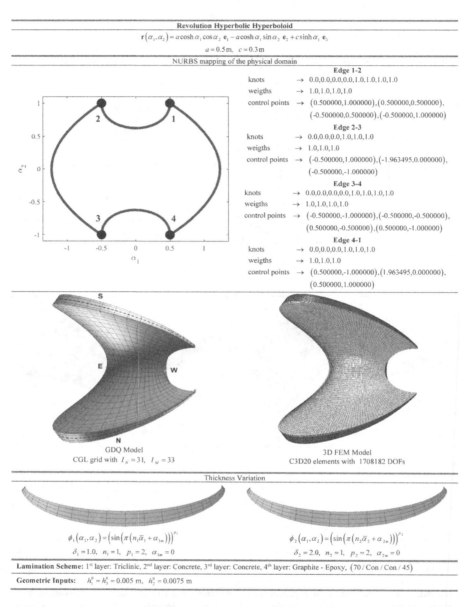

Revolution Hyperbolic Hyperboloid

$$\mathbf{r}(\alpha_1,\alpha_2)=a\cosh\alpha_1\cos\alpha_2\ \mathbf{e}_1-a\cosh\alpha_1\sin\alpha_2\ \mathbf{e}_2+c\sinh\alpha_1\ \mathbf{e}_3$$

$$a=0.5\,\text{m},\quad c=0.3\,\text{m}$$

NURBS mapping of the physical domain

Edge 1-2

knots → 0.0,0.0,0.0,0.0,1.0,1.0,1.0,1.0

weigths → 1.0,1.0,1.0,1.0

control points → $(0.500000,1.000000),(0.500000,0.500000),$
$(-0.500000,0.500000),(-0.500000,1.000000)$

Edge 2-3

knots → 0.0,0.0,0.0,1.0,1.0,1.0

weigths → 1.0,1.0,1.0

control points → $(-0.500000,1.000000),(-1.963495,0.000000),$
$(-0.500000,-1.000000)$

Edge 3-4

knots → 0.0,0.0,0.0,0.0,1.0,1.0,1.0,1.0

weigths → 1.0,1.0,1.0,1.0

control points → $(-0.500000,-1.000000),(-0.500000,-0.500000),$
$(0.500000,-0.500000),(0.500000,-1.000000)$

Edge 4-1

knots → 0.0,0.0,0.0,1.0,1.0,1.0

weigths → 1.0,1.0,1.0

control points → $(0.500000,-1.000000),(1.963495,0.000000),$
$(0.500000,1.000000)$

GDQ Model
CGL grid with $I_N=31,\ I_M=33$

3D FEM Model
C3D20 elements with 1708182 DOFs

Thickness Variation

$$\phi_1(\alpha_1,\alpha_2)=\left(\sin\left(\pi\left(n_1\bar{\alpha}_1+\alpha_{1m}\right)\right)\right)^{p_1}$$
$$\delta_1=1.0,\ n_1=1,\ p_1=2,\ \alpha_{1m}=0$$

$$\phi_2(\alpha_1,\alpha_2)=\left(\sin\left(\pi\left(n_2\bar{\alpha}_2+\alpha_{2m}\right)\right)\right)^{p_2}$$
$$\delta_2=2.0,\ n_2=1,\ p_2=2,\ \alpha_{2m}=0$$

Lamination Scheme: 1st layer: Triclinic, 2nd layer: Concrete, 3rd layer: Concrete, 4th layer: Graphite - Epoxy, $(70/\text{Con}/\text{Con}/45)$

Geometric Inputs: $h_1^0=h_3^0=0.005$ m, $h_2^0=0.0075$ m

FIGURE 1.5 Mechanical and geometric properties of a revolution hyperbolic hyperboloid of variable thickness and arbitrary shape laminated with generally anisotropic lamination scheme. NURBS description of the edges of the structure. Representation of the GDQ model developed with the present formulation and 3D FEM parabolic brick mesh employed for the calculation of the reference solution in terms of mode frequencies and shapes.

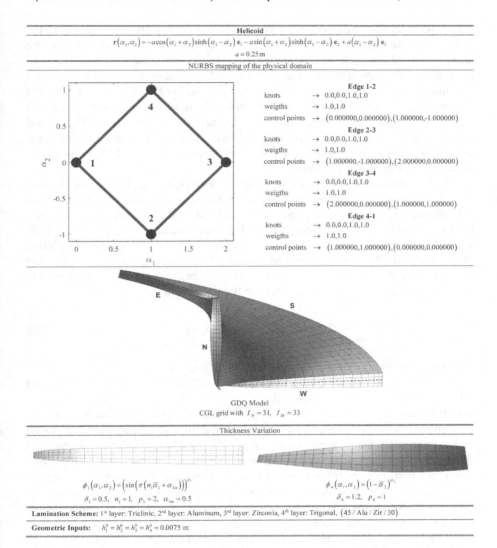

Helicoid

$$r(\alpha_1,\alpha_2) = -a\cos(\alpha_1+\alpha_2)\sinh(\alpha_1-\alpha_2)\,e_1 - a\sin(\alpha_1+\alpha_2)\sinh(\alpha_1-\alpha_2)\,e_2 + a(\alpha_1-\alpha_2)\,e_3$$

$$a = 0.25\,\text{m}$$

NURBS mapping of the physical domain

		Edge 1-2
knots	→	0.0,0.0,1.0,1.0
weigths	→	1.0,1.0
control points	→	$(0.000000,0.000000),(1.000000,-1.000000)$
		Edge 2-3
knots	→	0.0,0.0,1.0,1.0
weigths	→	1.0,1.0
control points	→	$(1.000000,-1.000000),(2.000000,0.000000)$
		Edge 3-4
knots	→	0.0,0.0,1.0,1.0
weigths	→	1.0,1.0
control points	→	$(2.000000,0.000000),(1.000000,1.000000)$
		Edge 4-1
knots	→	0.0,0.0,1.0,1.0
weigths	→	1.0,1.0
control points	→	$(1.000000,1.000000),(0.000000,0.000000)$

GDQ Model
CGL grid with $I_N = 31$, $I_M = 33$

Thickness Variation

$$\phi_1(\alpha_1,\alpha_2) = \left(\sin\left(\pi\left(n_1\bar{\alpha}_1 + \alpha_{1m}\right)\right)\right)^{p_1}$$

$$\delta_1 = 0.5, \quad n_1 = 1, \quad p_1 = 2, \quad \alpha_{1m} = 0.5$$

$$\phi_4(\alpha_1,\alpha_2) = \left(1 - \bar{\alpha}_2\right)^{p_4}$$

$$\delta_4 = 1.2, \quad p_4 = 1$$

Lamination Scheme: 1st layer: Triclinic, 2nd layer: Aluminum, 3rd layer: Zirconia, 4th layer: Trigonal, $(45/\text{Alu}/\text{Zir}/30)$

Geometric Inputs: $h_1^0 = h_2^0 = h_3^0 = h_4^0 = 0.0075\,\text{m}$

FIGURE 1.6 Mechanical and geometric properties of a helicoid of variable thickness and arbitrary shape laminated with generally anisotropic lamination scheme. NURBS description of the edges of the structure. Representation of the GDQ model developed with the present formulation and 3D FEM parabolic brick mesh employed for the calculation of the reference solution in terms of mode frequencies and shapes.

In all configurations, the first ten mode frequencies have been calculated employing Eqn. (1.84), taking into account a CGL two-dimensional computational grid with $I_N \times I_M = 31 \times 33$ defined from Eqn. (1.81). Based on the generalized approach of Eqn. (1.29) for the description of the displacement field, we report the results as provided from various higher-order theories starting from a power through-the-thickness kinematic expansion of Eqn. (1.33). For the same maximum kinematic expansion order N, the analyses have been performed systematically to check for the influence of the zigzag assumption within the displacement field components. We also provide a three-dimensional representation of the mode shapes as obtained from the present ESL approach. The accuracy of our formulation has been verified for a benchmark case, with respect to a reference 3D solution based on a finite element commercial package, employing 20-node brick elements. An excellent accordance among results has been always observed. Despite the higher-order ESL method following a two-dimensional formulation, it predicts a series of three-dimensional warping and coupling effects acting along the thickness of the shell that can be normally traced with refined models.

The first example consists of a rectangular plate characterized by a lamination scheme obtained from the superimposition of four layers of triclinic (87) and trigonal (88) materials with general orientations. A linear and two-wave sinusoidal shell thickness variation has been considered, following the approaches of Eqns. (1.8) and (1.10). The sets of knot, weights and control points for the NURBS description of the mapped edges are reported in Figure 1.2, accounting for two straight and two curved lines. The following equation [17] has been adopted for the geometric description of the reference surface within the ESL framework:

$$\mathbf{r}(\alpha_1, \alpha_2) = \alpha_1 \mathbf{e}_1 - \alpha_2 \mathbf{e}_2 \tag{1.92}$$

The first nine mode frequencies are reported in Table 1.1.

In the first case, a fully clamped (CCCC) layup has been considered. A perfect alignment between all the ESL simulations and the 3D FEM predictions is seen for both lower and higher frequencies even with classical FSDT and TSDT simulations. Accordingly, few discrepancies are pursued when the zigzag function is employed in the kinematic field. Apart from classical approaches, the best agreement is observable for a kinematic expansion order $N = 4$, with a zigzag function. Note also that a first-order approximation of each field variable component with the ED1 theory can get wrong results. On the other hand, the employment of the Murakami's function improves the ESL performances. Referring to the FFCC configuration, the same accuracy level is noticed: higher-order theories provide better results than lower-order ones. Accordingly, since lower frequencies can be predicted with an acceptable precision from the ED2 and EDZ2 theories, for a correct prediction of higher modes, the order of the kinematic expansion should be incremented up to $N = 4$. In Figure 1.7, the first nine mode shapes are reported for the structure at issue. They have been calculated employing the EDZ4 higher-order theory: note that the employment of a generally anisotropic lamination scheme and the presence of a thickness variation influence the modal response of the structure, especially for higher modes.

The next example consists of a cone, a singly curved structure defined by the following expression in terms of the principal coordinates α_1, α_2 [17]:

$$\mathbf{r}(\alpha_1, \alpha_2) = R_0(\alpha_1)\cos\alpha_2 \mathbf{e}_1 - R_0(\alpha_1)\sin\alpha_2 \mathbf{e}_2 + \alpha_1 \cos\varphi \mathbf{e}_3 \tag{1.93}$$

TABLE 1.1

Mode Frequencies of a Rectangular Plate of Variable Thickness with a Generally Anisotropic Lamination Scheme

Mode f[Hz]	3D FEM	FSDT	FSDTZ	TSDT	TSDTZ	Rectangular Plate							
						ED1	EDZ1	ED2	EDZ2	ED3	EDZ3	ED4	EDZ4
DOFs	1369062	5394	8091	10788	13485	5394	8091	8091	10788	10788	13485	13485	16182
						CCCC							
1	287.349	286.190	287.897	287.529	286.827	286.107	311.305	292.442	287.998	294.617	290.228	291.651	289.699
2	552.103	545.264	552.366	551.062	550.130	545.199	596.810	561.148	552.214	568.340	559.266	562.314	558.208
3	597.363	591.157	598.400	597.993	596.583	590.600	636.549	604.476	596.007	613.137	604.627	607.230	603.566
4	877.142	861.786	876.908	875.549	873.952	639.888	932.222	886.472	873.414	902.906	889.467	893.547	887.758
5	893.160	874.898	891.810	889.694	888.409	813.362	951.454	903.023	889.256	921.123	906.833	910.882	904.829
6	1010.470	995.254	1013.279	1012.415	1010.166	861.475	1064.692	1019.700	1006.053	1040.433	1026.425	1030.289	1024.499
7	1150.872	1160.994	1161.197	1153.984	1150.431	874.812	1187.196	1184.151	1179.922	1164.758	1161.118	1162.732	1160.292
8	1190.479	1187.206	1196.815	1192.959	1189.464	993.832	1225.685	1223.484	1208.230	1204.618	1200.980	1202.408	1200.190
9	1214.832	1198.306	1214.635	1212.702	1210.707	1074.150	1281.099	1230.069	1223.201	1255.552	1236.913	1241.655	1234.081
10	1255.718	1224.492	1252.253	1247.686	1245.583	1160.810	1315.746	1262.972	1244.889	1291.704	1272.873	1277.943	1270.138
						FFCC							
1	74.002	74.418	74.377	74.399	74.083	74.403	78.336	75.191	74.206	75.385	74.476	74.490	74.042
2	182.388	183.441	183.818	183.937	182.976	182.830	187.811	184.356	182.480	185.444	183.742	184.857	184.281
3	279.239	278.797	280.078	279.811	279.141	278.527	303.055	284.235	279.894	285.951	281.696	283.220	281.262
4	359.628	360.640	362.404	362.357	360.601	360.272	370.673	362.953	359.104	365.920	362.358	363.061	361.521
5	503.674	501.252	505.823	505.172	503.667	389.573	522.468	510.326	503.614	514.597	508.870	510.952	508.191
6	512.078	516.168	516.486	514.524	512.743	500.559	532.758	523.995	521.612	519.179	516.398	517.488	516.067

(Continued)

TABLE 1.1 *(Continued)*

Mode Frequencies of a Rectangular Plate of Variable Thickness with a Generally Anisotropic Lamination Scheme

Mode f[Hz]	Rectangular Plate												
	3D FEM	FSDT	FSDTZ	TSDT	TSDTZ	ED1	EDZ1	ED2	EDZ2	ED3	EDZ3	ED4	EDZ4
DOFs	1369062	5394	8091	10788	13485	5394	8091	8091	10788	10788	13485	13485	16182
7	605.433	601.145	606.843	606.455	604.957	515.935	648.211	613.376	604.542	620.402	611.626	614.649	610.764
8	642.990	639.671	645.183	644.597	642.328	598.029	677.653	649.996	641.432	656.552	648.298	651.284	647.560
9	677.994	681.827	681.902	680.209	678.499	610.341	701.018	701.304	699.180	686.884	685.161	685.732	685.006
10	739.228	735.972	744.150	743.686	740.751	639.014	762.422	745.342	736.984	755.558	747.540	750.389	747.123

Lamination Scheme: (60/30/45/10), $h_1^0 = h_2^0 = h_3^0 = h_4^0 = 0.0025$ m

Laminae Material Sequence: 1st layer: Triclinic, 2nd layer: Triclinic, 3rd layer: Trigonal, 4th layer: Triclinic

Thickness Variation: $\phi_2(\alpha_1, \alpha_2) = \left(\sin\left(\pi\left(n_2\bar{\alpha}_2 + \alpha_{2m}\right)\right)\right)^{p_2}$, $\delta_2 = 0.8$, $n_2 = 2$, $p_2 = 2$, $\alpha_{2m} = 0$

$\phi_3(\alpha_1, \alpha_2) = \left(1 - \bar{\alpha}_1\right)^{p_3}$, $\delta_3 = 1.0$, $p_3 = 3$

Computational Grid: CGL distribution with $I_N = 31$ and $I_M = 33$ discrete points

Aribtrarily-shaped rectangular plate with variable thickness - mode shapes
(FFCC)

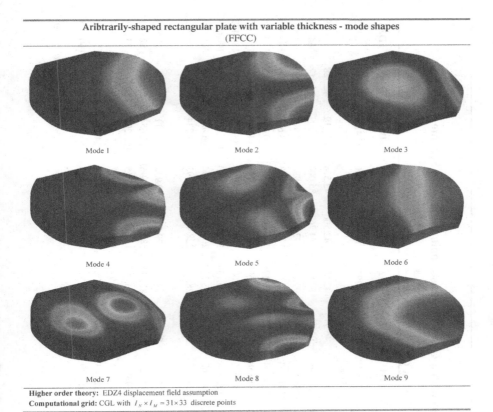

Mode 1   Mode 2   Mode 3

Mode 4   Mode 5   Mode 6

Mode 7   Mode 8   Mode 9

Higher order theory: EDZ4 displacement field assumption
Computational grid: CGL with $I_N \times I_M = 31 \times 33$ discrete points

FIGURE 1.7   Mode shapes of a rectangular plate of variable thickness and arbitrary geometry via the proposed ESL two-dimensional formulation. The employment of the EDZ4 axiomatic assumption allows to predict a series of stretching and warping effects acting along the thickness of the structure.

where $\varphi = 5°$. Moreover, $R_0(\alpha_1)$ is defined as:

$$R_0(\alpha_1) = R_b + \alpha_1 \sin\varphi \qquad (1.94)$$

where $R_b = 1.5$ m. As visible in Figure 1.3, a skew domain has been adopted for the mapping of the structure, characterized by some concave and convex opposite areas. In this case, the lamination scheme is characterized by generally oriented triclinic layers (87) in the core area, whereas the external parts consist of composite materials with mechanical properties from Eqns. (1.89) and (1.90). A power and a single-wave thickness variation has been considered. In Table 1.2, the mode frequencies are listed for the FCFC and CFCF boundary condition configurations. In the latter case, a stable level of accuracy is seen for the ED3 and ED4 theories, without any meaningful improvement coming from the employment of the zigzag function of Eqn. (1.32). Even though lower modes are predicted with a good level of accuracy,

TABLE 1.2

Mode Frequencies of a Skew Cone of Variable Thickness with a Generally Anisotropic Lamination Scheme

Mode f[Hz]	3D FEM	FSDT	FSDTZ	TSDT	TSDTZ	ED1	EDZ1	ED2	EDZ2	ED3	EDZ3	ED4	EDZ4
						Skew Cone							
DOFs	1062390	5394	8091	10788	13485	5394	8091	8091	10788	10788	13485	13485	16182
						FCFC							
1	42.099	42.337	42.420	42.354	42.351	39.247	46.713	42.569	42.506	42.574	42.519	42.528	42.489
2	84.063	84.565	84.840	84.665	84.658	77.746	91.044	84.868	84.762	85.000	84.909	84.917	84.846
3	91.902	92.600	92.835	92.684	92.678	88.851	99.992	92.898	92.790	92.994	92.902	92.908	92.847
4	103.364	103.562	103.698	103.579	103.576	101.368	106.471	103.689	103.642	103.733	103.696	103.688	103.665
5	168.118	169.493	170.143	169.738	169.720	164.254	185.057	170.123	169.882	170.453	170.241	170.273	170.115
6	184.230	184.286	184.599	184.337	184.325	181.858	189.400	184.603	184.515	184.755	184.685	184.653	184.599
7	188.798	188.287	188.690	188.504	188.492	186.576	194.017	188.504	188.354	188.598	188.478	188.512	188.423
8	198.938	198.994	199.428	199.236	199.226	197.338	203.852	199.139	199.004	199.260	199.154	199.176	199.097
9	203.735	205.206	205.983	205.525	205.507	201.562	219.544	205.738	205.515	206.194	206.002	206.021	205.879
10	226.767	228.217	228.812	228.418	228.400	223.117	237.678	228.667	228.471	228.897	228.734	228.733	228.602
						CFCF							
1	77.721	79.134	79.242	79.178	79.174	77.433	81.689	79.850	79.814	79.938	79.911	79.936	79.899
2	79.725	81.917	82.021	81.956	81.953	80.823	84.293	82.616	82.590	82.696	82.654	82.601	82.601
3	123.519	122.950	123.316	123.237	123.234	119.907	127.101	123.108	123.005	123.295	123.209	123.287	123.239
4	149.061	149.543	150.065	149.922	149.918	146.224	153.156	149.978	149.858	150.342	150.235	150.248	150.164
5	154.023	153.870	154.230	154.142	154.142	151.993	159.414	154.084	154.000	154.270	154.191	154.047	153.988
6	158.117	159.476	160.058	159.902	159.897	158.829	162.338	159.896	159.835	160.434	160.360	160.364	160.373

(Continued)

TABLE 1.2 (Continued)
Mode Frequencies of a Skew Cone of Variable Thickness with a Generally Anisotropic Lamination Scheme

Mode f[Hz]	3D FEM	FSDT	FSDTZ	TSDT	TSDTZ	ED1	EDZ1	ED2	EDZ2	ED3	EDZ3	ED4	EDZ4
					Skew Cone								
DOFs	1062390	5394	8091	10788	13485	5394	8091	8091	10788	10788	13485	13485	16182
7	169.536	171.186	171.913	171.707	171.701	169.592	177.161	171.522	171.420	172.141	172.054	172.146	172.079
8	215.990	217.504	218.455	218.147	218.134	212.183	232.200	217.940	217.682	218.512	218.282	218.346	218.224
9	243.053	242.704	244.091	243.759	243.752	240.795	251.303	243.019	242.809	244.092	243.917	243.987	243.875
10	248.316	246.832	248.195	247.890	247.884	244.867	255.182	247.201	247.013	248.287	248.103	247.994	247.872

Lamination Scheme: (0/45/45/30), $h_1^0 = h_4^0 = 0.02$ m, $h_2^0 = h_3^0 = 0.01$ m

Laminae Material Sequence: 1st layer: Graphite–Epoxy, 2nd layer: Graphite–Epoxy, 3rd layer: Triclinic, 4th layer: Graphite–Epoxy

Geometric Inputs: $R_b = 1.5$ m, $\varphi = 5$ deg

Thickness Variation: $\phi_2(\alpha_1,\alpha_2) = \left(\sin\left(\pi\left(n_2\bar{\alpha}_2 + \alpha_{2m}\right)\right)\right)^{p_2}$, $\delta_2 = 1.5$, $n_2 = 1$, $p_2 = 2$, $\alpha_{2m} = 0$

$\phi_3(\alpha_1,\alpha_2) = (1-\bar{\alpha}_1)^{p_3}$, $\delta_3 = 1.0$, $p_3 = 1$

Computational Grid: CGL distribution with $I_N = 31$ and $I_M = 33$ discrete points

better results are observed for higher frequencies with a precision of 0.1 Hz with respect to the 3D FEM-based reference solution. This time, the employment of the Murakami's function does not increase the accuracy of the simulation even with classic FSDT and TSDT theories, which means that the coupling between two adjacent laminae is predicted by a smooth displacement field variation. The FCFC layup is characterized by an excellent level of accuracy for both lower and higher modes when a higher-order displacement field is assumed. Apart from the ED1 and EDZ1 simulations, both the 3D FEM and ESL predictions are in perfect agreement, with a maximum discrepancy of 0.1 Hz. For these simulations, the EDZ4 has been adopted for the implementation of the first nine mode shapes as reported in Figure 1.8. Both in-plane and out-of-plane deformations are also plotted for each mode. Accordingly, in some cases, complex bending and torsional three-dimensional effects are noticed employing different wave numbers.

In Table 1.3, a free vibration analysis of a hyperbolic paraboloid has been performed. The geometry has been assessed within the ESL model starting from the

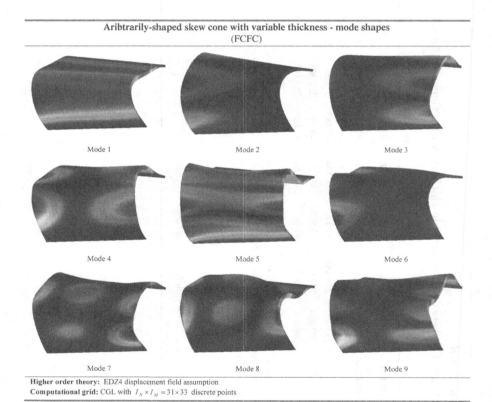

Aribtrarily-shaped skew cone with variable thickness - mode shapes (FCFC)

Mode 1   Mode 2   Mode 3

Mode 4   Mode 5   Mode 6

Mode 7   Mode 8   Mode 9

Higher order theory: EDZ4 displacement field assumption
Computational grid: CGL with $I_N \times I_M = 31 \times 33$ discrete points

FIGURE 1.8 Mode shapes of a skew cone of variable thickness and arbitrary geometry via the proposed ESL two-dimensional formulation. The employment of the EDZ4 axiomatic assumption allows to predict a series of stretching and warping effects acting along the thickness of the structure.

TABLE 1.3

Mode Frequencies of a Hyperbolic Paraboloid of Variable Thickness with a Generally Anisotropic Lamination Scheme

Hyperbolic Paraboloid

Mode f[Hz]	3D FEM	FSDT	FSDTZ	TSDT	TSDTZ	ED1	EDZ1	ED2	EDZ2	ED3	EDZ3	ED4	EDZ4
DOFs	729522	5394	8091	10788	13485	5394	8091	8091	10788	10788	13485	13485	16182
						CCFF							
1	57.548	58.518	57.842	57.895	57.851	58.021	59.694	58.097	57.028	57.807	57.395	57.509	56.880
2	194.641	201.242	195.503	197.496	197.321	194.450	199.175	199.927	192.332	197.892	194.413	197.677	194.966
3	271.353	278.955	274.888	274.752	274.179	262.331	276.697	275.893	271.145	275.207	272.931	274.610	272.797
4	409.120	426.688	412.038	416.862	416.285	403.295	421.249	423.267	405.401	418.584	410.043	417.140	410.164
5	432.593	453.641	435.009	441.594	441.006	431.915	441.698	449.598	426.738	443.167	432.764	441.138	432.599
6	453.363	477.407	458.520	464.740	463.997	467.686	466.948	472.353	450.509	465.701	455.791	463.986	455.718
7	568.132	607.500	572.714	584.954	583.900	558.933	582.154	600.196	561.264	587.147	569.786	584.365	569.869
8	624.766	633.639	626.952	627.691	626.928	602.869	637.594	631.572	621.508	630.984	626.222	629.675	625.807
9	700.481	746.286	702.870	721.604	720.483	701.128	704.190	740.997	686.222	724.239	699.420	720.659	699.717
10	745.209	800.785	749.638	768.579	767.226	754.633	768.947	793.551	733.446	773.921	747.614	769.315	746.252
						CFCF							
1	381.386	384.400	382.328	381.690	380.023	374.561	392.002	383.570	379.937	383.994	382.198	383.266	381.830
2	407.786	413.095	409.221	409.168	406.578	407.862	417.995	411.312	405.749	411.224	408.586	410.621	409.166
3	424.473	445.260	426.316	434.047	424.003	439.764	433.510	442.664	419.335	435.401	425.128	433.457	424.655
4	456.306	484.259	458.472	468.877	455.152	476.756	463.699	478.913	448.337	470.125	456.291	467.914	456.360
5	551.958	567.294	552.977	557.390	548.474	539.355	571.503	564.060	545.724	561.002	551.816	558.641	551.520
6	626.280	671.960	629.002	645.945	623.476	648.194	639.325	663.868	613.323	648.947	626.099	644.984	625.673

(Continued)

TABLE 1.3 *(Continued)*

Mode Frequencies of a Hyperbolic Paraboloid of Variable Thickness with a Generally Anisotropic Lamination Scheme

Mode f[Hz]	3D FEM	FSDT	FSDTZ	TSDT	TSDTZ	ED1	EDZ1	ED2	EDZ2	ED3	EDZ3	ED4	EDZ4
						Hyperbolic Paraboloid							
DOFs	729522	5394	8091	10788	13485	5394	8091	8091	10788	10788	13485	13485	16182
7	729.583	780.961	732.751	757.672	728.410	751.950	739.106	772.857	715.994	761.114	731.360	755.850	730.176
8	748.297	793.850	751.044	771.193	745.670	771.289	754.668	785.680	731.173	773.024	748.191	769.903	747.488
9	781.187	824.670	787.754	792.249	780.585	802.950	788.969	814.549	774.436	794.767	781.843	791.210	781.581
10	818.080	867.003	820.282	837.704	813.086	817.332	850.464	862.236	804.805	844.632	818.889	840.771	819.586

Lamination Scheme: (30/0/70), $h_1^0 = h_3^0 = 0.02$ m, $h_2^0 = 0.036$ m

Laminae Material Sequence: 1st layer: Triclinic, 2nd layer: Graphite–Epoxy, 3rd layer: Trigonal

Geometric Inputs: $k^{\alpha_1} = 3$, $k^{\alpha_2} = 2.5$

Thickness Variation: $\phi_i(\alpha_1, \alpha_2) = \left(\sin\left(\pi\left(n_i\bar\alpha_i + \alpha_{1m}\right)\right)\right)^{p_1}$, $\delta_i = 0.5$, $n_1 = 1$, $p_1 = 2$, $\alpha_{1m} = 0$

Computational Grid: CGL distribution with $I_N = 31$ and $I_M = 33$ discrete points

following parametrization of the reference surface $\mathbf{r}(\alpha_1, \alpha_2)$ taken from [17], setting $k^{\alpha_1} = 3$ and $k^{\alpha_2} = 2.5$:

$$\mathbf{r}(\alpha_1, \alpha_2) = \left(\frac{1}{2} k^{\alpha_1} \tan \alpha_1 + \frac{1}{4} k^{\alpha_2} \tan^2 \alpha_2 \sin \alpha_1 \right) \mathbf{e}_1 + \left(-\frac{1}{2} k^{\alpha_2} \tan^2 \alpha_2 \right) \mathbf{e}_2 +$$

$$+ \left(\frac{1}{4} k^{\alpha_1} \tan^2 \alpha_1 + \frac{1}{4} k^{\alpha_2} \tan^2 \alpha_2 \cos \alpha_1 \right) \mathbf{e}_3 \tag{1.95}$$

A curved mapping of the physical domain has been performed employing the generalized blending functions set of Eqn. (1.16), considering a NURBS description of the shell edges within the physical domain (see Figure 1.4). This time the stacking sequence consists of three layers, and accounts for a central core made of orthotropic Graphite–Epoxy (89), whereas the two external skins are made of a triclinic (87) and a trigonal (88) material applied at the bottom and top of the shell, respectively. A sinusoidal thickness variation has been applied along the α_1 principal direction, whereas a constant thickness is considered along the α_2 coordinate. Referring to the CCFF case, the first nine natural frequencies have been calculated employing ESL theories of different kinematic expansion orders, together with the Murakami's function of Eqn. (1.32). In such a case, the FSDT provides different results from the 3D FEM reference solutions, especially for higher modes. The accuracy improves a lot through the adoption of a zigzag function. A third-order expansion of the in-plane variables provides better results than FSDT, and the introduction of the kinematic field of Eqn. (1.32) for the $(N+1)$-th kinematic expansion order accounts for a high level of accuracy. If a linear description of the out-of-plane displacement field variable is assumed, erroneous results are provided, whereas a higher-order assumption of both in-plane and out-of-plane components accounts for a better agreement with the 3D FEM. The EDZ4 theory predicts exactly both lower- and higher-mode frequencies of the structure. In the second simulation group, two opposite edges of the shell at issue have been constrained (CFCF). Despite an increment of the displacement field expansion improves the accuracy of the simulations, the employment of the Murakami's zigzag function is essential for the correct prediction of eigenvalues referred to lower- and higher-vibration modes. Accordingly, the EDZ3 and EDZ4 theories are capable of predicting all the results provided by the 3D FEM with an excellent level of accuracy. Referring to classical theories where no-stretching effects are depicted, the introduction of the zigzag function improves a lot the accuracy of the numerical predictions, thus reducing the discrepancy from the 3D FEM under 1%. This means that for lower and higher vibrations, the adopted lamination scheme induces a severe coupling between adjacent laminae. Accordingly, the third-order expansion of in-plane components improves the accuracy of the simulation. Since it has been shown that the EDZ4 provides the best agreement with the FEM-based predictions, the first nine mode shapes have been calculated for the structure at issue by means of a fourth kinematic expansion order, together with the zigzag function of Eqn. (1.32). A three-dimensional representation of the modal eigenvectors is plotted in Figure 1.9. In particular, the employment of a higher-order theory along each principal direction provides through-the-thickness warping and

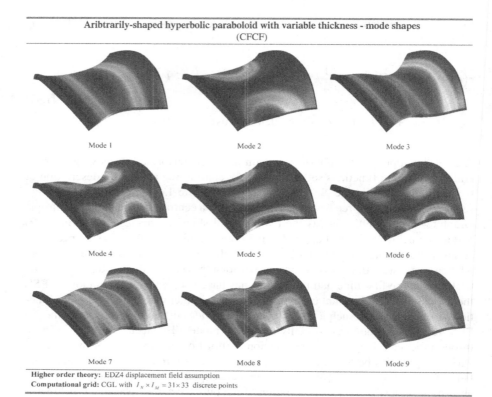

Aribtrarily-shaped hyperbolic paraboloid with variable thickness - mode shapes
(CFCF)

Mode 1   Mode 2   Mode 3

Mode 4   Mode 5   Mode 6

Mode 7   Mode 8   Mode 9

Higher order theory: EDZ4 displacement field assumption
Computational grid: CGL with $I_N \times I_M = 31 \times 33$ discrete points

FIGURE 1.9 Mode shapes of a hyperbolic paraboloid of variable thickness and arbitrary geometry via the proposed ESL two-dimensional formulation. The employment of the EDZ4 axiomatic assumption allows to predict a series of stretching and warping effects acting along the thickness of the structure.

stretching effects, especially when higher modes are considered, along with the bending vibrations.

In Table 1.4, we provide the results for another example, namely a revolution hyperbolic hyperboloid [17], whose geometric assessment has been performed within the ESL framework of Eqn. (1.3), according to the following expression, setting $a = 0.5$ m and $c = 0.3$ m:

$$\mathbf{r}(\alpha_1, \alpha_2) = a \cosh \alpha_1 \cos \alpha_2 \ \mathbf{e}_1 - a \cosh \alpha_1 \sin \alpha_2 \ \mathbf{e}_2 + c \sinh \alpha_1 \ \mathbf{e}_3 \qquad (1.96)$$

The shell thickness variation has been assessed employing a sinusoidal expression for both the principal directions of the structure. In the middle thickness of the structure, two isotropic layers of concrete (91) have been considered. The outer parts of the shell are made instead of a generally anisotropic triclinic layer (87) at the bottom surface level, whereas the top surface is made by an orthotropic Graphite–Epoxy

TABLE 1.4
Mode Frequencies of a Revolution Hyperbolic Hyperboloid of Variable Thickness with a Generally Anisotropic Lamination Scheme

Mode f[Hz]	3D FEM	FSDT	FSDTZ	TSDT	TSDTZ	ED1	EDZ1	ED2	EDZ2	ED3	EDZ3	ED4	EDZ4
						Revolution Hyperbolic Hyperboloid							
DOFs	1708182	5394	8091	10788	13485	5394	8091	8091	10788	10788	13485	13485	16182
						FFFC							
1	76.603	76.796	76.974	76.681	76.609	62.836	79.940	77.199	76.914	77.404	77.042	78.200	75.733
2	122.258	124.562	124.840	124.388	124.342	122.879	125.132	124.415	124.264	124.672	124.395	123.541	125.371
3	203.686	205.339	205.800	204.790	204.630	197.557	208.304	205.554	205.162	205.644	205.389	206.039	205.530
4	252.059	259.338	259.669	258.151	258.016	241.403	263.073	259.732	259.352	259.674	259.324	269.538	257.218
5	417.973	416.434	416.984	414.549	414.642	393.246	421.027	416.884	416.084	416.556	415.698	417.526	415.786
6	501.357	499.063	500.056	497.404	496.973	487.882	503.734	499.153	498.134	499.109	498.184	498.472	497.808
7	598.351	584.169	586.091	577.059	578.961	576.966	604.519	585.101	582.675	584.872	580.686	584.260	583.515
8	667.446	679.146	681.179	666.971	671.849	667.234	688.757	678.198	676.464	676.671	672.351	684.496	685.941
9	755.354	761.061	762.376	757.726	756.936	743.261	772.361	762.334	759.939	760.557	758.883	761.523	760.209
10	816.814	817.905	820.477	815.086	814.517	808.071	831.231	817.810	815.243	818.058	816.233	817.033	819.891
						CFCF							
1	167.997	171.583	171.747	171.102	171.036	165.464	172.657	171.569	171.363	171.433	171.280	171.204	170.958
2	202.818	207.013	207.285	206.356	206.301	177.503	211.197	207.639	207.357	207.719	207.457	207.302	207.192
3	296.704	296.950	297.615	296.523	296.333	289.939	298.884	297.075	296.597	297.313	296.984	297.023	296.587
4	302.968	302.532	303.169	302.014	301.822	294.783	306.550	303.222	302.681	303.311	302.889	302.753	302.506
5	516.469	519.490	520.619	518.275	517.995	503.205	523.555	518.946	518.102	519.351	518.725	518.839	518.431
6	542.553	541.159	542.384	540.387	539.936	523.035	548.368	541.590	540.224	541.910	540.882	541.092	540.385

(Continued)

TABLE 1.4 *(Continued)*

Mode Frequencies of a Revolution Hyperbolic Hyperboloid of Variable Thickness with a Generally Anisotropic Lamination Scheme

Mode f[Hz]	3D FEM	FSDT	FSDTZ	TSDT	TSDTZ	ED1	EDZ1	ED2	EDZ2	ED3	EDZ3	ED4	EDZ4
					Revolution Hyperbolic Hyperboloid								
DOFs	1708182	5394	8091	10788	13485	5394	8091	8091	10788	10788	13485	13485	16182
7	621.559	622.103	623.396	620.691	620.050	615.492	638.758	624.691	622.364	624.593	622.685	623.220	621.528
8	669.700	673.424	674.524	671.698	671.353	643.807	682.680	674.569	673.209	673.929	672.905	673.020	672.195
9	758.855	754.469	756.965	751.655	751.119	739.881	768.874	754.411	751.982	755.252	753.525	754.252	752.692
10	762.415	762.606	764.836	760.692	760.083	745.738	770.764	762.813	760.752	762.986	761.582	761.945	760.770

Lamination Scheme: (70/Con/Con/45), $h_1^0 = h_4^0 = 0.005$ m, $h_2^0 = h_3^0 = 0.0075$ m

Laminae Material Sequence: 1st layer: Triclinic, 2nd layer: Concrete, 3rd layer: Concrete, 4th layer: Graphite–Epoxy

Geometric Inputs: $a = 0.5$, $c = 0.3$

Thickness Variation: $\phi_1(\alpha_1, \alpha_2) = \left(\sin\left(\pi\left(n_1 \bar{\alpha}_1 + \alpha_{1m} \right) \right) \right)^{p_1}$, $\delta_1 = 1.0$, $n_1 = 1$, $p_1 = 2$, $\alpha_{1m} = 0$

$\phi_2(\alpha_1, \alpha_2) = \left(\sin\left(\pi\left(n_2 \bar{\alpha}_2 + \alpha_{2m} \right) \right) \right)^{p_2}$, $\delta_2 = 2.0$, $n_2 = 1$, $p_2 = 2$, $\alpha_{2m} = 0$

Computational Grid: CGL distribution with $I_N = 31$ and $I_M = 33$ discrete points

material, with engineering constants from Eqn. (1.89). In Figure 1.5, the interested reader can find all the useful information for the distortion of the domain based on NURBS curves, together with the lamination scheme features.

Referring to the FFFC case, a stable level of accuracy is seen for the same vibration mode, if the kinematic expansion order varies. Accordingly, the best predictions of the 3D FEM response are obtained for the fundamental frequency, especially in the case of ED4 and EDZ4 theories. Actually, for an accurate prediction of the first fundamental frequency, the ED1 should not be employed, whereas it provides acceptable results for higher modes.

In the CFCF configuration, the two concave edges of the structure have been fixed. In this case, the first fundamental vibration mode is predicted with the same level of accuracy, regardless of the choice of the displacement field assumption, whereas the second mode is very sensitive to the thickness function selection. If a linear relationship is taken for each field component, the zigzag assumption should be introduced for a proper definition of the vibration frequency. From the third mode on, a parabolic field variable expansion predicts the 3D FEM-based outcomes. Above all, for the structure at issue, an ESL assumption with $N = 4$ and the Murakami's function of Eqn. (1.32) is capable of providing lower- and higher-vibration frequencies.

The mode shapes of the shell at issue under a CFCF configuration have been collected in Figure 1.10. Note that the proposed approach correctly predicts the three-dimensional issues occurring in each vibration mode, taking into account a coupling between laminae as well as an in-plane stretching.

The last example of investigation consists of a mapped helicoid of variable thickness characterized by a lamination scheme embedding a central non-homogeneous core made of isotropic aluminium and zirconia (91) and two outer sheets composed by a triclinic (87) and trigonal (88) material. The physical domain-mapping information for the implementation via Eqn. (1.16) is reported in Figure 1.6, together with the thickness variation law-governing parameters, as well as the boundary lamination scheme. In this case, the generalized blending functions have been adopted with straight edges, as computed by Eqn. (1.12). The reference surface equation, expressed with principal coordinates, reads as follows [17]:

$$\mathbf{r}(\alpha_1,\alpha_2) = -a\cos(\alpha_1+\alpha_2)\sinh(\alpha_1-\alpha_2)\,\mathbf{e}_1$$
$$- a\sin(\alpha_1+\alpha_2)\sinh(\alpha_1-\alpha_2)\,\mathbf{e}_2 + a(\alpha_1-\alpha_2)\,\mathbf{e}_3 \qquad (1.97)$$

where $a = 0.25$ m is the featuring geometric parameter. The first ten mode frequencies have been calculated for the structure at issue, accounting for two different external constraint configurations. In the first configuration (FFFC), the structure is fixed only at the South edge, while keeping free the other sides. The mode frequencies have been calculated employing different kinematic field assumptions, starting from the classical approaches like the FSDT and TSDT, together with the power through-the-thickness expansion of the field variable even with the introduction of the Murakami's function. The results are reported in Table 1.5. All the simulations predict similar results, confirming the accuracy of the proposed

Aribtrarily-shaped revolution hyperbolic hyperboloid with variable thickness - mode shapes
(CFCF)

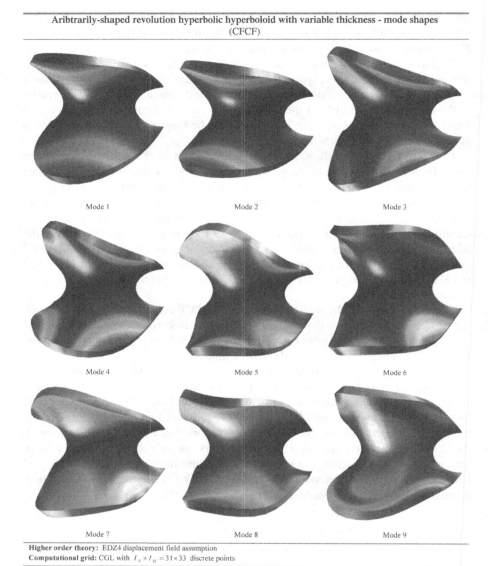

Higher order theory: EDZ4 displacement field assumption
Computational grid: CGL with $I_X \times I_M = 31 \times 33$ discrete points

FIGURE 1.10 Mode shapes of a revolution hyperbolic hyperboloid of variable thickness and arbitrary geometry via the proposed ESL two-dimensional formulation. The employment of the EDZ4 axiomatic assumption allows to predict a series of stretching and warping effects acting along the thickness of the structure.

methodology for the vibration behaviour of such geometry induced by the selected lamination scheme. For higher modes, a slight dependence of the natural frequency from the actual properties of the ESL axiomatic assumptions can be noticed. In particular, a slight decrease of the eigenvalue is seen as the order of the kinematic

TABLE 1.5

Mode Frequencies of a Helicoid of Variable Thickness with a Generally Anisotropic Lamination Scheme

Mode f[Hz]	FSDT	FSDTZ	TSDT	TSDTZ	ED1	EDZ1	ED2	EDZ2	ED3	EDZ3	ED4	EDZ4
						Helicoid						
DOFs	5394	8091	10788	13485	5394	8091	8091	10788	10788	13485	13485	16182
						FFFC						
1	32.009	31.982	31.715	31.723	31.910	34.641	32.324	32.210	31.889	31.842	31.827	31.787
2	88.867	88.855	88.043	87.871	88.195	95.579	89.560	89.199	88.674	88.490	88.602	88.513
3	122.696	122.287	121.858	121.809	116.766	124.158	122.673	122.225	122.305	122.024	122.303	122.089
4	150.873	150.298	149.450	149.295	145.215	152.333	150.228	149.611	149.705	149.447	149.629	149.423
5	285.070	284.103	282.283	281.769	281.119	290.358	283.972	282.711	282.526	282.001	282.254	281.850
6	374.358	374.216	371.808	371.087	363.932	391.578	375.430	373.720	373.183	372.251	372.493	371.906
7	456.933	456.115	453.201	452.616	442.148	474.210	458.126	456.177	455.060	454.100	454.667	453.986
8	518.987	518.238	514.988	513.935	513.850	539.990	519.657	517.199	516.227	515.106	515.499	514.909
9	623.213	621.988	615.788	614.956	587.468	649.183	622.651	620.212	617.966	616.813	617.198	616.564
10	730.215	727.855	725.634	724.906	676.630	731.480	728.976	726.303	728.266	726.681	728.182	726.620
						CFCC						
1	190.123	189.607	188.567	188.398	188.616	195.773	190.312	189.594	189.175	188.842	189.012	188.753
2	312.123	311.381	309.504	309.027	308.714	321.230	312.155	310.819	310.425	309.822	310.037	309.635
3	506.455	506.490	503.212	501.945	502.115	528.107	507.826	505.238	505.037	503.682	504.019	503.311
4	628.246	627.243	623.254	622.564	611.242	662.453	631.438	629.027	626.043	624.888	625.399	624.613
5	783.764	784.948	778.601	776.466	772.766	822.771	786.227	782.160	782.023	779.698	780.242	779.092
6	835.357	835.326	827.074	825.540	813.946	883.024	837.263	833.805	831.259	829.328	829.827	828.787

(Continued)

TABLE 1.5 *(Continued)*

Mode Frequencies of a Helicoid of Variable Thickness with a Generally Anisotropic Lamination Scheme

Mode f[Hz]	FSDT	FSDTZ	TSDT	TSDTZ	ED1	EDZ1	ED2	EDZ2	ED3	EDZ3	ED4	EDZ4
DOFs	5394	8091	10788	13485	5394	8091	8091	10788	10788	13485	13485	16182
7	1085.977	1087.032	1076.203	1073.878	1064.743	1149.316	1088.852	1084.243	1081.855	1079.100	1079.928	1078.410
8	1138.734	1143.460	1133.317	1128.906	1098.754	1187.157	1140.589	1133.855	1138.551	1134.065	1135.073	1132.953
9	1230.829	1232.023	1220.328	1217.811	1131.641	1309.308	1235.709	1230.421	1226.876	1223.824	1224.732	1223.062
10	1386.862	1386.185	1376.193	1373.782	1219.429	1431.362	1390.343	1384.765	1383.474	1380.169	1381.678	1379.430

(header: Helicoid)

Lamination Scheme: (45/Alu/Zir/30), $h_1^0 = h_2^0 = h_3^0 = h_4^0 = 0.0075$ m

Laminae Material Sequence: 1st layer: Triclinic, 2nd layer: Aluminium, 3rd layer: Zirconia, 4th layer: Trigonal

Geometric Inputs: $a = 0.25$

Thickness Variation: $\phi_1\left(\alpha_1, \alpha_2\right) = \left(\sin\left(\pi\left(n_1\bar{\alpha}_1 + \alpha_{1m}\right)\right)\right)^{p_1}$, $\delta_1 = 0.5$, $n_1 = 1$, $p_1 = 2$, $\alpha_{1m} = 0.5$

$\phi_4\left(\alpha_1, \alpha_2\right) = \tilde{\alpha}_2^{p_4}$, $\delta_4 = 1.2$, $p_4 = 1$

Computational Grid: CGL distribution with $I_N = 31$ and $I_M = 33$ discrete points

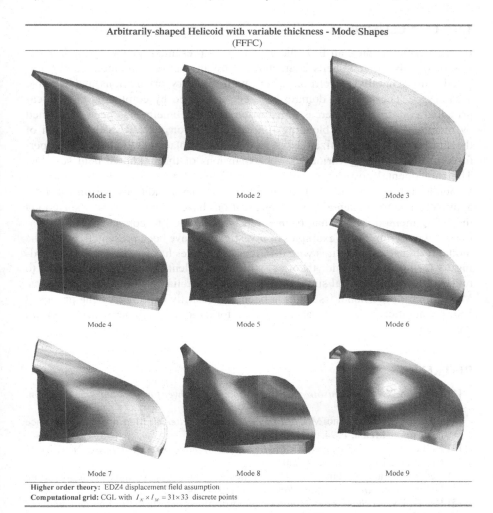

Arbitrarily-shaped Helicoid with variable thickness - Mode Shapes
(FFFC)

Mode 1   Mode 2   Mode 3

Mode 4   Mode 5   Mode 6

Mode 7   Mode 8   Mode 9

Higher order theory: EDZ4 displacement field assumption
Computational grid: CGL with $I_N \times I_M = 31 \times 33$ discrete points

FIGURE 1.11   Mode shapes of a helicoid of variable thickness and arbitrary geometry via the proposed ESL two-dimensional formulation. The employment of the EDZ4 axiomatic assumption allows to predict a series of stretching and warping effects acting along the thickness of the structure.

expansion gets higher. In Figure 1.11, we plot the corresponding mode shapes as computed by means of the EDZ4 higher-order theory.

Similar considerations can be repeated for a CFCC external edges configuration, except for the tenth natural frequency provided by the ED1 and EDZ1 theories. This proves the importance of using higher-order thickness functions within simulations.

1.9   CONCLUSIONS

In the present chapter, an ESL theory has been presented for the dynamic analysis of doubly curved shells characterized by an arbitrary geometry and laminated with generally anisotropic layers with arbitrary orientations and variable thicknesses. The physical domain has been described by employing curvilinear principal coordinates, whereas a distortion of the geometry has been performed by means of generalized blending functions based on a NURBS description of the shell edges. A generalized assessment of the displacement field has been performed, thus applying various power expansions of the unknown field variable. The fundamental governing equations have been derived from the Hamiltonian Principle, taking into account the natural boundary conditions within the ESL approach. The fundamental governing equations have been numerically solved via the GDQ method, which has been extensively applied to compute both derivatives and integrals. Some examples of investigation have been presented, where the modal response of structures with different geometries has been evaluated according to various higher-order theories, with an excellent accuracy with respect to refined three-dimensional simulations provided by reliable commercial packages. Many three-dimensional issues have been successfully captured by the proposed ESL formulation, proving its accuracy even for a reduced number of DOFs within the model.

REFERENCES

[1] Golfman, Y. *Hybrid anisotropic materials for wind power turbine blades*. Boca Raton: CRC Press, 2012.
[2] Pilato, L.A. and Michno, M.J. *Advanced composite materials*. Berlin: Springer Science & Business Media, 1994.
[3] Kollar, L.P. and Springer, G.S. *Mechanics of composite structures*. New York: Cambridge University Press, 2003.
[4] Reddy, J.N. and Robbins Jr., D.H. Theories and computational models for composite laminates. *Applied Mechanics Reviews*, 47 (1994), 147–169.
[5] Reddy, J.N. *Mechanics of laminated composite plates and shells: Theory and analysis*. Boca Raton: CRC Press, 1997.
[6] Liew, K.M., Pan, Z.Z. and Zhang L.W. An overview of layerwise theories for composite laminates and structures: Development, numerical implementation and application. *Composite Structures*, 216 (2019), 240–259.
[7] Li, D. Layerwise theories of laminated composite structures and their applications: A review. *Archives of Computational Methods in Engineering*, 28 (2021), 577–600.
[8] Thai, H.T. and Kim, S.E. A review of theories for the modeling and analysis of functionally graded plates and shells. *Composite Structures*, 128 (2015), 70–86.
[9] Kreja, I. A literature review on computational models for laminated composite and sandwich panels. *Open Engineering*, 1 (2011), 59–80.
[10] Tornabene, F., Viscoti, M. and Dimitri R. Generalized higher order layerwise theory for the dynamic study of anisotropic doubly-curved shells with a mapped geometry. *Engineering Analysis with Boundary Elements*, 134 (2022), 147–183.
[11] Tornabene, F. General higher-order layer-wise theory for free vibrations of doubly-curved laminated composite shells and panels. *Mechanics of Advanced Materials and Structures*, 23 (2016), 1046–1067.

[12] Tornabene, F., Fantuzzi, N., Bacciocchi, M. and Reddy, J.N. An equivalent layer-wise approach for the free vibration analysis of thick and thin laminated and sandwich shells. *Applied Sciences*, 7 (2017), 17.

[13] Tornabene, F. and Brischetto S. 3D capability of refined GDQ models for the bending analysis of composite and sandwich plates, spherical and doubly-curved shells. *Thin-Walled Structures*, 129 (2018), 94–124.

[14] Plagianakos, T.S. and Saravanos, D.A. Higher-order layerwise laminate theory for the prediction of interlaminar shear stresses in thick composite and sandwich composite plates. *Composite Structures*, 87 (2009), 23–35.

[15] Kim, H.S., Chattopadhyay, A. and Ghoshal, A. Dynamic analysis of composite laminates with multiple delamination using improved layerwise theory. *AIAA Journal*, 41 (2003), 1771–1779.

[16] Cho, M. and Kim, J.S. Higher-order zig-zag theory for laminated composites with multiple delaminations. *Journal of Applied Mechanics*, 68 (2001), 869–877.

[17] Tornabene, F. and Bacciocchi, M. Anisotropic doubly-curved shells. In: *Higher-order strong and weak formulations for arbitrarily shaped shell structures*. Bologna: Esculapio, 2018.

[18] Tornabene, F., Viola, E. and Fantuzzi N. General higher-order equivalent single layer theory for free vibrations of doubly-curved laminated composite shells and panels. *Composite Structures*, 104 (2013), 94–117.

[19] Reddy, J.N. An evaluation of equivalent-single-layer and layerwise theories of composite laminates. *Composite Structures*, 25 (1993), 21–35.

[20] Sánchez-Majano, A.R., Azzara, R., Pagani, A. and Carrera E. Accurate stress analysis of variable angle tow shells by high-order equivalent-single-layer and layer-wise finite element models. *Materials*, 14 (2021), 6486.

[21] Kraus, H. *Thin elastic shells*. New York: John Wiley & Sons, 1967.

[22] Draoui, A., Zidour, M., Tounsi, A. and Adim B. Static and dynamic behavior of nano-tubes-reinforced sandwich plates using (FSDT). *Journal of Nano Research*, 57 (2019), 117–135.

[23] Ferreira, A.J.M., Roque, C.M.C. and Jorge, R.M.N. Free vibration analysis of symmetric laminated composite plates by FSDT and radial basis functions. *Computer Methods in Applied Mechanics and Engineering*, 194 (2005), 4265–4278.

[24] Tornabene, F., Viola, E. and Inman, D.J. 2-D differential quadrature solution for vibration analysis of functionally graded conical, cylindrical shell and annular plate structures. *Journal of Sound and Vibration*, 328 (2009), 259–290.

[25] Tornabene, F. 2-D GDQ solution for free vibrations of anisotropic doubly-curved shells and panels of revolution. *Composite Structures*, 93 (2011), 1854–1876.

[26] Whitney, J.M. and Pagano, N.J. Shear deformation in heterogeneous anisotropic plates, ASME. *Journal of Applied Mechanics*, 37 (1970), 1031–1036.

[27] Katariya, P.V., Panda, S.K. and Mehar, K. Theoretical modelling and experimental verification of modal responses of skewed laminated sandwich structure with epoxy-filled softcore. *Engineering Structures*, 228 (2021), 111509.

[28] Reissner, E. The effect of transverse shear deformation on the bending of elastic plates. *ASME Journal of Applied Mechanics*, 12 (1945), A69–A77.

[29] Mindlin, R.D. Influence of rotatory inertia and shear on flexural motions of isotropic elastic plates. *ASME Journal of Applied Mechanics*, 18 (1951), 31–38.

[30] Reddy, J.N. A simple higher-order theory for laminated composite plates. *ASME Journal of Applied Mechanics*, 51 (1984), 745–752.

[31] Raghu, P., Rajagopal, A. and Reddy, J.N. Nonlocal nonlinear finite element analysis of composite plates using TSDT. *Composite Structures*, 185 (2018), 38–50.

[32] Reddy, J.N., A generalization of two-dimensional theories of laminated composite plates. *Communications in Applied Numerical Methods*, 3 (1987), 173–180.

[33] Washizu, K. *Variational methods in elasticity and plasticity.* Oxford: Pergamon Press, 1975.

[34] Cho, M. and Parmerter, R.R. Efficient higher order composite plate theory for general lamination configurations. *AIAA Journal*, 31 (1993), 1299–1306.

[35] Arya, H., Shimpi, R.P. and Naik, N.K. A zigzag model for laminated composite beams. *Composite Structures*, 56 (2002), 21–24.

[36] Shimpi, R.P. and Ghugal, Y.M. A new layerwise trigonometric shear deformation theory for two-layered cross-ply beams. *Composites Science and Technology*, 61 (2001), 1271–1283.

[37] Viola, E., Tornabene, F. and Fantuzzi N. General higher-order shear deformation theories for the free vibration analysis of completely doubly-curved laminated shells and panels. *Composite Structures*, 95 (2013), 639–666.

[38] Ramos, I.A., Mantari, J.L. and Zenkour, A.M. Laminated composite plates subject to thermal load using trigonometrical theory based on Carrera Unified Formulation. *Composite Structures*, 143 (2016), 324–335.

[39] Ghugal, Y.M. and Sayyad, A.S. A static flexure of thick isotropic plates using trigonometric shear deformation theory. *Journal of Solid Mechanics*, 2 (2010), 79–90.

[40] Murakami, H. Laminated composite plate theory with improved in-plane responses, ASME. *Journal of Applied Mechanics*, 53 (1986), 661–666.

[41] Toledano, A. and Murakami, H. A high-order laminated plate theory with improved in-plane responses. *International Journal of Solids Structures*, 23 (1987), 111–131.

[42] Toledano, A. and Murakami, H. A composite plate theory for arbitrary laminate configurations. *Journal of Applied Mechanics*, 54 (1987), 181–189.

[43] Gherlone, M. On the use of zigzag functions in equivalent single layer theories for laminated composite and sandwich beams: A comparative study and some observations on external weak layers. *Journal of Applied Mechanics*, 80 (2013), 061004.

[44] Tessler, A., Di Sciuva, M. and Gherlone, M. A consistent refinement of first-order shear deformation theory for laminated composite and sandwich plates using improved zigzag kinematics. *Journal of Mechanics of Materials and Structures*, 5 (2010), 341–367.

[45] Iurlaro, L., Gherlone, M.D., Di Sciuva, M. and Tessler, A. Assessment of the refined zigzag theory for bending, vibration, and buckling of sandwich plates: A comparative study of different theories. *Composite Structures*, 106 (2013), 777–792.

[46] Di Sciuva, M. Bending, vibration and buckling of simply supported thick multilayered orthotropic plates: An evaluation of a new displacement model. *Journal of Sound and Vibration*, 105 (1986), 425–442.

[47] Tornabene, F., Viscoti, M., Dimitri, R. and Reddy, J.N. Higher order theories for the vibration study of doubly-curved anisotropic shells with a variable thickness and isogeometric mapped geometry. *Composite Structures*, 267 (2021), 113829.

[48] Merdaci, S., Adda, H.M., Hakima, B., Dimitri, R. and Tornabene, F. Higher-order free vibration analysis of porous functionally graded plates. *Journal of Composites Science*, 5 (2021), 305.

[49] Tornabene, F., Viscoti, M. and Dimitri, R. Static analysis of anisotropic doubly-curved shells with arbitrary geometry and variable thickness resting on a Winkler-Pasternak support and subjected to general loads. *Engineering Analysis with Boundary Elements*, 140 (2022), 618–673.

[50] Kulikov, G.M. and Plotnikova, S.V. A sampling surfaces method and its application to three-dimensional exact solutions for piezoelectric laminated shells. *International Journal of Solids and Structures*, 50 (2013), 1930–1943.

[51] Khare, R.K., Kant, T. and Garg, A.K. Closed-form thermo-mechanical solutions of higher-order theories of cross-ply laminated shallow shells. *Composite Structures*, 59 (2003), 313–340.

[52] Tornabene, F., Viscoti, M. and Dimitri, R. Higher order formulations for doubly-curved shell structures with a honeycomb core. *Thin-Walled Structures*, 164 (2021), 107789.

[53] Arshid, H., Khorasani, M., Soleimani-Javid, Z., Dimitri, R. and Tornabene, F. Quasi-3D hyperbolic shear deformation theory for the free vibration study of honeycomb microplates with graphene nanoplatelets-reinforced epoxy skins. *Molecules*, 25 (2020), 5085.

[54] Tornabene, F., Viscoti, M., Dimitri, R. and Aiello, M.A. Higher-order modeling of anisogrid composite lattice structures with complex geometries. *Engineering Structures*, 244 (2021), 112686.

[55] Oden, J.T. and Reddy, J.N. *Variational methods in theoretical mechanics*. Berlin Heidelberg: Springer Science & Business Media, 2012.

[56] Jouneghani, F.Z., Dashtaki, P.M., Dimitri, R., Bacciocchi, M. and Tornabene, F. First-order shear deformation theory for orthotropic doubly-curved shells based on a modified couple stress elasticity. *Aerospace Science and Technology*, 73 (2018), 129–147.

[57] Tornabene, F., Brischetto, S., Fantuzzi, N. and Bacciocchi, M. Boundary conditions in 2D numerical and 3D exact models for cylindrical bending analysis of functionally graded structures. *Shock and Vibration*, 2016 (2016), 2373862.

[58] Brischetto, S., Tornabene, F., Fantuzzi, N. and Bacciocchi, M. Interpretation of boundary conditions in the analytical and numerical shell solutions for mode analysis of multilayered structures. *International Journal of Mechanical Sciences*, 122 (2017), 18–28.

[59] Tonti, E. The reason for analogies between physical theories. *Applied Mathematical Modelling*, 1 (1976), 37–50.

[60] Wu, C.P. and Liu, Y.C. A review of semi-analytical numerical methods for laminated composite and multilayered functionally graded elastic/piezoelectric plates and shells. *Composite Structures*, 147 (2016), 1–15.

[61] Li, H., Pang, F., Wang, X., Du, Y. and Chen, H. Free vibration analysis for composite laminated doubly-curved shells of revolution by a semi analytical method. *Composite Structures*, 201 (2018), 86–111.

[62] Shu, C., Chew, Y.T. and Richards, B.E. Generalized differential and integral quadrature and their application to solve boundary layer equations. *International Journal for Numerical Methods in Fluids*, 21 (1995), 723–733.

[63] Tornabene, F. and Dimitri, R. Generalized differential and integral quadrature: Theory and applications. In: *Mathematical Methods in Interdisciplinary Sciences*. Hoboken: John Wiley and Sons Inc, 2020.

[64] Shu, C. *Differential Quadrature and Its Application in Engineering*. Berlin Heidelberg: Springer Science & Business Media, 2012.

[65] Tornabene, F., Fantuzzi, N., Ubertini, F. and Viola, E. Strong formulation finite element method based on differential quadrature: A survey, ASME. *Applied Mechanics Reviews*, 67 (2015), 020801.

[66] Tornabene, F., Fantuzzi, N. and Bacciocchi, M. Strong and weak formulations based on differential and integral quadrature methods for the free vibration analysis of composite plates and shells: Convergence and accuracy. *Engineering Analysis with Boundary Elements*, 92 (2018), 3–37.

[67] Tornabene, F. Free vibration analysis of functionally graded conical, cylindrical shell and annular plate structures with a four-parameter power-law distribution. *Computer Methods in Applied Mechanics and Engineering*, 198 (2009), 2911–2935.

[68] Tornabene, F. Free vibrations of anisotropic doubly-curved shells and panels of revolution with a free-form meridian resting on Winkler–Pasternak elastic foundations. *Composite Structures*, 94 (2011), 186–206.

[69] Loy, C.T., Lam, K.Y. and Shu, C. Analysis of cylindrical shells using generalized differential quadrature. *Shock and Vibration*, 4 (1997), 193–198.

[70] Fazzolari, F.A., Viscoti, M., Dimitri, R. and Tornabene, F. 1D-Hierarchical Ritz and 2D-GDQ Formulations for the free vibration analysis of circular/elliptical cylindrical shells and beam structures. *Composite Structures*, 258 (2021), 113338.

[71] Tornabene, F., Liverani, A. and Caligiana, G. Laminated composite rectangular and annular plates: A GDQ solution for static analysis with a posteriori shear and normal stress recovery. *Composites Part B: Engineering*, 43 (2012), 1847–1872.

[72] Tornabene, F., Fantuzzi, N., Bacciocchi, M., Viola, E. and Reddy J.N. A numerical investigation on the natural frequencies of FGM sandwich shells with variable thickness by the local generalized differential quadrature method. *Applied Sciences*, 7 (2017), 131.

[73] Alinaghizadeh, F. and Shariati, M. Static analysis of variable thickness two-directional functionally graded annular sector plates fully or partially resting on elastic foundations by the GDQ method. *Journal of the Brazilian Society of Mechanical Sciences and Engineering*, 37 (2015), 1819–1838.

[74] Arefi, M., Bidgoli, E.M.R., Dimitri, R. and Tornabene, F. Free vibrations of functionally graded polymer composite nanoplates reinforced with graphene nanoplatelets. *Aerospace Science and Technology*, 81 (2018), 108–117.

[75] Tornabene, F., Liverani, A. and Caligiana, G. General anisotropic doubly-curved shell theory: A differential quadrature solution for free vibrations of shells and panels of revolution with a free-form meridian. *Journal of Sound and Vibration*, 331 (2012), 4848–4869.

[76] Tornabene, F., Dimitri, R. and Viola, E. Transient dynamic response of generally-shaped arches based on a GDQ-time-stepping method. *International Journal of Mechanical Sciences*, 114 (2016), 277–314.

[77] Viola, E., Dilena, M. and Tornabene, F. Analytical and numerical results for vibration analysis of multi-stepped and multi-damaged circular arches. *Journal of Sound and Vibration*, 299 (2007), 143–163.

[78] Piegl, L. and Tiller, W. *The NURBS book*. Berlin: Springer Science & Business Media, 1996.

[79] Dimitri, R., De Lorenzis, L., Scott, M.A., Wriggers, P., Taylor, R.L. and Zavarise G. Isogeometric large deformation frictionless contact using T-splines. *Computer Methods in Applied Mechanics and Engineering*, 269 (2014), 394–414.

[80] Hughes, T.J., Cottrell, J.A. and Bazilevs, Y. Isogeometric analysis: CAD, finite elements, NURBS, exact geometry and mesh refinement. *Computer Methods in Applied Mechanics and Engineering*, 194 (2005), 4135–4195.

[81] Tornabene, F., Fantuzzi, N. and Bacciocchi, M. *DiQuMASPAB: Differential quadrature for mechanics of anisotropic shells, plates, arches and beams*. Bologna: Esculapio, 2018.

2 Dynamical Problems of Functionally Graded Nonuniform Nanoplates under Thermal Field

Rahul Saini

CONTENTS

2.1   VIBRATION

Vibration is a phenomenon of oscillations of a particle, member, or a body about its equilibrium point. These oscillations are studied by the physical laws for the motions and forces of the system. The elastic bodies have the tendency to retain its equilibrium position during the forced disturbance and hence the vibrations occur. In general, it is visible in our daily life such as juicers, mixers, strings, motion of a tuning fork, the reed in a woodwind instrument or harmonica or the cone of a loudspeaker. However, vibrations may be weak or strong which can be disastrous such as earthquakes, winds, and tsunamis [1]. Thus, the knowledge of the theory of vibration is essential for a design engineer not only to prevent the undesirable

DOI: 10.1201/9781003328032-2

weak or strong vibrations and noise but also to increase the efficiency, efficacy, and life of the machines. People became interested in vibration when the first musical instrument, probably whistles or drums, was discovered. Consequently, vibration of elastic bodies has its beginning in the seventeenth century when Galileo observed that the frequency of a simple pendulum depends upon its length and documented that in *Discourses Concerning Two New Sciences*. The correct account of the vibration of strings was presented in his book *Harmonicorum Liber* by Mersenne in 1636 [2]. Eventually, the experimental and theoretical developments on this topic have been carried out by some of the pioneers like John Wallis, Robert Hook, Sir Isaac Newton, Sauveur, Brook Taylor, Daniel Bernoulli, Jean D Alembert, Leonard Euler, Joseph Fourier, Joseph Lagrange, Leibnitz [3], Chladni [4], Biot [5], Daniel Bernoulli [6], Euler [7], Coulomb [8], Cauchy [9], Saint-Venant [10], Poisson [11], Lamè, [12], Sophie Germain, Kirchhoff [13, 14], and Love [15].

Plates are two-dimensional structural elements in a plane with a thickness 'h' in normal direction which is much smaller than the in-plane dimensions. The plates are bounded by curved or straight lines and can have free, simply supported, and fixed boundary conditions, including uniform/nonuniform, in-plane forces, elastic supports, and elastic restraints, or in some cases even point supports. The plates can be visualized as a side-by-side combination of beams. As the joints of these beams have discontinuity which may cause delamination and cracks due to inter-lamina stresses, the two-dimensional plates are preferred over the years to avoid such ambiguity with an advantage of lighter weight and lower cost. A large number of structural components in engineering structures can be classified as plates – i.e., in civil engineering structures as floor and foundation slabs, lock-gates, thin retaining walls, bridge decks, and slab bridges. Plates are also indispensable in shipbuilding, aerospace industries, and the wings and a large part of the fuselage of an aircraft. Plates are also frequently parts of machineries and other mechanical devices. Due to these modern engineering applications of vibrations of plates, it is necessary to study the static and dynamic behaviours of plates preciously under different environmental and physical situations. In view of the various assumptions for the plates considered by the mentioned researchers, various displacement theories such as Kirchoff–Love plate theory [16, 17], Mindlin plate theory [18, 19], two-variable refined plate theory [20], Reddy's third-order shear deformation theory [21] are proposed by various researchers [22]. Among these, Kirchhoff–Love plate theory is the simplest and commonly used by the engineers and scientist to predict the behaviour of plates. This theory is based on some assumptions which are as follows:

- The thickness of the plate is small as compared to length of the plate.
- The normal stresses in the transverse direction of the middle plane of the plate are taken to be negligibly small.
- The middle surface of the plate remains unstrained during deformation.
- The normal to the undeformed middle surface remains straight and normal to the deformed middle surface.

In this regard, the mathematical theory of elasticity by Love [15] is one of the oldest and best books to develop the subject to the present level. After that, a number of books have appeared from time to time covering various aspects of bending and vibration of structural element.

2.2 NANOSTRUCTURES

The term *nanotechnology* was first introduced by Eric Drexler in the mid-1980s and further defined by American Ceramic Society as 'the creation, processing, characterization, and utilization of materials, devices, and systems with dimensions of the order of 0.1–100 nanometres (10^{-9} meter)'. In the last few decades, nanotechnology has become the topic of interest for engineers due to its wide applications in various technological situations such as invention of atomic precise materials, atomic and molecular configurations of materials, the electronic and logic devices with atomic or molecular level materials, modelling and simulation of nanomaterials based on physics and chemistry. The nanoscale devices and systems are extremely small in size and, therefore, exhibit different mechanical and thermal properties from the conventional ones. Many novel concepts and laws of normal scales are not true for nanoscale structures and, hence, modified and experimentally verified. Nanodevices provide higher thermal resistance, low weight, mechanical, optical, and electrical properties which encouraged the researchers to study their mechanical behaviour. As the local continuum theories do not consider the effect of size scale of the structure, therefore they cannot predict the behaviour of nanostructures. Hence, researchers have proposed various nanoscale theories such as strain gradient theory [19], couple stress theory [23], modified strain gradient theory [24], modified couple stress theory [20], nonlocal elasticity theory [25], and modified nonlocal elasticity theory [26]. Out of these, nonlocal elasticity theory [16–18, 21] is widely used by the researchers. According to this theory, the stress tensor at a reference point X in the domain of material is a function of the strains at all points in neighbourhood of X and can be expressed in the integral form as

$$\sigma_{ij}^{NL} = \int_V \beta\big(|X' - X|, \tau\big)\sigma_{ij}^L(X')dV(X')$$

$$\sigma_{ij}^L(X) = C(X) : \epsilon_{ij}$$

where, σ_{ij}^L and σ_{ij}^{NL} are the local and nonlocal stresses, $\beta(|X' - X|, \tau)$ is the nonlocal kernel function, $\tau = e_0 c/a$ being the length scale parameter, e_0 is the material constant, and c is an internal characteristics length; $C(X)$ is the elasticity tensor. The differential form of the nonlocal stress-strain relationship is given by

$$\big(1 - \mu\nabla^2\big)\sigma_{ij}^{NL}(X) = C(X) : \epsilon_{ij} \tag{2.1}$$

where, nonlocal parameter $\mu = (e_0 c)^2$ controls the size effect for nanostructures; ∇^2 is the Laplacian operator. The nonlocal parameter μ reveals the nonlocal effects, e.g., the effect of internal/cohesive forces on the response of nanostructures, which are prominent for nanostructures [27].

2.3 FUNCTIONALLY GRADED MATERIALS

Functionally graded materials (FGMs) or the advanced composite materials are heterogeneous composites fabricated by the mixture of two or more conventional materials at microstructural level to improve their overall performance. The Japanese material scientist invented the FGMs in 1984 which can be found in Koizumi [28] during an aerospace project. Since then, engineers and scientists are motivated to fabricate newer FGMs with superior characteristics to conventional materials. The advantage of continuous change in the material properties of these materials distinguishes them from the conventional composites, which have a mismatch of material properties at the interface due to the bonding of two distinct materials and leading to debonding, cracks, and residual stress. As the FGMs are tailored by mixing constituent materials, they have combined characteristics of their constituent materials. For example, in metal/ceramic FGMs, the toughness of a metal can be combined with the high thermal and corrosion resistance of ceramic. FGMs are mostly used in extremely challenging situations, i.e., metal-ceramic armour, thin-walled rotating blades, sensors, actuators, photodetectors, and dental implants. The various aspects of FGMs are published in past decades and summarized by various researchers in their review articles [29–33].

2.3.1 MICROMECHANICAL MODELS OF FUNCTIONALLY GRADED MATERIALS

To find the effective mechanical properties of functionally graded materials, the materials scientists gave a variety of micromechanical models [34–39] in terms of mechanical properties of constituent materials. Generally, these homogenization models are based on the approximation of shape and volume fraction of constituents in a domain. The comprehensive discussion of the homogenization models was presented by Shen [40]. Out of these, power-law function, Mori-Tanaka's scheme, and exponential function are widely used by the researchers. The mathematical expressions to compute the effective mechanical properties of the FGMs for these homogenization techniques are defined as:

i. **Power-law function** [41–43]:

$$E_z = E^m + \left\{ E^c - E^m \right\} \left(\frac{z}{h} + \frac{1}{2} \right)^n \tag{2.2a}$$

$$\rho_z = \rho^m + \left\{ \rho^c - \rho^m \right\} \left(\frac{z}{h} + \frac{1}{2} \right)^n \tag{2.2b}$$

$$\alpha_z = \alpha^m + \left\{ \alpha^c - \alpha^m \right\} \left(\frac{z}{h} + \frac{1}{2} \right)^n \tag{2.2c}$$

$$k_z = k^m + \left\{ k^c - k^m \right\} \left(\frac{z}{h} + \frac{1}{2} \right)^n \tag{2.2d}$$

where, superscripts c and m refer to ceramic and metal, respectively; E_z, ρ_z, α_z, and k_z are the Young's modulus, mass density, thermal expansion coefficient, and thermal conductivity, respectively, and non-homogeneity parameter n (≥ 0) controls the material distribution in the FG structure.

ii. **Mori-Tanaka's scheme** [44, 45]:

$$\frac{K_z - K^m}{K^c - K^m} = \frac{\left(\dfrac{z}{h} + \dfrac{1}{2}\right)^n}{1 + \left\{1 - \left(\dfrac{z}{h} + \dfrac{1}{2}\right)^n\right\}\left(\dfrac{K^c - K^m}{K^m + \dfrac{4}{3}G^m}\right)}, \quad K^c = \frac{E^c}{3(1 - v^c)}, \quad K^m = \frac{E^m}{3(1 - v_m)}$$

$$\frac{G_z - G^m}{G^c - G^m} = \frac{\left(\dfrac{z}{h} + \dfrac{1}{2}\right)^n}{1 + \left\{1 - \left(\dfrac{z}{h} + \dfrac{1}{2}\right)^n\right\}\left(\dfrac{G^c - G^m}{G^m + f^m}\right)}, \quad G^c = \frac{E^c}{2(1 + v^c)}, \quad G^m = \frac{E^m}{2(1 + v^m)}$$

$$f^m = \frac{G^m(9K^m + 8G^m)}{6(K^m + 2G^m)}, \quad E_z = \frac{9K_z G_z}{3K_z + G_z}, \quad \frac{k_z - k^m}{k^c - k^m} = \frac{\left(\dfrac{z}{h} + \dfrac{1}{2}\right)^n}{1 + \left\{1 - \left(\dfrac{z}{h} + \dfrac{1}{2}\right)^n\right\}\left(\dfrac{k^c - k^m}{3k^m}\right)}$$

$$\frac{\alpha_z - \alpha^m}{\alpha^c - \alpha^m} = \frac{\dfrac{1}{K_z} - \dfrac{1}{K^m}}{\dfrac{1}{K^c} - \dfrac{1}{K^m}}$$

iii. **Exponential function** [46]:

$$E_z = E^m e^{\frac{z}{h}\ln\frac{E^c}{E^m}}, \quad \rho_z = \rho^m e^{\frac{z}{h}\ln\frac{\rho^c}{\rho^m}}, \quad \alpha_z = \alpha^m e^{\frac{z}{h}\ln\frac{\alpha^c}{\alpha^m}}, \quad k_z = k^m e^{\frac{z}{h}\ln\frac{k^c}{k^m}}$$

2.3.2 THERMAL ANALYSIS OF FUNCTIONALLY GRADED MATERIALS

FGMs are commonly used as thermal barrier coatings (TBCs) at high temperature environment due to its ability to optimize temperature field, reduce thermal stress, and enhance thermal resistance of the material. For example, TBCs of SiC/C, ZrO_2/Ni, and ZrO_2/Y_2O_3 FGM were used for combustion chamber [47], rocket engine [48], and turbine blade [49], respectively. Such type of heating of structural components leads to variation in the mechanical properties of the material i.e., mechanical properties become functions of temperature. Extensive researches have been done to analyse the thermoelastic response of FGMs with various methods.

Recently, Swaminathan and Sangeetha [46] presented an extensive review of thermal analysis of FGM plates up to 2016. In view of this, the temperature-dependent mechanical properties of the material were developed by Cubberly [50], Munro [51], Chan [52], Touloukian [53] and expressed as,

$$P^i(T) = P_0^i \left(P_{-1}^i T^{-1} + 1 + P_1^i T^1 + P_2^i T^2 + P_3^i T^3 \right), \ i = m,c \qquad (2.3)$$

where, the experimental values of P_j ($j = -1, 0, 1, 2, 3$) refers for any mechanical property of ceramic and metal; T is the temperature within the plate. The experimental values of different ceramics and composites are reported in [54].

2.4 NONUNIFORM THICKNESS

In many practical situations, the appropriate variation in thickness of the plates helps the designer in reducing the weight and size of the plates and also increases the efficiency for vibration as compared to the plates of uniform thickness. This led to the study of vibrational characteristics of plates of variable thickness. The work up to 1965 dealing with the dynamic behaviour of plates of variable thickness with various geometries has been given in his monograph by Leissa [1969]. Thereafter, numerous studies dealing with the vibration of square and rectangular plates with various types of thickness variations have been carried out, such as linear [55], double linear [56], parabolic [57], exponential [58], quadratic [58], general [59], polynomial [60], arbitrary [61], and stepped [62].

Example: In view of the mentioned discussion and classifications, the examples are performed in order to show the formulation of mathematical model for the vibration analysis of nonuniform functionally graded annular and circular nanoplates. It covers the problems of uniform, linear, parabolic, and quadratic thickness variation of the annular and circular nanoplates under thermal environment for various boundary conditions. The effective temperature-dependent mechanical properties of the fabricated functionally graded material are obtained by power-law function. The stresses and strains of the plates are obtained in terms of displacement components of the plates using Hooke's law and Kirchhoff–Love plate theory. Further, the governing equation of the uniform, linear, parabolic, and quadratic nanoplates is derived from Hamilton's principle and Eringen's nonlocal theory. The mechanical properties for ceramic and metal constituents as ZrO_2 and 6061-T6Al, respectively, are reported in Table 2.1. The results are compared with the author's published research for varying values of different parameters with various combinations of boundary conditions.

2.5 MATHEMATICAL FORMULATION

Consider a functionally graded annular nanoplate with thickness h, inner radii b and outer radii a, and a circular plate ($b = 0$) (Figure 2.1). $z = 0$ is the mid-plane of the plates. The top and bottom surface of the plates are fully ceramic and metallic, respectively. A variable temperature T is distributed in the thickness direction of the plate. Based upon Kirchhoff–Love plate theory,

TABLE 2.1

Mechanical Properties for Ceramic and Metal

	Material	P_{-1}	P_0	P_1	P_2	P_3
$E(Pa)$	Ti-6Al-4V	0	122.7×10^9	-4.605×10^{-4}	0	0
	ZrO_2	0	132.2×10^9	-3.805×10^{-4}	-6.127×10^{-8}	0
$\rho \ (kg/m^3)$	Ti-6Al-4V	0	4420	0	0	0
	ZrO_2	0	3657	0	0	0
$\alpha \ (1/k)$	Ti-6Al-4V	0	7.43×10^{-6}	7.483×10^{-4}	-3.621×10^{-7}	0
	ZrO_2	0	13.3×10^{-6}	-1.421×10^{-3}	9.549×10^{-7}	0
$k \ (W/mK)$	Ti-6Al-4V	0	6.10	0	0	0
	ZrO_2	0	1.78	0	0	0

the displacement components for axisymmetric vibrations at the mid-plane of the plates are u_r and u_z, given by [63]:

$$u_r(r,z,t) = -z \frac{\partial w_0}{\partial r}, \ u_z(r,z,t) = w_0(r,t)$$

where, w_0 is the displacements of mid-plane.

The normal and shear, strains and stresses of the functionally graded nanoplates are as follows:

$$\epsilon_{rr} = -z \frac{\partial^2 w_0}{\partial r^2}, \ \epsilon_{\theta\theta} = -\frac{z}{r} \frac{\partial w_0}{\partial r}, \ \gamma_{\theta z} = \gamma_{rz} = \gamma_{r\theta} = 0 \quad (2.4)$$

$$\sigma_{rr} = \frac{E_z(T)}{1-v^2} \{\epsilon_{rr} + v\epsilon_{\theta\theta}\}, \ \sigma_{\theta\theta} = \frac{E_z(T)}{1-v^2} \{v\epsilon_{rr} + \epsilon_{\theta\theta}\}, \ \sigma_{\theta z} = \sigma_{rz} = \sigma_{r\theta} = 0 \quad (2.5)$$

Let us consider an element of a nanoplate with stresses at the mid-plane of the plate. These stresses vary in the z direction over thickness h of the plate. Then,

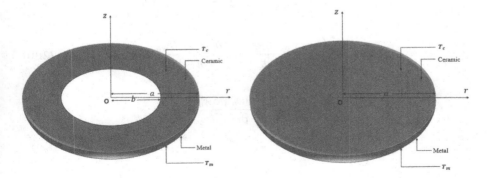

FIGURE 2.1 Geometry of the functionally graded annular and circular nanoplate subjected to thermal field.

the total potential and kinetic energies of the functionally graded nanoplates are [64, 65]:

$$U = \frac{1}{2} \int_b^a \int_{-h/2}^{h/2} (\sigma_{rr}\epsilon_{rr} + \sigma_{\theta\theta}\epsilon_{\theta\theta})\, rdz\, dr$$

$$V = \frac{1}{2} \int_b^a \int_{-h/2}^{h/2} \rho_z \left(\frac{\partial u_r}{\partial t} \frac{\partial u_r}{\partial t} + \frac{\partial u_z}{\partial t} \frac{\partial u_z}{\partial t} \right) rdz\, dr$$

The work done by the plate due to temperature profile can be written as:

$$W_T = -\frac{1}{2} \int_b^a \int_{-h/2}^{h/2} \frac{E_z(T)}{1-v} \alpha_z(T)(T - T_0) \frac{\partial w_0}{\partial r} \frac{\partial w_0}{\partial r}\, rdz\, dr$$

Taking the variations of U, V, and W_T

$$\delta U = \int_b^a \int_{-h/2}^{h/2} (\sigma_{rr}\delta\epsilon_{rr} + \sigma_{\theta\theta}\delta\epsilon_{\theta\theta})\, rdz\, dr = -\int_b^a \left(M_{rr} \frac{\partial^2 \delta w_0}{\partial r^2} + \frac{M_{\theta\theta}}{r} \frac{\partial \delta w_0}{\partial r} \right) rdr \qquad (2.6)$$

$$\delta V = \int_b^a \int_{-h/2}^{h/2} \rho(z) \left(\frac{\partial u}{\partial t} \frac{\partial \delta u}{\partial t} + \frac{\partial w}{\partial t} \frac{\partial \delta w}{\partial t} \right) rdz\, dr = \int_b^a \left(I_0 \frac{\partial w_0}{\partial t} \frac{\partial \delta w_0}{\partial t} + I_2 \frac{\partial^2 w_0}{\partial r \partial t} \frac{\partial^2 \delta w_0}{\partial r \partial t} \right) rdr$$

$$(2.7)$$

$$\delta W_T = -\int_b^a \int_{-h/2}^{h/2} \frac{E_z(T)}{1-v} \alpha_z(T)(T - T_0) \frac{\partial w_0}{\partial r} \frac{\partial \delta w_0}{\partial r}\, rdz\, dr \qquad (2.8)$$

Using the strain and stress relations from Eqns. (2.4) and (2.5) gives

$$\delta U = -\int_b^a \left(M_{rr} \frac{\partial^2 \delta w_0}{\partial r^2} + \frac{M_{\theta\theta}}{r} \frac{\partial \delta w_0}{\partial r} \right) rdr \qquad (2.9)$$

$$\delta V = \int_b^a \left(I_0 \frac{\partial w_0}{\partial t} \frac{\partial \delta w_0}{\partial t} + I_2 \frac{\partial^2 w_0}{\partial r \partial t} \frac{\partial^2 \delta w_0}{\partial r \partial t} \right) rdr \qquad (2.10)$$

$$\delta W_T = \int_b^a N_T \frac{\partial w_0}{\partial r} \frac{\partial \delta w_0}{\partial r}\, rdr \qquad (2.11)$$

where,

$$\begin{bmatrix} M_{rr} & M_{\theta\theta} \end{bmatrix} = \int_{-h/2}^{h/2} z \begin{bmatrix} \sigma_{rr} & \sigma_{\theta\theta} \end{bmatrix} dz$$

are the normal moment resultants of the plate,

$$\begin{bmatrix} I_0 & I_2 \end{bmatrix} = \int_{-h/2}^{h/2} \rho_z \begin{bmatrix} 1 & z^2 \end{bmatrix} dz$$

are the inertia of the plate, and

$$N_T = -\int_{-h/2}^{h/2} \frac{E_z(T)}{1-v} \alpha_z(T)(T-T_0) dz \tag{2.12}$$

T_0 is the reference temperature.

According to this principle, the actual dynamical path experienced by a mechanical system in motion is distinguished from all other admissible paths by the condition

$$\delta \int_{t_1}^{t_2} (V - U - W_T) dt = 0 \tag{2.13}$$

where, t_1 and t_2 are the initial and final values of time.

Substituting the expressions of energy variations from Eqns. (2.9)–(2.11) into $(\delta V - \delta U - \delta W_T)$, we get

$$\int_{t_1}^{t_2} \int_{b}^{a} \left(I_0 \frac{\partial w_0}{\partial t} \frac{\partial \delta w_0}{\partial t} + I_2 \frac{\partial^2 w_0}{\partial r \partial t} \frac{\partial^2 \delta w_0}{\partial r \partial t} + M_{rr} \frac{\partial^2 \delta w_0}{\partial r^2} + \frac{M_{\theta\theta}}{r} \frac{\partial \delta w_0}{\partial r} - N_T \frac{\partial w_0}{\partial r} \frac{\partial \delta w_0}{\partial r} \right) r dr\, dt = 0$$

$$\tag{2.14}$$

Integrating Eqn. (2.14) by parts leads to

$$\int_{b}^{a} \left(r I_0 \frac{\partial w_0}{\partial t} \delta w_0 + r I_2 \frac{\partial^2 w_0}{\partial r \partial t} \frac{\partial w_0}{\partial r} - r I_2 \frac{\partial^3 w_0}{\partial r \partial t^2} \frac{\partial \delta w_0}{\partial r} \right)_{t_1}^{t_2} dr +$$

$$\int_{t_1}^{t_2} \left[r M_{rr} \frac{\partial \delta w_0}{\partial r} - \left\{ \frac{\partial}{\partial r} (r M_{rr}) - M_{\theta\theta} \right\} \delta w_0 - r N_T \frac{\partial w_0}{\partial r} \delta w_0 \right]_{b}^{a} dt -$$

$$\int_{t_1}^{t_2} \int_{b}^{a} \left[r I_0 \frac{\partial^2 w_0}{\partial t^2} - I_2 \left(\frac{\partial^3 w_0}{\partial r \partial t^2} + r \frac{\partial^4 w_0}{\partial r^2 \partial t^2} \right) - \frac{\partial^2}{\partial r^2} (r M_{rr}) + \frac{\partial M_{\theta\theta}}{\partial r} - \frac{\partial}{\partial r} \left(r N_T \frac{\partial w_0}{\partial r} \right) \right] \delta w_0 dr\, dt = 0$$

$$\tag{2.15}$$

The above integrals will be independently zero, and hence equation of motion obtained as

$$\frac{\partial^2}{\partial r^2}(rM_{rr}) - \frac{\partial M_{\theta\theta}}{\partial r} + \frac{\partial}{\partial r}\left(rN_T\frac{\partial w_0}{\partial r}\right) = rI_0\frac{\partial^2 w_0}{\partial t^2} - I_2\frac{\partial^2}{\partial t^2}\left(\frac{\partial w_0}{\partial r} + r\frac{\partial^2 w_0}{\partial r^2}\right) \quad (2.16)$$

The coefficient I_2 involved in rotatory inertia term are often neglected in most of the books [64] due to its insignificant contribution in lowest frequencies of vibration of the plate. Accordingly, the equation of motion reduces to

$$\frac{\partial^2 M_{rr}}{\partial r^2} + \frac{2}{r}\frac{\partial M_{rr}}{\partial r} - \frac{1}{r}\frac{\partial M_{\theta\theta}}{\partial r} + \frac{1}{r}\frac{\partial}{\partial r}\left(rN_T\frac{\partial w_0}{\partial r}\right) = I_0\frac{\partial^2 w_0}{\partial t^2} \quad (2.17)$$

2.5.1 BOUNDARY CONDITIONS

The boundary conditions of the plates would be:

i. **Clamped edge:**

$$w = \frac{dw}{dr} = 0$$

ii. **Simply supported edge:**

$$w = M_{rr} = 0$$

2.5.2 ERINGEN'S NONLOCAL THEORY

For the present functionally graded nanoplates, the nonlocal stresses are defined in terms of displacements as

$$(1 - \mu\nabla^2)\sigma_{rr} = -z\frac{E_z(T)}{1-v^2}\left(\frac{\partial^2 w_0}{\partial r^2} + \frac{v}{r}\frac{\partial w_0}{\partial r}\right)$$

$$(1 - \mu\nabla^2)\sigma_{\theta\theta} = -z\frac{E_z(T)}{1-v^2}\left(v\frac{\partial^2 w_0}{\partial r^2} + \frac{1}{r}\frac{\partial w_0}{\partial r}\right)$$

The nonlocal normal moment resultants of functionally graded nanoplate can be expressed in terms of displacements as

$$(1 - \mu\nabla^2)\begin{bmatrix} M_{rr} \\ M_{\theta\theta} \end{bmatrix} = -D\left\{ v\begin{pmatrix} \dfrac{1}{r}\dfrac{\partial w_0}{\partial r} \\ \dfrac{\partial^2 w_0}{\partial r^2} \end{pmatrix} + \begin{pmatrix} \dfrac{\partial^2 w_0}{\partial r^2} \\ \dfrac{1}{r}\dfrac{\partial w_0}{\partial r} \end{pmatrix} \right\} \quad (2.18)$$

where,

$$D = \int\limits_{-h/2}^{h/2} z^2 E_z(T)\,dz$$

is the flexural rigidity of the plate.

The nonlocal governing equations of motion for functionally graded nanoplate under thermal environment are deduced by solving Eqns. (2.17) and (2.18) and can be written as:

$$
\begin{aligned}
&\left[D\left\{ \frac{\partial^4 w_0}{\partial r^4} + \frac{2}{r}\frac{\partial^3 w_0}{\partial r^3} - \frac{1}{r^2}\frac{\partial^2 w_0}{\partial r^2} + \frac{1}{r^3}\frac{\partial w_0}{\partial r} \right\} \right. \\
&\left. + \frac{\partial D}{\partial r}\left\{ 2\frac{\partial^3 w_0}{\partial r^3} + \frac{(2+v)}{r}\frac{\partial^2 w_0}{\partial r^2} - \frac{1}{r^2}\frac{\partial w_0}{\partial r} \right\} \right. \\
&\left. + \frac{\partial^2 D}{\partial r^2}\left\{ \frac{\partial^2 w_0}{\partial r^2} + \frac{v}{r}\frac{\partial w_0}{\partial r} \right\} - \frac{\partial w_0}{\partial r}\frac{\partial N_T}{\partial r} \right] \\
&+ \mu\left\{ N_T\left(\frac{\partial^4 w_0}{\partial r^4} + \frac{4}{r}\frac{\partial^3 w_0}{\partial r^3} + \frac{2}{r^2}\frac{\partial^2 w_0}{\partial r^2} \right) + \right. \\
&\left. \frac{\partial N_T}{\partial r}\left(3\frac{\partial^3 w_0}{\partial r^3} + \frac{8}{r}\frac{\partial^2 w_0}{\partial r^2} + \frac{2}{r^2}\frac{\partial w_0}{\partial r} \right) + \right. \\
&\left. \frac{\partial^2 N_T}{\partial r^2}\left(3\frac{\partial^2 w_0}{\partial r^2} + \frac{4}{r}\frac{\partial w_0}{\partial r} \right) + \frac{\partial w_0}{\partial r}\frac{\partial^3 N_T}{\partial r^3} \right\}
\end{aligned}
$$

$$
-N_T\left(\frac{\partial^2 w_0}{\partial r^2} + \frac{1}{r}\frac{\partial w_0}{\partial r} \right) = -I_0\frac{\partial^2}{\partial t^2}\left\{ w_0 - \mu\left(\frac{\partial^2 w_0}{\partial r^2} + \frac{1}{r}\frac{\partial w_0}{\partial r} \right) \right\} \tag{2.19}
$$

Assuming that the thickness of the nanoplates is varying radially and defined as $h = h_0 f(r)$, where,

$$
f(r) = 1 + \frac{\alpha r}{a} + \frac{\beta r^2}{a} \tag{2.20}
$$

If $\alpha \neq 0$, $\beta = 0 \Rightarrow$ the linear thickness variation of the plates; $\alpha = 0$, $\beta \neq 0 \Rightarrow$ the parabolic thickness variation of the plates; $\alpha \neq 0$, $\beta \neq 0 \Rightarrow$ the quadratic thickness variation of the plates. The positive values of these parameters represent that the thickness on the plates increases from inner radii to outer radii for annular plate (Figure 2.2) and centre to periphery for circular plate (Figure 2.2), and negative values show reverse impact. Using the harmonic solution along with nondimensional variables $x = r/a$, $H = h_0/a$, $w = aW(x)e^{i\omega t}$, $\gamma = \sqrt{\mu}/a$, and $\Omega^2 = \omega^2 a^4 \rho_0^c H/D^*$, Eqn. (2.19) is transformed to the nondimensional ordinary differential equation as:

$$
\begin{aligned}
&\left[B\left\{ \frac{d^4W}{dx^4} + \frac{2}{x}\frac{d^3W}{dx^3} - \frac{1}{x^2}\frac{d^2W}{dx^2} + \frac{1}{x^3}\frac{dW}{dx} \right\} \right. \\
&\left. + \frac{dB}{dx}\left\{ 2\frac{d^3W}{dx^3} + \frac{(2+v)}{x}\frac{d^2W}{dx^2} - \frac{1}{x^2}\frac{dW}{dx} \right\} \right. \\
&\left. + \frac{d^2B}{dx^2}\left\{ \frac{d^2W}{dx^2} + \frac{v}{x}\frac{dW}{dx} \right\} - \frac{dW}{dx}\frac{d\bar{N}_T}{dx} \right] \\
&+ \gamma^2\left\{ \bar{N}_T\left(\frac{d^4W}{dx^4} + \frac{4}{x}\frac{d^3W}{dx^3} + \frac{2}{x^2}\frac{d^2W}{dx^2} \right) + \right. \\
&\left. \frac{d\bar{N}_T}{dx}\left(3\frac{d^3W}{dx^3} + \frac{8}{x}\frac{d^2W}{dx^2} + \frac{2}{x^2}\frac{dW}{dx} \right) + \right. \\
&\left. \frac{d^2\bar{N}_T}{dx^2}\left(3\frac{d^2W}{dx^2} + \frac{4}{x}\frac{dW}{dx} \right) + \frac{dW}{dx}\frac{d^3\bar{N}_T}{dx^3} \right\}
\end{aligned}
$$

$$
-\bar{N}_T\left(\frac{d^2W}{dx^2} + \frac{1}{x}\frac{dW}{dx} \right) = \Omega^2\bar{I}_0\left\{ W - \gamma^2\left(\frac{d^2W}{dx^2} + \frac{1}{x}\frac{dW}{dx} \right) \right\} \tag{2.21}
$$

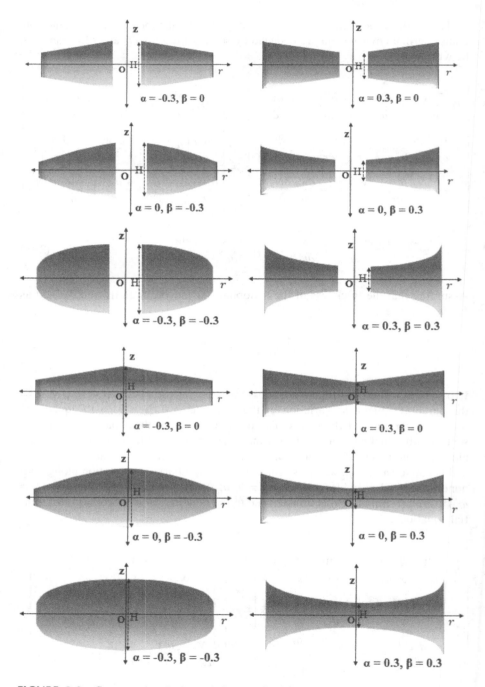

FIGURE 2.2 Cross-sectional view of the annular and circular nanoplates with linear, parabolic, and quadratic thickness.

where,

$$D = H^3 BD^*, \ I_0 = H\overline{I}_0 \rho_0^c, \ N_T = H\overline{N}_T D^*, \ D^* = \frac{E_0^c H^3}{12(1-v^2)}$$

2.6   SOLUTION TECHNIQUE

As the coefficients in dimensionless governing equations is variable, the closed-form solutions are not possible, except for the specific values of various parameters. Hence, an approximate solution has been obtained using generalized differential quadrature method [66]. According to this method, the domain of the plate is discretized into $x_1, x_2, \ldots, x_m$ points. The pth order derivative of a function $W(x)$ at point x_i is approximated by

$$\left\{ \frac{d^p W(x)}{dx^p} \right\}_{x=x_i} = \sum_{j=1}^{m} C_{ij}^p W_j, \ i = 1, 2, \ldots, m, \tag{2.22}$$

where, C_{ij}^p rare the weighting coefficients which can be obtained as

$$C_{ij}^1 = \begin{cases} \dfrac{M^1(x_i)}{M^1(x_j)(x_i - x_j)}, & i \neq j \\ -\displaystyle\sum_{\substack{j=1 \\ i \neq j}}^{m} C_{ij}^1, & i = j \end{cases}, \text{ where } i, j = 1, 2, \ldots, m \text{ and } M^1(x_i) = \prod_{\substack{k=1 \\ k \neq i}}^{m}(x_i - x_k)$$

$$C_{ij}^p = \begin{cases} p\left[C_{ij}^1 C_{ij}^{p-1} - \dfrac{C_{ij}^{p-1}}{(x_i - x_j)} \right], & i \neq j \\ -\displaystyle\sum_{\substack{j=1 \\ i \neq j}}^{m} C_{ij}^p, & i = j \end{cases} \text{ where, } p \geq 2, i, j = 1, 2, \ldots, m$$

The co-ordinates of m grid points were generated in accordance with the Chebyshev Gauss–Lobatto grid distribution and given by

$$x_i = \frac{1}{2}\left[1 - \cos\left(\frac{i-1}{m-1}\pi \right) \right], \ i = 1, 2, \ldots, m$$

Equation (2.21) is transformed and then discretized by approximating the derivatives of nondimensional transverse displacement W given in Eqn. (2.22) at any grid point x_i, $i = 1, 2, \ldots, m$, gives simultaneous equation in terms of m unknowns. The satisfaction of the resultant equation at all grid points provides a set of m-4 linear equations. These equations together with corresponding boundary conditions at the inner and outer peripheries provide m linear equations an eigen-value problem which is solved to compute the nondimensional frequencies of the annular nanoplates using MATLAB. However, for circular nanoplates, boundary and regularity conditions are incorporated at the outer periphery and centre of the plate, respectively.

2.7   RESULTS AND DISCUSSION

Free axisymmetric vibrations are presented for functionally graded circular and annular nanoplates to discuss the effect of size-dependency, non-homogeneity parameter, and temperature difference. The values of parameters are taken as nonlocal parameter $\gamma = 0, 0.1, 0.2$, temperature difference $\Delta T = 0, 200K$, non-homogeneity $n = 0, 0.5, 2$, radii ratio ε $(b/a) = 0.3, 0.5$.

2.7.1   CONVERGENCE AND COMPARISON

To choose an appropriate number of grid point m, a MATLAB code is run for increasing values of number of grid points $m = 10, 11, \ldots 21, 22$, to evaluate the value of frequency parameter for annular and circular nanoplates. As the grid point increases, a regular improvement in the values of frequency parameter is noticed for all boundary conditions. Being a complex nanostructure under thermal field and the inclusion of regularity conditions at the centre, the convergence in the values frequency parameter for the circular nanoplates is very slow. Consequently, 22 grid points are used for the whole analysis since further increase in m does not improve the result in the first and fourth decimal place for circular and annular nanoplates, respectively. In view of this, the values of frequency parameter with the number of grid points m are recorded in Tables 2.2–2.6 for specific annular nanoplates for different boundary conditions in the first three modes of vibration, as maximum deviations are observed for this data.

A comparison of results for uniform isotropic annular plates obtained by present method and other methods [2, 3] for clamped-clamped, clamped-simply supported, and simply supported-simply supported boundary conditions is presented in Table 2.7. Similarly, the values of frequency parameter for uniform isotropic/functionally graded circular plate in the absence of thermal environment, are compared with those available in the literature and reported in Tables 2.8–2.10 for clamped and simply-supported peripheral conditions.

To study the effect of thermal environment on the vibration behaviour of plates, two different approaches including the contribution of thermal field are available in various articles. In the first approach, the researchers obtained the displacements due to the thermal environment and then introduced thermal stresses in the equation of motion. However, in the second approach, the work done by the temperature is included straight forward in the Hamilton's principle for the plate. In this chapter, the second approach is adopted to analyse the influence of thermal gradient on the vibration behaviour of nonuniform annular and circular nanoplates. However, the first

TABLE 2.2
Convergence of Frequency Parameter for Annular Nanoplate with Clamped-Clamped Boundary Conditions $n = 2$, $\Delta T = 200$ K, $\gamma = 0.2$, $\varepsilon = 0.3$

	U			LV						PV						QV					
	$\alpha = 0$, $\beta = 0$			$\alpha = -0.3$, $\beta = 0$			$\alpha = 0.3$, $\beta = 0$			$\alpha = 0$, $\beta = -0.3$			$\alpha = 0$, $\beta = 0.3$			$\alpha = -0.3$, $\beta = -0.3$			$\alpha = -0.3$, $\beta = -0.3$		
m	I	II	III	I	II	III	I	II	III	I	II	III	I	II	III	I	II	III	I	II	III
10	26.1481	46.6046	69.3005	22.9302	40.9990	61.1443	28.9471	51.3637	75.2743	23.3219	42.1585	62.4872	28.6794	50.3115	73.2773	19.3588	34.8027	50.4894	31.2397	54.6535	78.2314
11	26.1483	46.6848	66.9714	22.9297	41.0224	59.2470	28.9435	51.4975	73.6690	23.3206	42.1622	61.3509	28.6732	50.5259	71.7803	19.3591	34.7817	51.5100	31.2276	54.8676	78.0636
12	26.1481	46.7149	67.3950	22.9296	41.0393	59.4407	28.9434	51.5020	74.3389	23.3204	42.1635	61.4402	28.6729	50.5162	72.7399	19.3590	34.7939	51.3886	31.2282	54.8077	79.1601
13	26.1479	46.7118	67.5805	22.9295	41.0392	59.5592	28.9432	51.4969	74.4103	23.3203	42.1636	61.4806	28.6728	50.5068	72.7544	19.3593	34.7929	51.3407	31.2280	54.8030	78.9289
14	**26.1479**	46.7106	67.5530	**22.9295**	41.0389	59.5534	**28.9432**	51.4972	74.3662	**23.3203**	42.1637	61.4801	**28.6728**	50.5085	72.6849	**19.3593**	34.7936	51.3471	**31.2280**	54.8068	78.8744
15	—	46.7107	67.5410	—	**41.0389**	59.5487	—	51.4973	74.3656	—	**42.1637**	61.4805	—	50.5087	72.6920	—	**34.7936**	51.3433	—	54.8067	78.9017
16	—	46.7108	67.5423	—	—	**59.5487**	—	**51.4973**	74.3676	—	—	61.4805	—	50.5086	72.6954	—	—	51.3436	—	54.8065	78.9025
17	—	**46.7108**	67.5428	—	—	—	—	—	74.3674	—	—	61.4804	—	**50.5086**	72.6945	—	—	51.3434	—	54.8066	78.9006
18	—	—	67.5428	—	—	—	—	—	74.3673	—	—	**61.4804**	—	—	72.6943	—	—	**51.3434**	—	**54.8066**	78.9007
19	—	—	67.5427	—	—	—	—	—	74.3673	—	—	—	—	—	72.6944	—	—	—	—	—	78.9008
20	—	—	**67.5427**	—	—	—	—	—	—	—	—	—	—	—	**72.6944**	—	—	—	—	—	**78.9008**
21	—	—	—	—	—	—	—	—	—	—	—	—	—	—	—	—	—	—	—	—	—
22	—	—	—	—	—	—	—	—	—	—	—	—	—	—	—	—	—	—	—	—	—

TABLE 2.3

Convergence of Frequency Parameter for Annular Nanoplate with Clamped-Simply Supported Boundary Conditions $n = 2$, $\Delta T = 200$ K, $\gamma = 0.2$, $\varepsilon = 0.3$

	U						LV			PV						QV					
	$\alpha=0$, $\beta=0$			$\alpha=-0.3$, $\beta=0$			$\alpha=0.3$, $\beta=0$			$\alpha=0$, $\beta=-0.3$			$\alpha=0$, $\beta=0.3$			$\alpha=-0.3$, $\beta=-0.3$			$\alpha=-0.3$, $\beta=-0.3$		
m	I	II	III	I	II	III	I	II	III	I	II	III	I	II	III	I	II	III	I	II	III
10	17.4896	38.7966	65.1184	16.0590	34.5787	58.0285	18.7390	42.3658	68.5017	16.7053	35.6693	55.4379	18.1596	41.1476	66.6152	14.9201	29.9189	44.6768	19.3358	44.5240	69.4097
11	17.4781	38.8279	58.5192	16.0537	34.4619	52.6477	18.7111	42.6274	63.4688	16.7011	35.6245	54.4476	18.1276	41.6619	61.1511	14.9002	29.9377	45.3637	19.2754	45.2199	65.8657
12	17.4824	38.8810	58.5998	16.0551	34.4632	52.0656	18.7204	42.6601	64.5011	16.7024	35.6178	54.0631	18.1366	41.7015	62.9252	14.8949	29.9511	45.4487	19.2939	45.1373	68.7619
13	17.4806	38.8807	58.9988	16.0545	34.4695	52.1303	18.7167	42.6476	64.8914	16.7018	35.6184	54.0399	18.1329	41.6662	63.4146	14.8924	29.9552	45.4885	19.2872	45.1084	68.6908
14	17.4812	38.8783	58.9932	16.0547	34.4699	52.1896	18.7179	42.6465	64.7650	16.7020	**35.6184**	54.0601	18.1342	41.6665	63.1467	14.8917	29.9574	45.4907	19.2895	45.1150	68.3714
15	17.4810	38.8784	58.9662	16.0546	34.4698	52.1921	18.7175	42.6472	64.7482	16.7019	—	54.0639	18.1338	41.6683	63.1318	14.8914	29.9581	45.4912	19.2887	45.1157	68.4169
16	17.4811	38.8784	58.9664	**16.0546**	**34.4698**	52.1890	18.7176	42.6471	64.7558	**16.7019**	—	54.0636	18.1339	41.6681	63.1505	14.8914	29.9584	45.4912	19.289	45.1153	68.4315
17	17.4810	38.8785	58.9676	—	—	52.1886	18.7176	**42.6471**	64.7560	—	—	54.0635	**18.1339**	41.6680	63.1497	14.8913	29.9585	45.4911	19.2889	**45.1153**	68.4270
18	**17.4810**	38.8785	**58.9676**	—	—	52.1886	**18.7176**	—	64.7557	—	—	54.0634	—	**41.6680**	63.1488	**14.8913**	**29.9585**	45.4910	**19.2889**	—	68.4268
19	—	38.8785	—	—	—	52.1887	—	—	**64.7557**	—	—	**54.0634**	—	—	63.1489	—	—	**45.4910**	—	—	68.4270
20	—	38.8784	—	—	—	**52.1887**	—	—	—	—	—	—	—	—	63.1490	—	—	—	—	—	—
21	—	**38.8784**	—	—	—	—	—	—	—	—	—	—	—	—	**63.1490**	—	—	—	—	—	**68.4270**
22	—	—	—	—	—	—	—	—	—	—	—	—	—	—	—	—	—	—	—	—	—

TABLE 2.4

Convergence of Frequency Parameter for Annular Nanoplate with Simply Supported-Simply Supported Boundary Conditions $n = 2$, $\Delta T = 200$ K, $\gamma = 0.2$, $\varepsilon = 0.3$

	U						LV						PV						QV					
	$\alpha=0,\ \beta=0$			$\alpha=-0.3,\ \beta=0$			$\alpha=0.3,\ \beta=0$			$\alpha=-0.3,\ \beta=0$			$\alpha=0,\ \beta=-0.3$			$\alpha=0,\ \beta=0.3$			$\alpha=-0.3,\ \beta=-0.3$			$\alpha=-0.3,\ \beta=-0.3$		
m	I	II	III	I	II	III	I	II	III	I	II	III	I	II	III	I	II	III	I	II	III	I	II	III
10	12.6606	32.4558	53.9450	11.2561	28.6051	47.3406	13.8339	35.8004	58.0737	—	—	—	11.7607	29.7123	48.1583	13.3900	34.8246	56.2660	9.9295	24.8958	39.1401	14.4562	37.9894	59.4679
11	12.6609	32.4909	51.5690	11.2579	28.6054	45.6739	13.8298	35.8817	56.6796	—	—	—	11.7621	29.7109	47.5317	13.3853	34.9894	54.9104	9.8938	24.9097	39.9185	14.4441	38.1370	59.8367
12	12.6609	32.5064	51.8262	11.2581	28.6083	45.7523	13.8304	35.8775	57.2188	—	—	—	11.7623	29.7088	47.5567	13.3857	34.9702	55.8230	9.8998	24.9109	39.7630	14.4466	38.0762	60.8906
13	12.6610	32.5053	51.9840	11.2584	28.6085	45.8163	13.8299	35.8755	57.2574	—	—	—	11.7625	29.7087	47.5546	13.3854	34.9643	55.7912	9.8945	24.9104	39.8349	14.4453	38.0765	60.5703
14	**12.6610**	32.5046	51.9682	**11.2584**	**28.6085**	45.8159	13.8300	35.8754	57.2245	—	—	—	11.7625	29.7086	47.5549	13.3855	34.9655	55.7267	9.8953	24.9121	39.8246	14.4456	38.0783	60.5315
15	—	32.5047	51.9598	—	—	45.8154	13.8299	35.8754	57.2259	—	—	—	11.7626	29.7086	47.5559	13.3854	34.9657	55.7377	9.8946	24.9118	39.8305	14.4455	38.0782	60.5597
16	—	**32.5047**	51.9604	—	—	45.8153	13.8299	**35.8754**	57.2272	—	—	—	**11.7626**	**29.7086**	**47.5559**	**13.3854**	34.9656	55.7404	9.8947	24.9122	39.8297	**14.4455**	38.0781	60.5591
17	—	—	51.9607	—	—	45.8152	13.8299	—	57.2269	—	—	—	—	—	—	—	**34.9656**	55.7394	9.8946	24.9121	39.8301	—	**38.0781**	60.5577
18	—	—	**51.9607**	—	—	**45.8152**	13.8299	—	**57.2269**	—	—	—	—	—	—	—	—	55.7393	**9.8946**	24.9122	**39.8301**	—	—	60.5578
19	—	—	—	—	—	—	13.8300	—	—	—	—	—	—	—	—	—	—	55.7394	—	**24.9122**	—	—	—	60.5579
20	—	—	—	—	—	—	13.8299	—	—	—	—	—	—	—	—	—	—	**55.7394**	—	—	—	—	—	60.5579
21	—	—	—	—	—	—	**13.8299**	—	—	—	—	—	—	—	—	—	—	—	—	—	—	—	—	**60.5579**
22	—	—	—	—	—	—	—	—	—	—	—	—	—	—	—	—	—	—	—	—	—	—	—	—

TABLE 2.5

Convergence of Frequency Parameter for Circular Nanoplate with Clamped Boundary Conditions $n = 2$, $\Delta T = 200$ K, $\gamma = 0.2$

	U			LV						PV						QV					
	$\alpha=0$ $\beta=0$			$\alpha=-0.3$ $\beta=0$			$\alpha=0.3$ $\beta=0$			$\alpha=0$ $\beta=-0.3$			$\alpha=0$ $\beta=0.3$			$\alpha=-0.3$ $\beta=-0.3$			$\alpha=-0.3$ $\beta=-0.3$		
m	I	II	III	I	II	III	I	II	III	I	II	III	I	II	III	I	II	III	I	II	III
10	7.2	21.1	35.9	5.6	18.3	32.5	8.7	23.8	38.8	6	19.3	33.1	8.9	23.4	37.7	4.2	16.1	28.3	10.2	25.5	40.6
11	7.2	21.1	35.6	5.6	18.4	31.5	8.7	23.8	39.2	6	19.4	33.2	8.9	23.5	38.3	4.3	16.1	28.3	10.3	25.6	41.4
12	7.3	21.2	35.5	5.7	18.5	31.7	–	23.9	39.3	6	19.4	33.2	8.9	23.5	38.5	4.3	16.2	28.4	10.3	25.6	41.4
13	7.3	21.2	35.6	5.7	18.5	31.8	–	23.9	39.3	6.1	19.4	33.2	9	23.5	38.5	–	16.2	28.4	–	25.6	41.4
14	–	21.2	35.6	–	18.6	31.8	–	–	39.4	6.1	19.4	33.3	9	23.6	38.5	–	–	28.4	–	25.7	41.5
15	–	21.3	35.7	–	18.6	31.9	–	–	39.4	–	19.5	33.3	–	23.6	38.5	–	–	28.4	–	25.7	41.5
16	–	21.3	35.7	–	–	31.9	–	–	–	–	19.5	33.3	–	–	38.6	–	–	28.5	–	–	41.5
17	–	–	35.7	–	–	–	–	–	–	–	–	33.3	–	–	38.6	–	–	28.5	–	–	41.5
18	–	–	35.8	–	–	–	–	–	–	–	–	33.3	–	–	–	–	–	–	–	–	41.6
19	–	–	35.8	–	–	–	–	–	–	–	–	33.4	–	–	–	–	–	–	–	–	41.6
20	–	–	–	–	–	–	–	–	–	–	–	33.4	–	–	–	–	–	–	–	–	–
21	–	–	–	–	–	–	–	–	–	–	–	–	–	–	–	–	–	–	–	–	–
22	–	–	–	–	–	–	–	–	–	–	–	–	–	–	–	–	–	–	–	–	–

TABLE 2.6

Convergence of Frequency Parameter for Circular Nanoplate with Simply Supported Boundary Conditions $n = 2$, $\Delta T = 200$ K, $\gamma = 0.2$

| | U | | | LV | | | | | | PV | | | | | | QV | | | | | |
| | $\alpha = 0$ $\beta = 0$ | | | $\alpha = -0.3$ $\beta = 0$ | | | $\alpha = 0.3$ $\beta = 0$ | | | $\alpha = 0$ $\beta = -0.3$ | | | $\alpha = 0$ $\beta = 0.3$ | | | $\alpha = -0.3$ $\beta = -0.3$ | | | $\alpha = -0.3$ $\beta = -0.3$ | | |
m	I	II	III	I	II	III	I	II	III	I	II	III	I	II	III	I	II	III	I	II	III
10	3.6	16.5	30.6	3.1	14.7	27.8	4	18.1	32.8	3.2	15.3	28.6	3.9	17.5	31.4	2.6	13.3	24.8	4.3	19.1	33.2
11	**3.6**	16.5	30.5	3.1	14.8	27.6	**4**	18.2	33.1	3.2	15.3	28.6	**3.9**	17.7	32	2.6	13.3	24.8	4.2	19.2	34.8
12	–	16.6	30.7	3.1	14.8	27.6	–	18.2	33.4	3.2	15.4	28.6	–	17.6	32.6	2.6	13.3	24.8	4.3	19.2	35.1
13	–	**16.6**	30.7	3.1	14.8	27.6	–	18.2	33.3	**3.3**	15.4	28.7	–	17.7	32.5	2.7	13.3	24.9	**4.3**	19.3	34.9
14	–	–	30.7	3.1	14.8	27.7	–	18.2	33.4	3.3	15.4	28.7	–	17.7	32.5	**2.7**	13.3	24.9	–	19.3	35
15	–	–	30.8	3.2	14.9	27.7	–	18.3	33.4	–	15.4	28.7	–	–	32.5	–	13.4	24.9	–	19.3	35
16	–	–	**30.8**	**3.2**	**14.9**	27.7	–	**18.3**	33.5	–	15.4	28.7	–	17.7	32.6	–	**13.4**	24.9	–	19.3	35
17	–	–	–	–	–	27.7	–	–	**33.5**	–	15.5	28.8	–	–	32.6	–	–	25	–	19.4	35
18	–	–	–	–	–	27.8	–	–	–	**15.5**	**28.8**		–	–	**32.6**	–	–	**25**	–	**19.4**	35.1
19	–	–	–	–	–	**27.8**	–	–	–	–	–	–	–	–	–	–	–	–	–	–	**35.1**
20	–	–	–	–	–	–	–	–	–	–	–	–	–	–	–	–	–	–	–	–	–
21	–	–	–	–	–	–	–	–	–	–	–	–	–	–	–	–	–	–	–	–	–
22	–	–	–	–	–	–	–	–	–	–	–	–	–	–	–	–	–	–	–	–	–

TABLE 2.7

Comparison of Frequency Parameter for Uniform Isotropic Annular Plate

	$b/a \rightarrow$	0.1	0.2	0.3	0.4
CC	Present	24.6289	30.8414	39.3983	51.7036
	[67]	24.629	30.841	39.398	51.704
	[68]	24.634	–	39.395	–
CS	Present	16.5746	21.0565	27.3789	36.7015
	[67]	16.575	21.057	27.379	36.701
SS	Present	13.8742	16.1636	20.2187	27.6976
	[67]	13.874	16.164	20.219	26.698

TABLE 2.8

Comparison of Frequency Parameter for Uniform Isotropic Circular Plate

Ref.\Modes	C-plate			S-plate		
	I	II	III	I	II	III
Present	10.2158	39.7711	89.1041	4.9351	29.7200	74.1561
[69]	10.2159	39.7766	89.1708	4.9352	29.7222	74.1938
[70]	10.216	39.771	89.103	4.935	29.720	74.156
[71]	10.2158	39.7711	89.1041	4.935	29.7200	74.1560
[72]	10.2158	39.7711	89.1041	4.935	29.7200	74.1560
[73]	10.2158	39.7711	89.1041	4.9351	29.72	74.1561
[74]	10.216	39.773	–	4.935	29.736	–
[75]	10.215	39.771	89.104	4.935	29.720	74.156
[76]	10.215	39.771	89.104	4.935	29.720	74.156

TABLE 2.9

Comparison of Frequency Parameter for Uniform Functionally Graded Circular Plate

n	Ref.\Modes	C-plate			S-plate		
		I	II	III	I	II	III
1	Present	8.4988	33.0865	74.1277	4.1057	24.7247	61.6921
	[73]	8.4988	33.0865	74.1277	4.1057	24.7247	61.6921
	[74]	8.500	33.093	–	4.106	24.742	–
	[75]	8.498	33.086	74.127	4.105	24.724	61.692
	[76]	8.498	33.086	74.127	4.105	24.724	61.692
2	Present	8.1236	31.6258	70.8551	3.9244	23.6332	58.9685
	[73]	8.1236	31.6258	70.8551	3.9244	23.6332	58.9685

(Continued)

TABLE 2.9 *(Continued)*
Comparison of Frequency Parameter for Uniform Functionally Graded Circular Plate

n	Ref.\Modes	C-plate			S-plate		
		I	**II**	**III**	**I**	**II**	**III**
	[74]	8.125	31.634	–	3.925	23.651	–
	[75]	8.123	31.625	70.855	3.924	23.633	58.968
	[76]	8.123	31.625	70.855	3.924	23.633	58.968
3	Present	7.9112	30.7988	69.0024	3.8218	23.0152	57.4266
	[73]	7.9112	30.7988	69.0024	3.8218	23.0152	57.4266
	[75]	7.911	30.798	69.002	3.821	23.015	57.426
	[76]	7.911	30.798	69.002	3.821	23.015	57.426
4	Present	7.7335	30.1074	67.4532	3.7360	22.4985	56.1372
	[73]	7.7335	30.1074	67.4532	3.736	22.4985	56.1370
	[75]	7.733	30.107	67.453	3.736	22.498	56.137
	[76]	7.733	30.107	67.453	3.736	22.498	56.137
5	Present	7.5738	29.4854	66.0598	3.6588	22.0338	54.9777
	[73]	7.5738	29.4854	66.0598	3.6588	22.0338	54.9777
	[74]	7.576	29.496	–	3.659	22.052	–
	[75]	7.573	29.485	66.059	3.658	22.033	54.977
	[76]	7.573	29.485	66.059	3.658	22.033	54.977

TABLE 2.10
Comparison of Frequency Parameter for Nonuniform Functionally Graded Circular Plate

α	β	n	Ref./Modes	C-plate			S-plate		
				I	**II**	**III**	**I**	**II**	**III**
0.5	0	0	Present	14.3021	51.3480	112.6360	6.2927	37.7423	93.0342
			[77]	14.302	51.349	112.64	6.2928	37.743	93.042
			[78]	14.3022	51.3487	112.6399	6.2928	37.7427	93.0367
			[72]	14.3021	51.3480	112.6360	6.2927	37.7423	93.0342
−0.5	0	0	Present	6.1504	27.3002	63.0611	3.5498	21.2386	53.4404
			[77]	6.1504	27.300	63.062	3.5498	21.239	53.441
			[78]	6.1504	27.3004	63.0618	3.5498	21.2387	53.4409
			[72]	6.1504	27.3002	63.0611	3.5498	21.2386	53.4404
0.1	0	1	Present	9.1762	35.0520	78.1579	4.3311	26.0778	64.9168
			[73]	9.1762	35.0520	78.1579	4.3311	24.2748	64.9168
		3	Present	8.5418	32.6284	72.7539	4.0316	24.2748	60.4283
			[73]	8.5418	32.6284	72.7539	4.0316	24.2748	60.4283

(Continued)

TABLE 2.10 *(Continued)*
Comparison of Frequency Parameter for Nonuniform Functionally Graded Circular Plate

α	β	n	Ref./Modes	C-plate			S-plate		
				I	**II**	**III**	**I**	**II**	**III**
−0.1	0	1	Present	7.8223	31.0941	70.0212	3.8798	23.3582	58.4115
			[73]	7.8223	31.0941	70.0212	3.8798	23.3582	58.4115
		3	Present	7.2815	28.9442	65.1798	3.6116	21.7432	54.3728
			[73]	7.2815	28.9442	65.1798	3.6116	21.7432	54.3728
0	0.5	0	Present	13.7310	48.1833	105.2133	5.8537	34.729	86.1120
			[78]	13.7310	48.1833	105.2133	5.8537	34.729	86.1120
	−0.5	0	Present	6.6320	30.0152	69.8624	4.0392	23.887	59.9533
			[78]	6.6320	30.0152	69.8624	4.0392	23.887	59.9533
0.5	0.5	0	Present	17.8331	59.2181	127.3436	7.2492	42.4881	103.9770
			[78]	17.8332	59.2193	127.3516	7.2492	42.4887	103.9816

approach is adopted by the author to analyse circular plates in his published work for the circular plates with uniform [79], linear [42], parabolic [41], and quadratic [80] thickness variation [81] and the results are compared with those obtained by the present approach and recorded in Tables 2.11 and 2.12 for clamped and simply supported circular plates, respectively. The effect of various parameters is compared for both approaches, and it is found that the values of frequency parameter is higher for the present approach as compared to the earlier ones for both boundary conditions, keeping all parameters fixed, which shows that the contribution of thermal stress is more in the first approach as compared to the second ones. This effect is more pronounced for higher values of volume fraction index and taper parameters.

2.7.2   PARAMETRIC DISCUSSION

2.7.2.1   Annular Nanoplates

The numerical results for the nonuniform annular nanoplates with CC, CS, and SS-boundary conditions are presented in Tables 2.13–2.15 for the first three modes of vibration. Due to edge fixity conditions of the nanoplates, the values of frequency parameter are found to be in the order of $\Omega_{CC} > \Omega_{CS} > \Omega_{SS}$. The volume fraction index, nonlocal parameter, and temperature difference reduce the stiffness of the functionally graded nanoplate. Thereafter, the value of frequency parameter decreases with the increasing values of volume fraction index, nonlocal parameter, and temperature difference. However, it increases with the increase in radii ratio of the functionally graded annular nanoplates. In view of thickness variations, the values of frequency parameter are found to be in the order of $\Omega_{UV} < \Omega_{PV} < \Omega_{LV} < \Omega_{QV}$, when the nanoplate becomes thicker and thicker towards the outer periphery while it is $\Omega_{UV} > \Omega_{PV} > \Omega_{LV} > \Omega_{QV}$, when the nanoplate becomes thinner and thinner towards the outer periphery.

TABLE 2.11

Comparison of Frequency Parameter for Different Approaches of Thermal for Nonuniform Functionally Graded Clamped Circular Plate $\Delta T = 200$ K

		U			LV						PV						QV		
		$\alpha = 0$			$\alpha = -0.5$			$\alpha = 0.5$			$\alpha = 0$			$\alpha = 0$			$\alpha = 0.5$		
		$\beta = 0$			$\beta = 0$			$\beta = 0$			$\beta = -0.5$			$\beta = 0.5$			$\beta = 0.5$		
n		I	II	III	I	II	III	I	II	III	I	II	III	I	II	III	I	II	III
0	a	8.4544	34.6408	78.5269	5.0831	26.9049	64.1711	11.2946	40.9382	90.0656	5.3655	28.1491	67.4982	11.2693	39.7997	87.1113	13.8652	45.4034	97.3421
	b	8.4466	34.6117	78.4625	4.3916	23.1864	54.9585	12.2701	45.0771	99.5638	4.9283	25.7097	61.1291	11.7277	42.2143	92.9079	15.4975	52.1510	112.7180
1	a	7.9484	32.4117	73.3977	4.8303	25.2236	60.0311	10.5867	38.2768	84.1565	5.0808	26.3701	63.1215	10.5680	37.2189	81.4034	12.9822	42.4411	90.9460
	b	7.9447	32.3931	73.3539	4.2057	21.7522	51.4303	11.4980	42.1534	93.0487	4.6950	24.0997	57.1838	10.9942	39.4834	86.8364	14.5044	48.7543	105.3293
2	a	7.7555	31.7157	71.8663	4.6833	24.6527	58.7484	10.3483	37.4708	82.4160	4.9363	25.7849	61.7856	10.3271	36.4313	79.7155	12.6980	41.5536	89.0707
	b	7.7507	31.6947	71.8178	4.0638	21.2550	50.3258	11.2409	41.2637	91.1185	4.5497	23.5601	55.9678	10.7457	38.6459	85.0304	14.1902	47.7332	103.1516
3	a	7.6426	31.3454	71.0715	4.5848	24.3356	58.0685	10.2164	37.0491	81.5199	4.8430	25.4648	61.0836	10.1925	36.0174	78.8445	12.5444	41.0922	88.1079
	b	7.6372	31.3225	71.0194	3.9635	20.9768	49.7386	11.1001	40.7982	90.1237	4.4512	23.2630	55.3267	10.6084	38.2058	84.0975	14.0223	47.2027	102.0326
4	a	7.5642	31.1011	70.5552	4.5120	24.1212	57.6212	10.1274	36.7737	80.9407	4.7750	25.2503	60.6240	10.1013	35.7464	78.2807	12.4421	40.7920	87.4865
	b	7.5582	31.0768	70.5001	3.8872	20.7880	49.3515	11.0056	40.4943	89.4801	4.3776	23.0631	54.9062	10.5159	37.9178	83.4931	13.9115	46.8578	101.3101
5	a	7.5052	30.9230	70.1821	4.4551	23.9626	57.2956	10.0616	36.5742	80.5233	4.7225	25.0925	60.2904	10.0336	35.5498	77.8741	12.3670	40.5750	87.0393
	b	7.4988	30.8975	70.1247	3.8270	20.6480	49.0694	10.9361	40.2740	89.0162	4.3200	22.9156	54.6006	10.4476	37.7087	83.0572	13.8306	46.6084	100.7900

TABLE 2.12

Comparison of Frequency Parameter for Different Approaches of Thermal for Nonuniform Functionally Graded Simply Supported Circular Plate $\Delta T = 200$ K

| | | U | | | LV | | | | | | PV | | | | | | QV | | |
| | | α = 0 β = 0 | | | α = -0.5 β = 0 | | | α = 0.5 β = 0 | | | α = 0 β = -0.5 | | | α = 0 β = 0.5 | | | α = 0.5 β = 0.5 | | |
n		I	II	III	I	II	III	I	II	III	I	II	III	I	II	III	I	II	III
0	a	3.1459	25.5135	65.1079	1.7171	20.5315	54.2181	4.1558	29.7707	74.1578	2.4949	22.2246	58.0188	3.8228	28.2317	70.9368	4.7525	32.2541	79.2225
	b	3.0959	25.4614	65.0232	0.7000	17.4764	46.1792	4.6578	32.8180	82.0171	1.8423	20.0038	52.1285	4.1791	30.0685	75.7955	5.6620	37.1323	91.8334
1	a	3.0551	23.9031	60.8752	1.8135	19.3011	50.7559	3.9634	27.8572	69.3064	2.4599	20.8588	54.2845	3.6703	26.4270	66.3050	4.5107	30.1689	74.0292
	b	3.0333	23.8743	60.8231	1.1174	16.4608	43.2623	4.4452	30.7243	76.6772	1.9458	18.8039	48.8042	4.0085	28.1626	70.8718	5.3653	34.7449	85.8382
2	a	2.9246	23.3715	59.5933	1.6565	18.8336	49.6505	3.8346	27.2578	67.8648	2.3343	20.3733	53.1194	3.5370	25.8526	64.9206	4.3765	29.5268	72.4956
	b	2.9023	23.3416	59.5382	0.8963	16.0579	42.3164	4.3078	30.0621	75.0777	1.7989	18.3617	47.7525	3.8749	27.5494	69.3879	5.2176	34.0048	84.0554
3	a	2.8241	23.0803	58.9226	1.5098	18.5603	49.0550	3.7455	26.9382	67.1188	2.2319	20.0976	52.4996	3.4404	25.5436	64.2018	4.2877	29.1877	71.7048
	b	2.7982	23.0473	58.8628	0.6166	15.8181	41.8030	4.2134	29.7076	74.2474	1.6613	18.1065	47.1892	3.7794	27.2180	68.6148	5.1226	33.6130	83.1339
4	a	2.7448	22.8849	58.4846	1.3817	18.3705	48.6595	3.6785	26.7271	66.6349	2.1492	19.9090	52.0909	3.3662	25.3386	63.7346	4.2222	28.9650	71.1931
	b	2.7149	22.8489	58.4204	0.1757	15.6496	41.4602	4.1424	29.4727	73.7078	1.5415	17.9301	46.8160	3.7065	26.9974	68.1111	5.0533	33.3553	82.5366
5	a	2.6809	22.7410	58.1672	1.2696	18.2280	48.3700	3.6259	26.5731	66.2857	2.0813	19.7686	51.7929	3.3073	25.1885	63.3970	4.1713	28.8031	70.8242
	b	2.6469	22.7022	58.0991	*	15.5222	41.2083	4.0866	29.3010	73.3176	1.4378	17.7979	46.5430	3.6486	26.8355	67.7465	4.9998	33.1675	82.1055

Notes:

[a] Present results.

[b] Results from ref. [2].

* Value does not exist due to buckling

TABLE 2.13
Numerical Values of Frequency Parameter of Functionally Graded Annular Nanoplate in First Mode

				U	LV		PV		QV	
				$\alpha = 0$	$\alpha = -0.3$	$\alpha = 0.3$	$\alpha = 0$	$\alpha = 0$	$\alpha = -0.3$	$\alpha = 0.3$
$\Delta T \downarrow$	$\varepsilon \downarrow$	$\gamma \downarrow$	$n \downarrow$	$\beta = 0$	$\beta = 0$	$\beta = 0$	$\beta = -0.3$	$\beta = 0.3$	$\beta = -0.3$	$\beta = 0.3$
CC										
0	0.3	0.1	0.5	36.2238	32.2285	39.7607	32.6006	39.4500	27.8865	42.6824
			2	34.7191	30.8898	38.1091	31.2464	37.8114	26.7282	40.9095
		0.2	0.5	28.6410	25.5101	31.4581	25.8309	31.2194	22.2895	33.8271
			2	27.4514	24.4505	30.1514	24.7580	29.9226	21.3637	32.4220
	0.5	0.1	0.5	65.1564	57.1370	72.2515	58.2233	71.3452	48.8896	77.8502
			2	62.4500	54.7637	69.2504	55.8049	68.3817	46.8589	74.6166
		0.2	0.5	46.0403	40.4074	51.0733	41.2233	50.4498	34.9064	55.0988
			2	44.1279	38.7290	48.9518	39.5110	48.3543	33.4565	52.8101
200	0.3	0.1	0.5	35.0194	30.9598	38.5644	31.3753	38.2359	26.4689	41.4591
			2	33.5571	29.6623	36.9569	30.0618	36.6415	25.3519	39.7324
		0.2	0.5	27.2970	23.9464	30.2094	24.3519	29.9285	20.2371	32.5916
			2	26.1479	22.9295	28.9432	23.3203	28.6728	19.3593	31.2280
	0.5	0.1	0.5	63.1591	55.0742	70.2271	56.2012	69.3138	46.5573	75.7632
			2	60.5256	52.7706	67.3033	53.8523	66.4274	44.5964	72.6111
		0.2	0.5	43.9575	37.9472	49.1448	38.8754	48.4765	31.5200	53.2042
			2	42.1089	36.3361	47.0871	37.2288	46.4452	30.1485	50.9810
CS										
0	0.3	0.1	0.5	24.0813	22.4959	25.5339	23.2503	24.7889	21.4441	26.1718
			2	23.0811	21.5615	24.4733	22.2845	23.7593	20.5534	25.0847
		0.2	0.5	19.2158	17.9540	20.3977	18.5841	19.8023	17.2809	20.9417
			2	18.4176	17.2083	19.5505	17.8122	18.9798	16.5631	20.0719
	0.5	0.1	0.5	44.1170	40.0468	47.8225	41.6199	46.4068	37.1756	49.9220
			2	42.2846	38.3834	45.8361	39.8912	44.4792	35.6315	47.8484
		0.2	0.5	31.5300	28.6156	34.2089	29.7731	33.2086	26.7768	35.7633
			2	30.2204	27.4270	32.7879	28.5364	31.8292	25.6645	34.2778
200	0.3	0.1	0.5	23.1349	21.4205	24.6485	22.2140	23.8945	20.1380	25.3145
			2	22.1654	20.5182	23.6184	21.2801	22.8951	19.2829	24.2577
		0.2	0.5	18.2508	16.7689	19.5377	17.4427	18.9293	15.5698	20.1322
			2	17.4810	16.0546	18.7176	16.7019	18.1339	14.8913	19.2889
	0.5	0.1	0.5	42.6396	38.4257	46.3872	40.0203	44.9809	35.1688	48.5014
			2	40.8588	36.8143	44.4538	38.3441	43.1053	33.6818	46.4817
		0.2	0.5	30.0664	26.8151	32.8903	28.0190	31.8825	24.0510	34.5126
			2	28.8011	25.6751	31.5126	26.8308	30.5460	23.0011	33.0700
SS										
0	0.3	0.1	0.5	17.1834	15.6135	18.5502	16.1793	18.0043	14.2511	19.2615
			2	16.4696	14.9650	17.7797	15.5073	17.2565	13.6592	18.4614
		0.2	0.5	14.0618	12.7996	15.1795	13.2727	14.7355	11.7891	15.7754

(Continued)

TABLE 2.13 *(Continued)*

Numerical Values of Frequency Parameter of Functionally Graded Annular Nanoplate in First Mode

				U	LV		PV		QV	
				$\alpha = 0$	$\alpha = -0.3$	$\alpha = 0.3$	$\alpha = 0$	$\alpha = 0$	$\alpha = -0.3$	$\alpha = 0.3$
$\Delta T\downarrow$	$\varepsilon\downarrow$	$\gamma\downarrow$	$n\downarrow$	$\beta = 0$	$\beta = 0$	$\beta = 0$	$\beta = -0.3$	$\beta = 0.3$	$\beta = -0.3$	$\beta = 0.3$
			2	13.4777	12.2680	14.5490	12.7214	14.1234	11.2995	15.1201
	0.5	0.1	0.5	30.2737	26.7474	33.3569	27.6459	32.5243	23.3916	35.3564
			2	29.0162	25.6364	31.9713	26.4976	31.1733	22.4200	33.8878
		0.2	0.5	22.2953	19.7198	24.5660	20.3998	23.9584	17.3815	26.0565
			2	21.3692	18.9007	23.5456	19.5525	22.9633	16.6595	24.9742
200	0.3	0.1	0.5	16.3473	14.6399	17.7851	15.2609	17.2194	13.0788	18.5248
			2	15.6584	14.0177	17.0390	14.6147	16.4960	12.5158	17.7490
		0.2	0.5	13.2219	11.7643	14.4384	12.2887	13.9754	10.3530	15.0792
			2	12.6610	11.2584	13.8299	11.7626	13.3854	9.89460	14.4455
	0.5	0.1	0.5	29.0787	25.3891	32.2247	26.3351	31.3842	21.6903	34.2435
			2	27.8600	24.3178	30.8785	25.2262	30.0722	20.7624	32.8150
		0.2	0.5	21.1098	18.2488	23.5102	18.9876	22.8858	15.2299	25.0570
			2	20.2177	17.4673	22.5228	18.1772	21.9237	14.5547	24.0075

TABLE 2.14

Numerical Values of Frequency Parameter of Functionally Graded Annular Nanoplate in Second Mode

				U	LV		PV		QV	
				$\alpha = 0$	$\alpha = -0.3$	$\alpha = 0.3$	$\alpha = 0$	$\alpha = 0$	$\alpha = -0.3$	$\alpha = 0.3$
$\Delta T\downarrow$	$\varepsilon\downarrow$	$\gamma\downarrow$	$n\downarrow$	$\beta = 0$	$\beta = 0$	$\beta = 0$	$\beta = -0.3$	$\beta = 0.3$	$\beta = -0.3$	$\beta = 0.3$
CC										
0	0.3	0.1	0.5	80.2197	71.5759	87.8438	73.2155	86.3642	63.0252	93.3593
			2	76.8876	68.6029	84.1950	70.1744	82.7769	60.4073	89.4815
		0.2	0.5	51.0485	45.5057	55.8776	46.5523	54.8965	39.8209	59.2851
			2	48.9281	43.6156	53.5566	44.6186	52.6163	38.1669	56.8226
	0.5	0.1	0.5	129.8404	113.9800	143.7957	116.8227	141.2896	98.1153	154.0828
			2	124.4473	109.2456	137.8229	111.9702	135.4209	94.0399	147.6827
		0.2	0.5	75.3776	66.1132	83.4509	67.7187	81.9433	56.4139	89.2915
			2	72.2467	63.3670	79.9846	64.9059	78.5396	54.0707	85.5826
200	0.3	0.1	0.5	77.7746	69.0708	85.3693	70.7531	83.8838	60.2702	90.8234
			2	74.5321	66.1835	81.8148	67.7979	80.3901	57.7377	87.0440
		0.2	0.5	48.7608	42.8555	53.7481	44.0253	52.7183	36.3651	57.1978
			2	46.7108	41.0389	51.4973	42.1637	50.5086	34.7936	54.8066
	0.5	0.1	0.5	126.0519	110.1481	139.9078	113.0289	137.4151	93.8776	150.0663

(Continued)

TABLE 2.14 *(Continued)*

Numerical Values of Frequency Parameter of Functionally Graded Annular Nanoplate in Second Mode

ΔT	ε	γ	n	U $\alpha = 0$ $\beta = 0$	LV $\alpha = -0.3$ $\beta = 0$	LV $\alpha = 0.3$ $\beta = 0$	PV $\alpha = 0$ $\beta = -0.3$	PV $\alpha = 0$ $\beta = 0.3$	QV $\alpha = -0.3$ $\beta = -0.3$	QV $\alpha = 0.3$ $\beta = 0.3$
			2	120.8005	105.5475	134.0863	108.3111	131.6961	89.9343	143.8256
		0.2	0.5	72.0613	62.2052	80.3778	63.9739	78.8236	51.1178	86.2936
			2	69.0331	59.5670	77.0143	61.2670	75.5228	48.8984	82.6894
CS										
0	0.3	0.1	0.5	65.5388	59.1503	71.2699	60.9468	69.7055	53.7569	75.0606
			2	62.8165	56.6934	68.3096	58.4153	66.8101	51.5240	71.9428
		0.2	0.5	42.5066	38.2522	46.2884	39.3605	45.3017	34.3893	48.8148
			2	40.7410	36.6633	44.3657	37.7256	43.4200	32.9609	46.7872
	0.5	0.1	0.5	107.3787	94.9762	118.4038	97.8812	115.9234	83.6994	126.1428
			2	102.9185	91.0312	113.4857	93.8155	111.1083	80.2228	120.9032
		0.2	0.5	63.4578	56.0177	70.0245	57.6355	68.5871	48.8073	74.6442
			2	60.8220	53.6909	67.1159	55.2415	65.7382	46.7800	71.5437
200	0.3	0.1	0.5	63.4720	56.9793	69.2104	58.8050	67.6478	51.2502	72.9780
			2	60.8241	54.5952	66.3275	56.3467	64.8290	49.0929	69.9402
		0.2	0.5	40.5852	35.9963	44.5114	37.1919	43.4913	31.3141	47.0840
			2	38.8784	34.4698	42.6471	35.6184	41.6680	29.9585	45.1153
	0.5	0.1	0.5	104.1926	91.7011	115.1644	94.6248	112.7045	79.9315	122.8241
			2	99.8507	87.8690	110.3715	90.6734	108.0129	76.5703	117.7155
		0.2	0.5	60.6575	52.6903	67.4389	54.4264	65.9686	44.1427	72.1307
			2	58.1083	50.4552	64.6167	52.1229	63.2059	42.2240	69.1179
SS										
0	0.3	0.1	0.5	54.3633	48.6643	59.5011	50.3117	58.0950	44.0064	62.9206
			2	52.1052	46.6430	57.0296	48.2219	55.6819	42.1785	60.3070
		0.2	0.5	35.6299	31.8600	39.0156	32.9265	38.1018	28.7150	41.2740
			2	34.1499	30.5366	37.3950	31.5588	36.5191	27.5222	39.5596
	0.5	0.1	0.5	88.1813	77.5171	97.6732	79.9577	95.6643	67.9400	104.4693
			2	84.5185	74.2972	93.6161	76.6365	91.6907	65.1179	100.130
		0.2	0.5	52.3932	46.0238	58.0416	47.4493	56.8477	40.1732	62.0740
			2	50.2169	44.1121	55.6307	45.4785	54.4864	38.5045	59.4956
200	0.3	0.1	0.5	52.5484	46.7426	57.7028	48.4247	56.2927	41.7833	61.1044
			2	50.3539	44.7836	55.2974	46.3975	53.9450	40.0204	58.5592
		0.2	0.5	33.9337	29.8781	37.4455	31.0232	36.4976	26.0422	39.7414
			2	32.5047	28.6085	35.8754	29.7086	34.9656	24.9122	38.0781
	0.5	0.1	0.5	85.4692	74.7084	94.9284	77.1772	92.9288	64.6947	101.6589
			2	81.9053	71.5833	90.9761	73.9515	89.0586	61.9697	97.4292
		0.2	0.5	50.0038	43.1914	55.8357	44.7223	54.6068	36.2148	59.9261
			2	47.9004	41.3569	53.4976	42.8275	52.3182	34.6376	57.4217

TABLE 2.15

Numerical Values of Frequency Parameter of Functionally Graded Annular Nanoplate in Third Mode

				U	LV		PV		QV	
				$\alpha = 0$	$\alpha = -0.3$	$\alpha = 0.3$	$\alpha = 0$	$\alpha = 0$	$\alpha = -0.3$	$\alpha = 0.3$
$\Delta T\downarrow$	$e\downarrow$	$\gamma\downarrow$	$n\downarrow$	$\beta = 0$	$\beta = 0$	$\beta = 0$	$\beta = -0.3$	$\beta = 0.3$	$\beta = -0.3$	$\beta = 0.3$
CC										
0	0.3	0.1	0.5	126.6306	113.1532	138.5002	116.3078	135.6422	100.4219	146.5574
			2	121.3708	108.4532	132.7473	111.4768	130.0080	96.2507	140.4698
		0.2	0.5	73.7353	65.8939	80.6442	67.7488	78.9649	58.5351	85.3226
			2	70.6726	63.1569	77.2945	64.9347	75.6850	56.1037	81.7785
	0.5	0.1	0.5	194.4296	170.8143	215.2218	175.5462	211.0656	147.8047	230.1688
			2	186.3536	163.7192	206.2821	168.2545	202.2986	141.6653	220.6083
		0.2	0.5	106.8736	93.9124	118.3090	96.5365	116.0322	81.4479	126.5558
			2	102.4344	90.0116	113.3948	92.5267	111.2126	78.0648	121.2991
200	0.3	0.1	0.5	122.9305	109.4201	134.7191	112.6067	131.8736	96.3638	142.6760
			2	117.8091	104.8515	129.1127	107.9083	126.3840	92.3227	136.7411
		0.2	0.5	70.5052	62.1814	77.6166	64.1919	75.8737	53.6569	82.3426
			2	67.5427	59.5487	74.3673	61.4804	72.6944	51.3434	78.9008
	0.5	0.1	0.5	188.8877	165.2684	209.4979	170.0279	205.3774	141.7246	224.2449
			2	181.0216	158.3701	200.7830	162.9352	196.8322	135.7787	214.9211
		0.2	0.5	102.2350	88.4798	113.9875	91.3132	111.6432	73.9944	122.3205
			2	97.9403	84.7302	109.2185	87.4523	106.9686	70.7868	117.2119
CS										
0	0.3	0.1	0.5	110.4562	99.1553	120.4867	102.2925	117.7199	89.3132	127.0213
			2	105.8682	95.0367	115.4820	98.0436	112.8302	85.6034	121.7452
		0.2	0.5	64.3810	57.7734	70.2219	59.6038	68.5946	51.9782	73.9919
			2	61.7068	55.3737	67.3051	57.1281	65.7454	49.8192	70.9185
	0.5	0.1	0.5	170.1870	149.9874	188.0409	154.5499	184.0962	131.1633	200.5711
			2	163.1180	143.7574	180.2302	148.1304	176.4494	125.7152	192.2400
		0.2	0.5	93.2131	82.1656	102.9678	84.6801	100.7853	71.9136	109.7782
			2	89.3413	78.7527	98.6908	81.1628	96.5990	68.9266	105.2184
200	0.3	0.1	0.5	107.1861	95.8186	117.1662	98.9758	114.4162	85.5814	123.6318
			2	102.7195	91.8164	112.2895	94.8447	109.6526	81.9895	118.4885
		0.2	0.5	61.5541	54.4965	67.5851	56.4485	65.9108	47.5436	71.4119
			2	58.9676	52.1887	64.7557	54.0635	63.1490	45.4910	68.4270
	0.5	0.1	0.5	165.3039	145.0644	183.0177	149.6392	179.1120	125.6475	195.3920
			2	158.4192	139.0082	175.4038	143.3958	171.6591	120.3732	187.2675
		0.2	0.5	89.1658	77.3973	99.2098	80.0770	96.9791	65.2286	106.1117
			2	85.4201	74.1170	95.0591	76.6907	92.9187	62.3983	101.6802
SS										
0	0.3	0.1	0.5	97.2009	86.9659	106.3202	89.8909	103.7633	78.2624	112.2497
			2	93.1635	83.3536	101.9040	86.1571	99.4533	75.0116	107.5872
		0.2	0.5	56.7984	50.7962	62.1170	52.4978	60.6124	45.5834	65.5412

(Continued)

TABLE 2.15 *(Continued)*
Numerical Values of Frequency Parameter of Functionally Graded Annular Nanoplate in Third Mode

				U	LV		PV		QV	
				$\alpha = 0$	$\alpha = -0.3$	$\alpha = 0.3$	$\alpha = 0$	$\alpha = 0$	$\alpha = -0.3$	$\alpha = 0.3$
$\Delta T\downarrow$	$e\downarrow$	$\gamma\downarrow$	$n\downarrow$	$\beta = 0$	$\beta = 0$	$\beta = 0$	$\beta = -0.3$	$\beta = 0.3$	$\beta = -0.3$	$\beta = 0.3$
			2	54.4392	48.6863	59.5368	50.3172	58.0947	43.6900	62.8188
	0.5	0.1	0.5	148.9009	130.8794	164.8404	134.9238	161.3917	114.1547	176.1087
			2	142.7160	125.4431	157.9935	129.3195	154.6880	109.4131	168.7937
		0.2	0.5	81.4763	71.5994	90.1900	73.7970	88.2911	62.3155	96.3215
			2	78.0920	68.6254	86.4438	70.7317	84.6238	59.7271	92.3206
200	0.3	0.1	0.5	94.2555	83.9527	103.3347	86.8998	100.7900	74.8888	109.2032
			2	90.3262	80.4441	99.0324	83.2709	96.5923	71.7432	104.6590
		0.2	0.5	54.2414	47.8430	59.7287	49.6554	58.1787	41.6289	63.2008
			2	51.9607	45.8152	57.2269	47.5559	55.7394	39.8301	60.5579
	0.5	0.1	0.5	144.5699	126.5040	160.3911	130.5664	156.9716	109.2517	171.5206
			2	138.5474	121.2209	153.7175	125.1172	150.4388	104.6632	164.3877
		0.2	0.5	77.8865	67.3788	86.8553	69.7285	84.9086	56.4576	93.0642
			2	74.6134	64.5214	83.2205	66.7785	81.3524	54.0063	89.1766

The above effects increase with the increase in the number of modes of vibration of the functionally graded annular nanoplates.

2.7.2.2 Circular Nanoplates

For circular plates, the results are shown in Table 2.16 for C and S boundary conditions and the first three modes of vibration. The values of frequency parameter for circular nanoplates are smaller as compared to annular nanoplates. Similar to the annular nanoplate, the value of frequency parameter for circular nanoplate decreases with the increase in volume fraction index, nonlocal parameter, and temperature difference for both boundary conditions. However, in view of thickness variations, these are found to be in the order of $\Omega_{UV} < \Omega_{PV} \leq \Omega_{LV} < \Omega_{QV}$ when the nanoplate becomes thicker and thicker towards the outer periphery, while for rest of the cases it remains the same for annular nanoplates.

2.8 CONCLUSIONS

The vibrational behaviour of functionally graded nonuniform annular and circular nanoplates under thermal environment has been studied on the basis of Eringen's nonlocal elasticity theory and classical plate theory by utilizing generalized differential quadrature method. The effect of size-dependency, radii ratio, ceramic/metallic

TABLE 2.16

Numerical Values of Frequency Parameter of Functionally Graded Circular Nanoplate

			U		LV				PV				QV			
			$\alpha = 0$		$\alpha = -0.3$		$\alpha = 0.3$		$\alpha = 0$		$\alpha = 0$		$\alpha = -0.3$		$\alpha = 0.3$	
			$\beta = 0$		$\beta = 0$		$\beta = 0$		$\beta = -0.3$		$\beta = 0.3$		$\beta = -0.3$		$\beta = 0.3$	
$\Delta T\downarrow$	$\gamma\downarrow$	$n\downarrow$	C	S	C	S	C	S	C	S	C	S	C	S	C	S
Mode I																
0	0.1	0.5	8.8	4.3	7.1	3.8	10.4	4.7	7.2	4.0	10.4	4.6	5.3	3.4	11.9	5.0
		2	8.4	4.1	6.8	3.7	10.0	4.5	6.9	3.8	10.0	4.4	5.1	3.3	11.4	4.8
	0.2	0.5	8.1	4.0	6.5	3.5	9.5	4.4	6.6	3.7	9.5	4.3	4.9	3.2	10.9	4.6
		2	7.7	3.8	6.2	3.4	9.1	4.2	6.3	3.5	9.1	4.1	4.7	3.1	10.5	4.4
200	0.1	0.5	8.4	3.7	6.6	3.2	10.0	4.2	6.7	3.4	10.0	4.1	4.8	2.8	11.5	4.5
		2	8.0	3.6	6.3	3.0	9.5	4.0	6.4	3.2	9.5	3.9	4.6	2.6	11.0	4.3
	0.2	0.5	7.8	3.7	6.2	3.2	9.3	4.1	6.2	3.3	9.2	4.0	4.4	2.7	10.7	4.4
		2	7.5	3.5	5.9	3.0	8.9	3.9	6.0	3.2	8.9	3.8	4.2	2.6	10.2	4.2
Mode II																
0	0.1	0.5	30.7	23.2	27.1	21.0	33.9	25.3	27.7	21.7	33.4	24.6	23.6	19.3	36.3	26.7
		2	29.4	22.2	26.0	20.1	32.5	24.3	26.6	20.8	32.0	23.6	22.6	18.5	34.8	25.6
	0.2	0.5	23.2	17.8	20.6	16.1	25.6	19.5	21.0	16.7	25.2	18.9	18.0	14.9	27.4	20.5
		2	22.2	17.1	19.7	15.4	24.5	18.6	20.1	16.0	24.1	18.1	17.3	14.3	26.2	19.7
200	0.1	0.5	29.7	22.3	26.1	20.0	32.9	24.5	26.7	20.8	32.3	23.7	22.4	18.2	35.3	25.8
		2	28.4	21.4	25.0	19.1	31.5	23.5	25.5	19.9	31.0	22.7	21.4	17.4	33.8	24.7
	0.2	0.5	22.4	17.2	19.7	15.3	24.8	18.9	20.1	15.9	24.4	18.3	16.8	13.8	26.6	20.0
		2	21.5	16.4	18.8	14.7	23.8	18.1	19.3	15.3	23.4	17.5	16.0	13.2	25.5	19.1
Mode III																
0	0.1	0.5	59.3	50.0	53.5	45.5	64.4	54.1	54.9	47.1	63.1	52.7	48.2	42.1	67.8	56.5
		2	56.8	47.9	51.3	43.6	61.7	51.9	52.6	45.1	60.5	50.5	46.2	40.4	65.0	54.2
	0.2	0.5	38.9	33.1	35.2	30.1	42.2	35.8	36.1	31.1	41.4	34.8	31.7	27.8	44.4	37.4
		2	37.3	31.7	33.7	28.9	40.5	34.3	34.6	29.9	39.6	33.4	30.3	26.7	42.6	35.8
200	0.1	0.5	57.5	48.4	51.7	43.9	62.6	52.5	53.1	45.5	61.3	51.1	46.2	40.3	66.0	54.9
		2	55.1	46.4	49.5	42.0	60.0	50.3	50.9	43.6	58.7	49.0	44.2	38.6	63.2	52.7
	0.2	0.5	37.4	31.9	33.5	28.7	40.8	34.6	34.5	29.8	39.9	33.7	29.5	25.8	43.0	36.2
		2	35.9	30.5	32.1	27.5	39.1	33.2	33.0	28.5	38.3	32.3	28.2	24.7	41.2	34.7

constituent, thermal field, and taper parameters of the nanoplate has been discussed for corresponding boundary conditions and concluded that:

- The values of frequency parameter are found in the order of $\Omega_{CC} > \Omega_{CS} > \Omega_{SS}$ for annular nanoplates and $\Omega_C > \Omega_S$ for circular nanoplates.
- Due to the increase in surrounding temperature, the stiffness of the plate decreases and hence the fundamental frequency decreases.
- As the radii ratio of the annular nanoplates increases, the value of the dimensionless frequency increases.

- The increasing contribution of the metallic constituent in the plate leads to the fall off the value of frequency parameter Ω for all the three plates, for the same set of the values of other parameters.
- The thickness variations have significant effect on the frequency parameter of the nanoplates.
- The functionally graded nanoplates become more size-dependent as the nonlocal parameters increase and diminish the effect of other parameters.

REFERENCES

[1] S. Chakraverty. *Vibration of plates*. Boca Raton: CRC Press, 2009.

[2] R. Saini. *Numerical solution of some vibration problems of FGM circular plates under thermal environment*. Roorkee: Indian Institute of Technology, 2020.

[3] W. Soedel. *Vibrations of shells and plates*, third edition. Boca Raton, Florida: CRC Press, 2004.

[4] E. Chladni. *Entdeckungen über die Theorie des Klanges*. Leipzig: Weidmann und Reich, 1787.

[5] J. Biot. *Traite de Physique experimentale et Mathematique*. Paris: Deterville, 1816.

[6] D. Bernoulli. *Letters to Euler*. Basel, 1735. Referencia extraida de Vibration of shells and plates. Werner Soedel.

[7] L. Euler. *Methodus inveniendi lineas curvas maximi minimive proprietate gaudentes*. Berlin, 1744.

[8] C. Coulomb. *Recherches theoriques et experimentales sur la force torsion et sur l'elasticite des fils de metal*. Paris, 1784.

[9] A. Cauchy. *Exercices de mathematiques*. Paris, 1827.

[10] Saint-Venant. Memoir sur les vibrations tournantes des verges elastiques. *Comptes Rendus*, 28(1849): 69.

[11] S. Poisson. *Sur l'equilibre et le mouvement des corps elastiques*. Paris, 1829.

[12] G. Lamè. *Lecons sur la theorie mathematique de l'elasticite des corps solides*. Paris, 1852.

[13] G. Kirchhoff. Uber das gleichgewicht and bewegung einer elastischen scheibe. Journal Fur die Reine. *Angenwandte Math*, 40 (1850): 51–88.

[14] G. Kirchhoff. Uber die schwingungen einer kreisformigen elastischen scheibe, Ann. *Der Phys. Und Chemie*, 81 (1850): 258–264.

[15] A.E.H. Love. *A treatise on the mathematical theory of elasticity*. New York: Dover Publications, 1944.

[16] Z. Bin Shen, H.L. Tang, D.K. Li, and G.J. Tang. Vibration of single-layered graphene sheet-based nanomechanical sensor via nonlocal Kirchhoff plate theory. *Comput. Mater. Sci.*, 61 (2012): 200–205. https://doi.org/10.1016/j.commatsci.2012.04.003.

[17] J. Awrejcewicz, G. Sypniewska-Kamińska, and O. Mazur. Analysing regular nonlinear vibrations of nano/micro plates based on the nonlocal theory and combination of reduced order modelling and multiple scale method. *Mech. Syst. Signal Process*, 163 (2022): 108132. https://doi.org/10.1016/j.ymssp.2021.108132.

[18] M.S. Sari, M. Al-Rbai, and B.R. Qawasmeh. Free vibration characteristics of functionally graded Mindlin nanoplates resting on variable elastic foundations using the nonlocal elasticity theory. *Adv. Mech. Eng.*, 10 (2018): 1–17. https://doi.org/10.1177/1687814018813458.

[19] P.T. Thang, P. Tran, and T. Nguyen-Thoi. Applying nonlocal strain gradient theory to size-dependent analysis of functionally graded carbon nanotube-reinforced composite nanoplates. *Appl. Math. Model.*, 93 (2021): 775–791. https://doi.org/10.1016/j.apm.2021.01.001.

[20] Z. Shafiei, S. Sarrami-Foroushani, F. Azhari, and M. Azhari. Application of modified couple-stress theory to stability and free vibration analysis of single and multi-layered graphene sheets. *Aerosp. Sci. Technol.*, 98 (2020): 105652. https://doi.org/10.1016/j.ast.2019.105652.

[21] E. Allahyari and A. Kiani. Employing an analytical approach to study the thermo-mechanical vibration of a defective size-dependent graphene nanosheet. *Eur. Phys. J. Plus.*, 133 (2018): 223. https://doi.org/10.1140/epjp/i2018-12058-2.

[22] C.M. Wang, J.N. Reddy and K.H. Lee. *Shear deformable beams and plates: Relationship with classical solutions*. The Boulevard, Langford Lane, UK: Elsevier, 2000.

[23] C.P. Wu and H.X. Hu. A unified size-dependent plate theory for static bending and free vibration analyses of micro- and nano-scale plates based on the consistent couple stress theory. *Mech. Mater.*, 162 (2021): 104085. https://doi.org/10.1016/j.mechmat.2021.104085.

[24] M. Goodarzi, M. Mohammadi, M. Khooran, and F. Saadi. Thermo-mechanical vibration analysis of FG circular and annular nanoplate based on the visco-pasternak foundation. *J. Solid Mech.*, 8 (2016): 788–805.

[25] A.C. Eringen. *Nonlocal continuum field theories*. New York: Springer, n.d.

[26] H. Salehipour, A.R. Shahidi, and H. Nahvi. Modified nonlocal elasticity theory for functionally graded materials. *Int. J. Eng. Sci.*, 90 (2015): 44–57. https://doi.org/10.1016/j.ijengsci.2015.01.005.

[27] E. Kröner. Elasticity theory of materials with long range cohesive forces. *Int. J. Solids Struct.*, 3 (1967): 731–742. https://doi.org/10.1016/0020-7683(67)90049-2.

[28] M. Koizumi. The concept of FGM. *Ceram. Trans. Func. Grad. Mater.*, 34 (1993): 3–10.

[29] Y. Tanigawa. Some basic thermoelastic problems for nonhomogeneous structural materials. *Appl. Mech. Rev.*, 48 (1995): 287–300. https://doi.org/10.1115/1.3005103.

[30] A.J. Markworth, K.S. Ramesh, and W.P. Parks. Modelling studies applied to functionally graded materials. *J. Mater. Sci.*, 30 (1995): 2183–2193. https://doi.org/10.1007/BF01184560.

[31] T. Fuchiyama and N. Noda, Analysis of thermal stress in a plate of functionally gradient material. *JSAE Rev.*, 16 (1995): 263–268. https://doi.org/10.1016/0389-4304(95)00013-W.

[32] N. Noda. Thermal stresses in functionally graded materials. *J. Therm. Stress.*, 22 (1999): 477–512. https://doi.org/10.1080/014957399280841.

[33] G.H. Paulino, Z.H. Jin, and R.H. Dodds. 2.13 – Failure of functionally graded materials. *Compr. Struct. Integr.*, 2 (2007): 607–644. https://doi.org/10.1016/B0-08-043749-4/02101-7.

[34] J.R. Zuiker. Functionally graded materials: Choice of micromechanics model and limitations in property variation. *Compos. Eng.*, 5 (1995): 807–819. https://doi.org/10.1016/0961-9526(95)00031-H.

[35] T. Reiter, G.J. Dvorak, and V. Tvergaard. Micromechanical models for graded composite materials. *J. Mech. Phys. Solids.*, 45 (1997): 1281–1302. https://doi.org/10.1016/S0022-5096(97)00007-0.

[36] M.M. Gasik. Micromechanical modelling of functionally graded materials. *Comput. Mater. Sci.*, 13 (1998): 42–55. https://doi.org/10.1016/s0927-0256(98)00044-5.

[37] V. Birman and L.W. Byrd. Modeling and analysis of functionally graded materials and structures. *Appl. Mech. Rev.*, 60 (2007): 195–216. https://doi.org/10.1115/1.2777164.

[38] A.H. Akbarzadeh, A. Abedini, and Z.T. Chen. Effect of micromechanical models on structural responses of functionally graded plates. *Compos. Struct.*, 119 (2015): 598–609. https://doi.org/10.1016/j.compstruct.2014.09.031.

[39] J.H. Kim and G.H. Paulino. An accurate scheme for mixed-mode fracture analysis of functionally graded materials using the interaction integral and micromechanics models. *Int. J. Numer. Methods Eng.*, 58 (2003): 1457–1497. https://doi.org/10.1002/nme.819.

[40] H.S. Shen. Functionally graded materials: Nonlinear analysis of plates and shells, 2016.

[41] R. Lal and R. Saini. Vibration analysis of functionally graded circular plates of variable thickness under thermal environment by generalized differential quadrature method. *J. Vib. Control*, 26(1–2) (2020): 73–87. https://doi.org/10.1177/1077546319876389.

[42] R. Lal and R. Saini. On radially symmetric vibrations of functionally graded nonuniform circular plate including non-linear temperature rise. *Eur. J. Mech. A/Solids*, 77 (2019): 103796. https://doi.org/10.1016/j.euromechsol.2019.103796.

[43] R. Lal and R. Saini. On the high-temperature free vibration analysis of elastically supported functionally graded material plates under mechanical in-plane force via GDQR. *J. Dyn. Syst. Meas. Control*, 141 (2019): 101003. https://doi.org/10.1115/1.4043489.

[44] A.H. Sofiyev. Review of research on the vibration and buckling of the FGM conical shells. *Compos. Struct.*, 211 (2019): 301–317. https://doi.org/10.1016/j.compstruct.2018.12.047.

[45] F.A. Fazzolari. Natural frequencies and critical temperatures of functionally graded sandwich plates subjected to uniform and non-uniform temperature distributions. *Compos. Struct.*, 121 (2015): 197–210. https://doi.org/10.1016/j.compstruct.2014.10.039.

[46] K. Swaminathan and D.M. Sangeetha. Thermal analysis of FGM plates – A critical review of various modeling techniques and solution methods. *Compos. Struct.*, 160 (2017): 43–60. https://doi.org/10.1016/j.compstruct.2016.10.047.

[47] Y. Tada. *Space and aerospace vehicle components*. Hellertown-Pennsylvania, 1995.

[48] Y. Kuroda, K.K.A. Moro, and M. Togawa. Evaluation tests of ZrO2/Ni functionally gradient materials for regeneratively cooled thrust engine applications. In: *Proc. Second Int'l Symp. FGM'92, Ceramic Transactions 34*. Westerville, Ohio: American Ceramic Society, 1991: pp. 289–296.

[49] G.W. Goward, D.A. Grey, and R.C. Krutenat. US: 1994. Patent No. 4 248 940.

[50] W.H. Cubberly. *Metals handbook*, ninth edition. New York: SME, 1989.

[51] R.G. Munro. Evaluated material properties for a sintered α-alumina. *J. Am. Ceram. Soc.* 80 (1997): 1919–1928. https://doi.org/10.1111/j.1151-2916.1997.tb03074.x.

[52] S.K. Chan, Y. Fang, M. Grimsditch, Z. Li, M.V. Nevitt, W.M. Robertson, and E.S. Zouboulis. Temperature dependence of the elastic moduli of monoclinic zirconia. *J. Am. Ceram. Soc.*, 74 (1991): 1742–1744. https://doi.org/10.1111/j.1151-2916.1991.tb07177.x.

[53] Y.S. Touloukian. *Thermophysical properties of matter*. Purdue University, 1973.

[54] P. Malekzadeh and A. Alibeygi Beni. Free vibration of functionally graded arbitrary straight-sided quadrilateral plates in thermal environment. *Compos. Struct.*, 92 (2010): 2758–2767. https://doi.org/10.1016/j.compstruct.2010.04.011.

[55] P. Laura, R. Gutierrez, R. Carnicer, and H. Sanzi. Free vibrations of a solid circular plate of linearly varying thickness and attached to a Winkler foundation. *J. Sound Vib.*, 144 (1991): 149–161.

[56] B. Singh and V. Saxena. Axisymmetric vibration of a circular plate with double linear thickness. *J. Sound Vib.*, 179(5) (1995): 879–897.

[57] T. Lenox and H. Conway. An exact, closed form, solution for the flexural vibration of a thin annular plate having a parabolic thickness variation. *J. Sound Vib.*, 68 (1980): 231–239.

[58] E. Efraim and M. Eisenberger. Exact vibration analysis of variable thickness thick annular isotropic and FGM plates. *J. Sound Vib.*, 299 (2007): 720–738.

[59] D.S. Chehil and S.S. Dua. Buckling of rectangular plates with general variation in thickness. *J. Appl. Mech.*, (1973): 745–751.

[60] M. Eisenberger and M. Jabareen. Axisymmetric vibrations of circular and annular plates with variable thickness. *Int. J. Struct. Stab. Dyn.*, 1 (2001): 195–206.

[61] B. Singh and V. Saxena. Transverse vibration of a quarter of a circular plate with variable thickness. *J. Sound Vib.*, 183 (1995): 49–67. https://doi.org/10.1006/jsvi.1995.0238.

[62] F. Ju, H. Lee, and K. Lee. Free vibration of plates with stepped variations in thickness on non-homogeneous elastic foundations. *J. Sound Vib.*, 183 (1995): 533–545.

[63] O.A. Bauchau and J.I. Craig. *Structural Analysis*. Canada: Springer, 2009.

[64] J.N. Reddy. *Theory and Analysis of Elastic Plates and Shells*, 2008. https://doi.org/10.1002/zamm.200890020.

[65] J.N. Reddy. *Mechanics of laminated composite plates and shells: theory and analysis*, second edition. New York: CRC Press, 2003. https://doi.org/10.1007/978-1-4471-0095-9.

[66] C. Shu. *Differential quadrature and its applications in engineering*. London: Springer, 2000.

[67] J.B. Han and K.M. Liew. Axisymmetric free vibration of thick annular plates. *Int. J. Mech. Sci.*, 41 (1999): 1089–1109. https://doi.org/10.1016/S0020-7403(98)00057-5.

[68] M.H. Amini, M. Soleimani, A. Altafi, and A. Rastgoo. Effects of geometric nonlinearity on free and forced vibration analysis of moderately thick annular functionally graded plate. *Mech. Adv. Mater. Struct.*, 20 (2013): 709–720. https://doi.org/10.1080/15376494.2012.676711.

[69] G.C. Paradoen. Asymmetric vibration and stability of circular plates. *Comp. and Struct.*, 9 (1977): 89–95.

[70] S. Azimi. Free vibration of circular plates with elastic edge supports using the receptance method. *J. Sound Vib.*, 120 (1988): 19–35.

[71] U.S. Gupta and A.H. Ansari. Free vibration of polar orthotropic circular plates of variable thickness with elastically restrained edge. *J. Sound Vib.*, 213 (1998): 429–445.

[72] S. Sharma, R. Lal, and N. Singh. Asymmetric vibrations of non-homogenous circular plates of variable thickness. *International J. Contemp. Math.*, 3 (2012): 127–135.

[73] R. Lal and N. Ahlawat. Buckling and vibration of functionally graded non-uniform circular plates resting on Winkler Foundation. *Lat. Am. J. Solids Struct.*, 52 (2015): 2231–2258.

[74] K.K. Pradhan and S. Chakraverty. Free vibration of functionally graded thin elliptic plates with various edge supports. *Struct. Eng. Mech.*, 53 (2015): 337–354.

[75] K.K. Żur. Quasi-Green's function approach to free vibration analysis of elastically supported functionally graded circular plates. *Compos. Struct.*, 183 (2018): 600–610. https://doi.org/10.1016/j.compstruct.2017.07.012.

[76] K.K. Żur. Free-vibration analysis of discrete-continuous functionally graded circular plate via the Neumann series method. *Appl. Math. Model.*, 73 (2019): 166–189. https://doi.org/10.1016/j.apm.2019.02.047.

[77] B. Singh and V. Saxena. Axisymmetric vibration of a circular plate with double linear thickness. *J. Sound Vib.*, 179 (1995): 879–897.

[78] U.S. Gupta, R. Lal, and S. Sharma. Thermal effect on axisymmetric vibrations of non-uniform polar orthotropic circular plates with elastically restrained edge. *Proc. 2nd* Int. Congress Comput. Mech. Simul., 2006.

[79] R. Lal and R. Saini, Vibration analysis of FGM circular plates under non-linear temperature variation using generalized differential quadrature rule. *Appl. Acoust.*, 158 (2020): 107027. https://doi.org/10.1016/j.apacoust.2019.107027.

[80] R. Lal and R. Saini. Thermal effect on radially symmetric vibrations of temperature-dependent FGM circular plates with nonlinear thickness variation. *Mater. Res. Express.*, 6 (2019): 0865f1. https://doi.org/10.1088/2053-1591/ab24ee.

[81] R. Saini. Numerical solution of some vibration problems of FGM circular plates under thermal environment. Roorkee: Indian Institute of Technology, 2020.

3 Effect of External Resistances on Energy Harvesting Behaviour of Porous Functionally Graded Magneto-Electro-Elastic Beam

Arjun Siddharth Mangalasseri, Vinyas Mahesh, Vishwas Mahesh, Sriram Mukunda, Sathiskumar A. Ponnusami, and Dineshkumar Harursampath

CONTENTS

3.1 INTRODUCTION

Multifunctional materials have been in the research spotlight for several years now. They possess unique coupling characteristics which enable them to be used in various applications. Of these, magneto-electro-elastic materials display three-way coupling in electric, magnetic and elastic fields. This permits their utilization in different devices, including actuators, sensors and energy harvesters. Several research works have analysed MEE composites [1–3]. Vinyas and Kattimani [4] carried out

DOI: 10.1201/9781003328032-3

the hygrothermal analysis of MEE plates using finite element analysis. A detailed review of properties, modelling characteristics and analysis methods of MEE materials was looked into by Vinyas [5]. Different research works covered the analysis of MEE plates subjected to various loading conditions and containing different circuit implementations [6–8].

Specially engineered materials called functionally graded materials (FGMs) have a varying composition profile. They address the ever-changing need to develop efficient materials for efficient production. Udupa et al. [9] described the classification, properties and various concepts related to FGMs. They focused on research and applications related to functionally graded materials, focusing on a CNT-reinforced Al FG powder produced by powder metallurgy technique. Naebe and Shirvanimoghaddam [10] presented a detailed review of the properties, fabrication methods and challenges associated with FGMs. They also looked into the prospects of these materials from a scientific and application perspective. Several researchers have also focused on functionally graded MEE structures, carrying out analyses for a wide variety of applications. Huang et al. [11] developed an analytical method to evaluate the behaviour of magneto-electro-elastic beams. A series of analytical solutions for different cases such as tension, pure bending, shear force at free end for MEE cantilever beams are obtained. Tang et al. [12] carried out a study of 2D FG-nanobeams under the influence of MEE fields. The effect of the Pasternak foundation coefficient, length to thickness ratio, MEE loading conditions and material distributions are investigated. The GDQ method is used to discretize the model of the nanobeams. The influence of porosities created during the sintering process of producing FGMs is of grave importance. They affect the performance of the material and, in turn, the structure in question. This was researched by several works [13–17], considering various structures comprising of porous functionally graded magneto-electro-elastic materials.

The conversion of ambient energy into utilizable forms has been the talking point of the research community for several years, especially with the advent of sustainable practices in engineering. Several engineers have focused on this area, researching the domain of energy harvesting. Heshmati and Amini [18] developed a functionally graded piezoelectric energy harvester, operating using the vibrations of travelling oscillators. Using finite element formulations, a parametric study is done to assess the influence of material distribution, mass ratio, damping ratio, etc. Qi [19], in his work, developed an analytical model for functionally graded flexo-electric energy harvester. The device was tuned to obtain the optimum working frequency. Motlagh et al. [20] presented an electro-mechanical approach for FG-panels with piezo patches and validated the results with finite element analyses. Here too, voltage and power output were maximized through material and geometry variation. Optimization of the energy harvesting system ensures enhanced performance characteristics. This leads to better utilization of the energy, thus enabling sustainable practices. Gomez et al. [21] investigated the optimum load resistance ratio considering other system factors such as the actual temperature of the thermoelectric materials. A thermoelectric energy harvesting system is considered, and the work is carried out using numerical methods and later validated experimentally. Wei et al. [22] simulated and fabricated an impact-driven piezoelectric energy harvester based on human motion. They determined the variation in voltage and power produced for change in walking speed and external resistances. Such a device finds application in different wearable devices. In this regard, several other works [23–25] have also focused on modelling and experimentation of energy harvesting devices and optimizing their performance based on the dependent factors such as resistance and geometry of the device.

From the literature review, it is clear that no work has focused on evaluating the resistance effects on porous functionally graded MEE materials for energy harvesting purposes. Although Shirbani et al. [26] looked into the effect of resistance on the harvester's performance, the effect of porosity factors, substrate material, porosity distribution and the influence of using multiphase MEE materials was not covered. This chapter makes the first attempt to this end.

3.2 DESIGN AND MATERIALS

3.2.1 ENERGY HARVESTER CONFIGURATION

The pictorial representation of the proposed unimorph P-FGMEE energy harvesting system is shown in Figure 3.1. The basic configuration consists of a unimorph cantilever beam, fixed at one end. The beam has a length L, a cross-sectional width b, and thickness h, and these parameters are considered along x-, y- and z-axes of the Cartesian coordinate system. The cantilever beam essentially consists of two distinct layers – the top layer of P-FGMEE and an inactive substrate layer at the bottom.

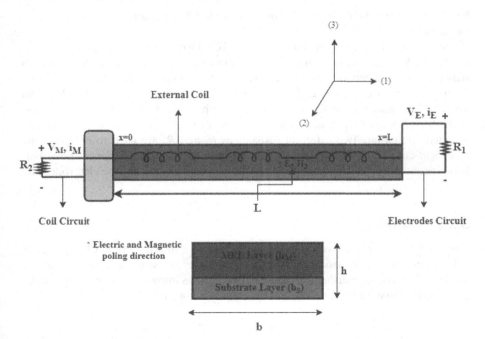

FIGURE 3.1 MEE energy harvesting system.

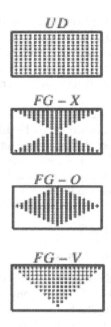

FIGURE 3.2 Porosity patterns [17].

Figure 3.2 shows the different functionally graded distribution patterns considered in the chapter.

Further, h_S and h_M are the substrate and MEE layer thickness, respectively. The upper and lower surfaces of the layers are covered by thin, continuous electrodes which extract the generated electric potential. The electrode material covering the layer is assumed to be perfectly conductive. This is to ensure that the electrical potential generated is uniform across the length of the beam. Meanwhile, a coil with 'N' turns is wrapped across the thickness of the beam to harvest the induced magnetic potential. Two different resistances R_1 and R_2 are used to complete the electrical circuit to obtain the harvested power, and the effect of these on the performance is studied. They are connected to the circuit formed by the electrodes and coil, respectively.

3.2.2 MATERIAL

The proposed energy harvesting system consists of a multifunctional composite of MEE material with porosities. The energy harvester is assumed to operate in 3-1 mode in this chapter. Further, the effect of porosities is introduced in the functionally graded MEE material through the modified power-law given by [16]:

$$M_{fg}(z) = M_l + (M_u - M_l)\left\{\left(\frac{z}{h}\right) + \left(\frac{1}{2}\right)\right\}^n - (M_u + M_l)(p/2)V_{por} \qquad (3.1)$$

TABLE 3.1
MEE Material Properties for BaTiO$_3$–CoFe$_2$O$_4$ [16]

Material Property	Material Constants	Piezomagnetic (F)	Piezoelectric (B)
Elastic Constant (GPa)	$C_{11}=C_{22}$	286	166
Piezoelectric Constant (C/m^2)	e_{31}	0	–4.4
Dielectric Constant (10^{-9} C^2/ Nm^2)	η_{33}	0.093	12.6
Magnetic Permeability (10^{-4} Ns^2 /C^2)	μ_{33}	1.57	0.1
Piezomagnetic Constant (N/Am)	q_{31}	580	0
Magneto-electric constant (10^{-12} Ns/ VC)	m_{33}	0	0
Density (kg/m^3)	ρ	5300	5800

Here, $M_{fg}(z)$ is the generalized representation of the FG material coefficient, such as, $[e]$, $[h]$, $[f]$, $[g]$, $[C]$, $[\mu]$. M_l and M_u represent the material property at the bottom and top layers, respectively. n, p and V_{por} are the gradient index, porosity volume and porosity distribution. The different porosity distributions considered for the analysis are as follows [16]:

1. Even Porosity: $V_{por} = 1$
2. Uneven Porosity:

$$\text{X distribution: } V_{por} = 4\frac{|z|}{h}$$

$$\text{V distribution: } V_{por} = \left(1+2\frac{|z|}{h}\right)$$

$$\text{O distribution: } V_{por} = 2\left(1-2\frac{|z|}{h}\right) \tag{3.2}$$

In Eqn. (3.2), the terms z and h are the point of interest along the z-axis and the overall thickness of the MEE layer in the cantilever beam, respectively. This chapter considers two forms of functionally graded patterns, namely, 'B' rich bottom and 'F' rich bottom [16]. The material properties of the MEE material used in this study are mentioned in Table 3.1.

3.3 MATHEMATICAL MODELLING

The lumped parameter model and the frequency response function for the base excited P-FGMEE energy harvester beam system is presented in this section. To this end, a single degree of freedom mass-spring-damper setup, as illustrated in Shirbani et al. [26], has been considered in this work as well. In the time domain, the energy balance equations of P-FGMEE energy harvesters can be expressed as:

$$\int F\dot{Z}\,dt = \frac{1}{2}M_{eq}Z^2 + C_{eq}\int \dot{Z}^2\,dt + \frac{1}{2}K_{eq}Z^2 + \int \theta_{EM}V_E\dot{Z}\,dt + \int \theta_{MM}V_M\dot{Z}\,dt \tag{3.3}$$

$$\int (\theta_{EM} \dot{Z}(t) + \lambda_{ME} \dot{V}_M) V_E \, dt = \int V_E i_E \, dt + \frac{1}{2} C_M V_E^2 \tag{3.4}$$

$$\int (\theta_{MM} \dot{Z} + \lambda_{EM} \dot{V}_E) V_M \, dt = \int V_M i_M \, dt + \frac{1}{2} \frac{L_C}{R_2^2} V_E^2 \tag{3.5}$$

The natural frequency of the P-FGMEE energy harvester in the beam form can be expressed using the equivalent lumped parameters as follows:

$$\omega_n = \sqrt{\frac{K_{eq}}{M_{eq}}} = \sqrt{\frac{25EI}{2(\rho_S h_S + \rho_M h_M) L^3}} \tag{3.6}$$

The frequency response functions related to the output parameters of the P-FGMEE energy harvester such as total power, electric voltage, magnetic voltage, and relative displacements can be mentioned as follows:

The coupling effects produced by the P-FGMEE material, i.e., the electrical and magnetic stiffness and damping, tend to affect the system's natural frequency. The frequency is given by Eqn. (3.6) is modified to include the coupling effects and is thus given as:

$$\omega_{tr} = \sqrt{\frac{K_{eq} + K_{ele} + K_{mag}}{M_{eq}}} = \omega_n \sqrt{1 + \frac{K_{ele}}{K_{eq}} + \frac{K_{mag}}{K_{eq}}} \tag{3.7}$$

ω_{tr} is referred to as the true resonant frequency of the system. The true resonant frequency of the system thus models the dynamic behaviour of the system. It is also used while studying the effect of resistances on R_1 and R_2 on the harvester output performance, i.e. $\omega = \omega_{tr}$ while varying R_1 and R_2.

The frequency response functions related to the output parameters of the energy harvester such as total power, electric voltage, magnetic voltage and relative displacements can be mentioned as follows:

$$\frac{Z}{\bar{Y} e^{j\omega_a t}} = f_Z = \frac{M_{eq} \omega_a^2}{\left(K_{eq} + K_{ele} + K_{mag} - M_{eq} \omega_a^2\right) + j\omega_a (C_{eq} + C_{ele} + C_{mag})} \tag{3.8}$$

$$\frac{V_E(t)}{\omega_a^2 \bar{Y} e^{j\omega_a t}} = \frac{f_{V_E}}{\omega_a^2} \tag{3.9}$$

$$\frac{V_M(t)}{\omega_a^2 \bar{Y} e^{j\omega_a t}} = \frac{f_{V_M}}{\omega_a^2} \tag{3.10}$$

$$\frac{P_E(t)}{(\omega_a^2 \bar{Y} e^{j\omega_a t})^2} = \frac{|f_{V_E}|^2}{2R_1 \omega_a^2} \tag{3.11}$$

$$\frac{P_M(t)}{\left(\omega_a^2 \overline{Y} e^{j\omega_a t}\right)^2} = \frac{|f_{V_M}|^2}{2R_2\omega_a^2} \tag{3.12}$$

$$P_{ME} = P_E + P_M \tag{3.13}$$

The various lumped parameters and other quantities used to arrive at Eqns. (3.8)–(3.13) are shown in Appendix.

3.4   RESULTS AND DISCUSSION

In this section, the effect of the external resistances used on the output parameters for various parameters of the P-FGMEE material is studied. For this set of results, N is fixed as 1, and the true resonant frequency of the system ω_{tr} is used.

3.4.1   EFFECT OF POROSITY DISTRIBUTION – B RICH BOTTOM FGM

The variation in electric potential V_E for various substrates is shown in Figure 3.3(a)–(d). The study is carried out for three different values of R_2: 10°, 10^2 and $10^6 \ \Omega$. Here, the

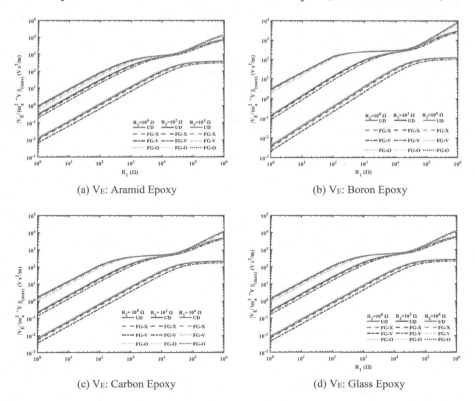

(a) V_E: Aramid Epoxy

(b) V_E: Boron Epoxy

(c) V_E: Carbon Epoxy

(d) V_E: Glass Epoxy

FIGURE 3.3   Variation in electric potential V_E for (a) Aramid epoxy, (b) Boron epoxy, (c) Carbon epoxy and (d) Glass epoxy.

porosity distribution serves as the control parameter. For all substrates, it is seen that an increase in coil resistance R_2 causes a significant increase in the value of electric potential. With respect to porosity distributions, the largest electric potential variation is for FG-O distribution, followed by UD, FG-X and FG-V. The maximum electric potential value $|V_E|_{max}$ occurs at $R_1 = R_2 = 10^6 \ \Omega$, for FG-O porosity distribution, at 14,531 V s^2/m for Aramid epoxy, 8684.3 V s^2/m for Boron epoxy, 12,423 V s^2/m for Carbon epoxy and 13,717 V s^2/m for Glass epoxy substrate. It is also observed that for $R_1 > 10^5 \Omega$, potential values are close together for $R_2 = 10^2$ and $10^6 \ \Omega$.

The variation in magnetic potential V_M for various substrates is shown in Figure 3.4(a)–(d). As seen in the figures, there is an increase in the magnetic potential, with an increase in external coil resistant R_2. From $R_2 = 10°$ to $R_2 = 10^2$, there is a drastic increase in the potential value, which reduces on further increase in R_2 to 10^6. For all substrates, FG-O porosity distribution produces the highest value, followed by UD, FG-X and FG-V. With an increase in R_1, there is a continuous decrease in V_M until a certain R_1 value, beyond which it increases. This can be associated with the increase in magnetic damping coefficient C_{mag}, with an increase in R_1. This effect is more pronounced for higher values of R_2, such as $10^2 \ \Omega$ and $10^6 \ \Omega$. It can be

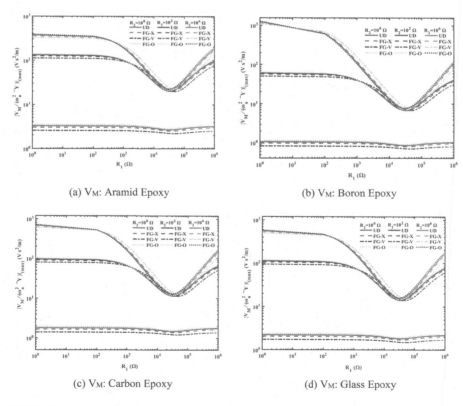

(a) V$_M$: Aramid Epoxy   (b) V$_M$: Boron Epoxy

(c) V$_M$: Carbon Epoxy   (d) V$_M$: Glass Epoxy

FIGURE 3.4   Variation in magnetic potential V_M for (a) Aramid epoxy, (b) Boron epoxy, (c) Carbon epoxy and (d) Glass epoxy.

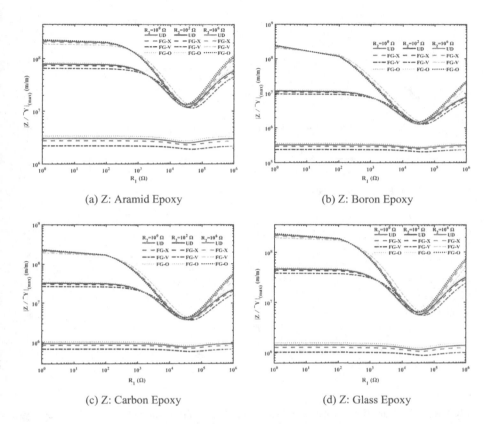

(a) Z: Aramid Epoxy

(b) Z: Boron Epoxy

(c) Z: Carbon Epoxy

(d) Z: Glass Epoxy

FIGURE 3.5 Variation in relative displacement Z for (a) Aramid epoxy, (b) Boron epoxy, (c) Carbon epoxy and (d) Glass epoxy.

deduced that there is a slight shift in the minimum point towards the right, in order of porosity distribution – FG-O, UD, FG-X, FG-V. The maximum magnetic potential value, $|V_M|_{max}$, occurs at $R_1 = 10° \, \Omega, R_2 = 10^6 \Omega$, for FG-O distribution. The maximum values are $- 382.8798$ V s²/m for Aramid epoxy substrate, 1285.2 V s²/m for Boron epoxy, 715.8241 V s²/m for Carbon epoxy and 581.4125 V s²/m for Glass epoxy.

The variation in normalized displacement with resistance R_1 for different values of R_2 and porosity distributions is shown in Figure 3.5(a)–(d). The trends of the graphs are similar to those of magnetic potential V_M. The displacement increases with an increase in the value of R_2, maximum value is seen for the FG-O distribution. Here a slight shift in the minimum point towards the right is observed, in decreasing order of porosity distribution. The maximum normalized displacement value $|Z/\bar{Y}|_{max}$ occurs at $R_1 = 10° \, \Omega, R_2 = 10^6 \, \Omega$, for FG-O porosity distribution. The corresponding values for Aramid, Boron, Carbon and Glass epoxy substrates are 2.3054 × 10⁸, 2.4630 × 10⁸, 2.3660 × 10⁸ and 2.3536 × 10⁸, respectively.

The study of total harvested power P_{ME} is shown in Figure 3.6(a)–(d). As the value of R_2 is increased from $10° \, \Omega$ to $10^2 \, \Omega$, there is a significant increase in the power

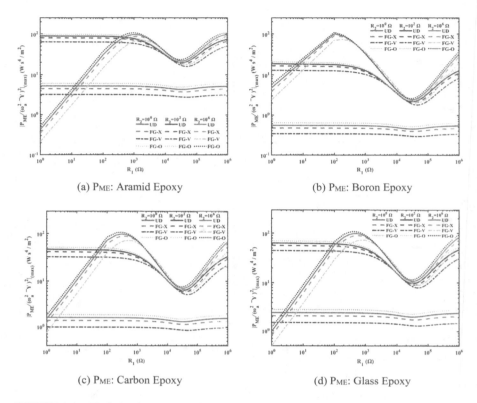

(a) P_{ME}: Aramid Epoxy

(b) P_{ME}: Boron Epoxy

(c) P_{ME}: Carbon Epoxy

(d) P_{ME}: Glass Epoxy

FIGURE 3.6   Variation in total harvested power P_{ME} for (a) Aramid epoxy, (b) Boron epoxy, (c) Carbon epoxy and (d) Glass epoxy.

generated. On increasing R_2 to 10^6 Ω, a unique variation is observed. The power generated starts at a value below the graphs of $R_2 = 10^0$ Ω (between 10^0 Ω and 10^2 Ω for Boron epoxy), increases to the highest value, decreases to a minimum value, and further increases up to $R_2 = 10^6$ Ω. The maximum value of total harvested power occurs for FG-O porosity distribution, for $R_2 = 10^6$ Ω, at varying R_1 values. The corresponding power and R_1 values are 105.6249 W s⁴/m² at 972401 Ω for Aramid epoxy, 109.6823 W s⁴/m² at 101 Ω for Boron epoxy, 106.6109 W s⁴/m² at 301 Ω for Carbon epoxy and 108.5298 W s⁴/m² at 401Ω for Glass epoxy.

3.4.2   EFFECT OF POROSITY INDEX

This section discusses the effect of variation of porosity index on the output values. The coil resistance R_2 and porosity index p form the control parameters. The study is carried out for three different $R_2 - 10^0, 10^2$ and 10^6 Ω and four values of porosity index $-$ 0, 0.2, 0.3 and 0.4. Here the gradation is fixed as F rich bottom, the substrate as Aramid epoxy, and the plots are obtained for two different porosity distributions $-$ FG-O and FG-X.

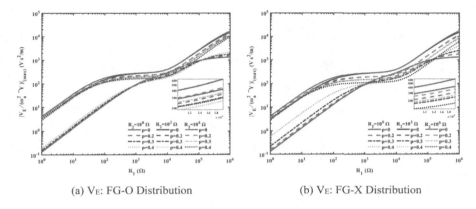

(a) V_E: FG-O Distribution   (b) V_E: FG-X Distribution

FIGURE 3.7 Variation of electric potential V_E with porosity index for (a) FG-O and (b) FG-X.

The variation in V_E is shown in Figure 3.7(a) and (b). It is observed that increasing R_1 and R_2 leads to an increase in the value of potential. However, the increase in potential as R_2 is increased from 10^2 to 10^6 Ω is less pronounced, and a slight overlap in the values is evident from the figures. For $R_2 = 10°$ Ω, $p = 0.4$ produces the highest potential, which decreases with a decrease in porosity index. For higher values of R_2 (10^2 and 10^6 Ω), there is a reverse in the trend followed. $p = 0$ shows the highest value of potential, which decreases with an increase in porosity index. The maximum electric potential value $|V_E|_{max}$, occurs for $R_1 = R_2 = 10^6$ Ω, for porosity index $p = 0$. The corresponding value is 17431 V s²/m for all porosity distributions.

Figure 3.8(a) and (b) represents the plots for variation of magnetic potential V_M with respect to resistance R_1. V_M continuously decreases as R_1 increases until R_1

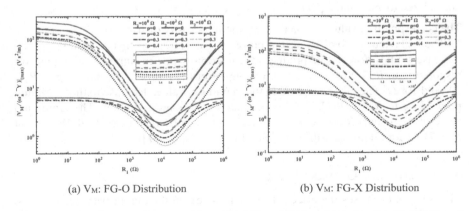

(a) V_M: FG-O Distribution   (b) V_M: FG-X Distribution

FIGURE 3.8 Variation of magnetic potential V_M with porosity index for (a) FG-O and (b) FG-X.

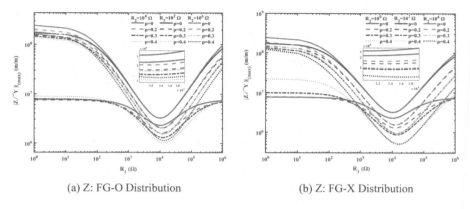

(a) Z: FG-O Distribution   (b) Z: FG-X Distribution

FIGURE 3.9 Variation of relative displacement Z with porosity index for (a) FG-O and (b) FG-X.

reaches a value slightly ahead of 10^4 Ω, beyond which it increases again. V_M also clearly increases with an increase in the value of R_2. With respect to the porosity index, it is observed that $p = 0$ produces the largest output potential, which decreases with an increase in the index to 0.2, 0.3 and 0.4. This is valid for all porosity distributions. This is probably because, with an increase in porosity index, there is a decrease in material properties, causing a decrease in the output value. The maximum value of magnetic potential $|V_M|_{max}$ is observed for $R_1 = 10°$ Ω, $R_2 = 10^6$ Ω, for $p = 0$. The corresponding value is 224.5293 V s^2/m for all porosity distributions.

Normalized displacement varies with resistance R_1, as seen in Figure 3.9(a) and (b). The trends followed are similar to those in the case of V_M. Normalized displacement increases with R_2 and decreases with R_1 until the minimum point is slightly ahead of 10^4 Ω. Beyond this, the value continues to increase. Porosity index $p = 0$ produces the largest displacement and $p = 0.4$ the lowest. The maximum value of normalized displacement $|Z/\bar{Y}|_{max}$ is observed for $R_1 = 10°$ Ω, $R_2 = 10^6$ Ω, for $p = 0$. The corresponding value is 2.4347×10^8 for all porosity distributions.

The behaviour of total harvested power P_{ME} is similar to that seen in the previous case studies. $R_2 = 10^6$ Ω produces the maximum power output. The graphs corresponding to $R_2 = 10^6$ Ω begins at a value lower than those corresponding to $R_2 = 10°$ and 10^6 Ω. They then increase to the maximum value, decrease to the minimum point close to $R_1 = 10^4$ Ω, and further increase beyond this point. The maximum value of P_{ME} occurs for $p = 0$, $R_2 = 10^6$ Ω. The value observed is 159.0718 W s^4/m^2 at $R_1 = 71$ Ω, for all porosity distributions (Figure 3.10[a] and [b]).

3.4.3 EFFECT OF GRADIENT INDEX

This section discusses the effect of gradient index on the output parameters of the energy harvesting system. For this particular study, Aramid epoxy substrate is used, with a P-FGMEE layer on top. The porosity volume is fixed as 0.2 and the plots

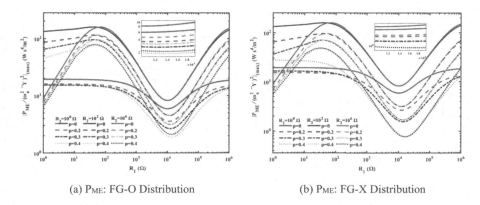

(a) P_{ME}: FG-O Distribution   (b) P_{ME}: FG-X Distribution

FIGURE 3.10   Variation of total harvested power P_{ME} with porosity index for (a) FG-O and (b) FG-X.

are obtained for two porosity distributions, FG-O and FG-X, and different gradient indexes with $R_2 = 10^6\ \Omega$.

Electric voltage V_E varies as shown in Figure 3.11(a) and (b). V_E continuously increases with an increase in R_1. At $R_1 = 10^0\ \Omega$, $n = 0.2$ has larger V_E for F rich gradation, while $n = 0.8$ has larger V_E for B rich gradation. As R_1 crosses $10^2\ \Omega$, a change in the trend is witnessed. For F rich bottom, it is seen that an increase in the value of potential occurs with an increase in gradient index. The maximum potential value is observed for $n = 0.8$ and $R_1 = 10^6\ \Omega$ and the values are 14500 V s²/m for FG-O, 12827 V s²/m for FG-X. A reverse trend is seen for B rich gradation, with $n = 0.2$ having the highest value and $n = 0.8$ having the least.

Magnetic potential V_M varies with resistance R_1 as shown in Figure 3.12(a) and (b). The values decrease with an increase in R_1, until a value of $10^4\ \Omega$ is reached. Beyond this, the value of V_M starts increasing again. For B rich gradation,

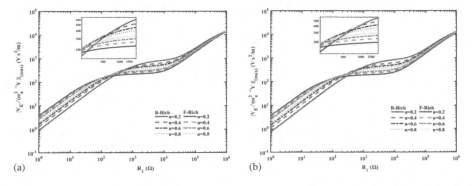

FIGURE 3.11   Variation of electric potential V_E with gradient index for (a) FG-O and (b) FG-X.

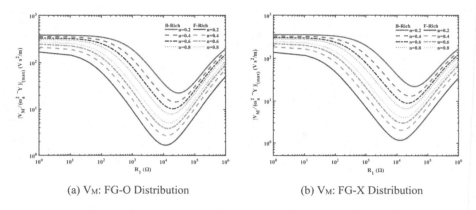

(a) V$_M$: FG-O Distribution   (b) V$_M$: FG-X Distribution

FIGURE 3.12   Variation of magnetic potential V_M with gradient index for (a) FG-O and (b) FG-X.

gradient-index $n = 0.2$ produces the highest value of potential whereas, for F rich gradation, this trend is reversed – $n = 0.8$ produces the largest value of V_M. It is also observed that an increase in gradient index causes a shift of the minimum point of the curves to the right for F rich bottom. A similar shift is observed for B rich gradation, with a decrease in gradient index. The normalized displacement value $\left| Z/\overline{Y} \right|$ varies with R_I, as seen in Figure 3.13(a) and (b). The pattern of variation of the plots is similar to that seen in the case of V_M. The maximum values for FG-O and FG-X are 2.3054×10^8 and 2.1095×10^8, respectively. The effect of gradient index on total harvested power P_{ME} is depicted in Figure 3.14(a) and (b). It is seen that although F rich gradation produces higher output power initially, at $R_2 = 10^6\ \Omega$, a greater value is produced for B rich gradation. The graphs for P_{ME} first increase

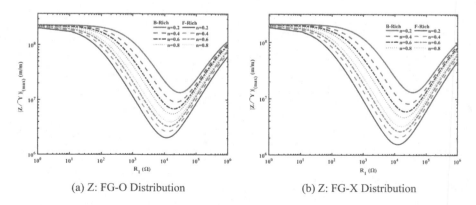

(a) Z: FG-O Distribution   (b) Z: FG-X Distribution

FIGURE 3.13   Variation of relative displacement Z with gradient index for (a) FG-O and (b) FG-X.

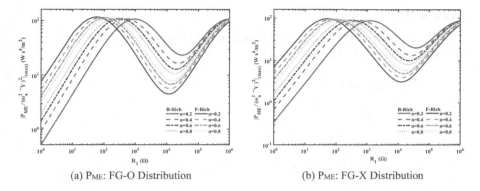

(a) P_{ME}: FG-O Distribution   (b) P_{ME}: FG-X Distribution

FIGURE 3.14 Variation of total harvested power P_{ME} with gradient index for (a) FG-O and (b) FG-X.

until a certain value, decrease until $R_1 = 10^4 \, \Omega$, and then further increases beyond this point. Maximum output power is produced by F rich bottom FGM, for $n = 0.2$ at different values of R_1 as 115.1115 W s⁴/m² at 61 Ω for FG-O and 95.8395 W s⁴/m² at 51 Ω for FG-X.

3.5   CONCLUSIONS

This chapter covers the energy harvesting performance of a magneto-electro-elastic energy harvester with respect to changing material properties. Here, porous functionally graded MEE materials are considered. In particular, the influence of porosity on the performance with respect to changing external load resistances is explored. This forms a basis for tuning of energy harvester performance for various applications. The following conclusions can be drawn effectively:

1. FG-O porosity distribution shows the highest value for all output parameters. The values increase with an increase in R_1 and R_2, and the maximum value is observed for $R_2 = 10^6 \, \Omega$.
2. The effect of porosity index is studied and explained. It is observed that $p = 0$ displays the largest value, and $p = 0.4$ the least.
3. The effect of gradient index on the output parameters is studied and assessed. Maximum output power is produced by F rich bottom FGM, for $n = 0.2$.

APPENDIX

The various lumped parameter formulation used to obtain the natural frequency of the system in Eqn. (3.10) are given as:

$$M_{eq} = \frac{24}{100}(\rho_s h_s + \rho_M h_M)bL$$

$$K_{eq} = \frac{3(EI)_{eq}}{L^3}$$

$$EI_{eq} = b\left[E_T^S \left[\frac{h_s^3}{12} + h_s \left(\bar{n} - \frac{h_s}{2} \right)^2 \right] + E_{11}^M \left[\frac{h_M^3}{12} + h_M \left(\frac{h_M}{2} + h_S - \bar{n} \right)^2 \right] \right]$$

$$\bar{n} = \frac{h_S^2 + h_M^2 r + 2h_s h_M r}{2(h_s + r h_M)}$$

$$r = \frac{E_T^S}{E_{11}^M}$$

$$\theta_{EM} = -e_{31}^M \frac{bh_{Mc}}{L}$$ (A.1)

$$h_{Mc} = \frac{h_M + 2h_s - 2\bar{n}}{2}$$

$$\theta_{MM} = q_{31}^M \frac{Nbh_{Mc}}{R_2 L}$$

$$C_M = \frac{h_{33}^M bL}{h_M}$$

$$L_C = \mu_{33}^M \frac{N^2 bL}{h_M}$$

$$\lambda_{EM} = \lambda_{ME} = g_{33}^M \frac{NbL}{R_2 h_M}$$

The different harmonic output parameters used to obtain FRFs in Eqns. (3.11)–(3.15) are represented as $Z = \bar{Z}e^{j\omega_a t}, V_E = \bar{V}_E e^{j\omega_a t}$ and $V_M = \bar{V}_M e^{j\omega_a t}$, where $\bar{Z}, \bar{V}_E, \bar{V}_M$ are the amplitudes of displacement of the mass, electric and magnetic potential and are given by:

$$\bar{Z} = \frac{M_{eq}\omega_a^2 \bar{Y}}{\left(K_{ele} + K_{eq} + K_{mag} - M_{eq}\omega_a^2 \right) + j\omega_a (C_{eq} + C_{ele} + C_{mag})} = f_Z \bar{Y}$$

$$\bar{V}_E = \frac{K_{ele} + j\omega_a C_{ele}}{\theta_{EM}} \bar{Z} = \frac{K_{ele} + j\omega_a C_{ele}}{\theta_{EM}} f_Z \bar{Y} = f_{V_E} \bar{Y}$$ (A.2)

$$\bar{V}_M = \frac{K_{mag} + j\omega_a C_{mag}}{\theta_{MM}} \bar{Z} = \frac{K_{mag} + j\omega_a C_{mag}}{\theta_{MM}} f_Z \bar{Y} = f_{V_M} \bar{Y}$$

f_Z, f_{V_E}, f_{V_M} are the transfer functions between relative displacement of the mass and the harvested voltages over input base excitation. K_{eq}, K_{ele}, K_{mag} and C_{eq}, C_{ele}, C_{mag} are the electric and magnetic stiffness and damping terms, given by:

$$K_{ele} = \omega_a \theta_{EM} \operatorname{Im}(-f_E)$$

$$K_{mag} = \omega_a \theta_{MM} \operatorname{Im}(-f_M)$$

$$C_{ele} = \theta_{EM} \operatorname{Re}(f_E)$$

$$C_{mag} = \theta_{MM} \operatorname{Re}(f_M)$$

The parameters f_E, f_M are given as:

$$f_E = \frac{1}{c_1 + c_2 x_1}$$

$$f_M = \frac{1}{c_3 + c_4 x_2}$$

$$c_1 = \frac{1}{\theta_{EM}} \left(\frac{1}{R_1} + j\omega_a C_M \right)$$

$$c_2 = \frac{-j\omega_a \lambda_{ME}}{\theta_{EM}}$$

$$c_3 = \frac{R_2 + j\omega_a L_C}{\theta_{MM} R_2^2}$$ (A.3)

$$c_4 = \frac{-j\omega_a \lambda_{EM}}{\theta_{MM}}$$

$$x_1 = \frac{1}{x_2} = \frac{c_1 - c_4}{c_3 - c_2}$$

The expression for electric and magnetic power harvested is a function of the generated voltage and resistance value used. It is given as:

$$P_E = \frac{|V_E^2|}{2R_1} = \frac{|f_{V_E}|^2 |\bar{Y}|^2}{2R_1}$$

$$= \frac{|\bar{Y}|^2}{2R_1} \left| \frac{K_{ele}^2 + \omega_a^2 C_{ele}^2}{\theta_{EM}^2} \right| \left| \frac{\left(M_{eq} \omega_a^2 \right)^2}{\left(K_{eq} + K_{ele} + K_{mag} - M_{eq}\omega_a^2 \right)^2 + \omega_a^2 (C_{eq} + C_{ele} + C_{mag})^2} \right|$$

$$P_M = \frac{|V_M^2|}{2R_2} = \frac{|f_{V_M}|^2 |\bar{Y}|^2}{2R_2}$$

$$= \frac{|\bar{Y}|^2}{2R_2} \left| \frac{K_{mag}^2 + \omega_a^2 C_{mag}^2}{\theta_{MM}^2} \right| \left| \frac{\left(M_{eq} \omega_a^2 \right)^2}{\left(K_{eq} + K_{ele} + K_{mag} - M_{eq}\omega_a^2 \right)^2 + \omega_a^2 (C_{eq} + C_{ele} + C_{mag})^2} \right|$$

(A.4)

ACKNOWLEDGEMENTS

The financial support by the Royal Society through Newton International Fellowship (NIF\R1\212432) is sincerely acknowledged by the authors Vinyas Mahesh and Sathiskumar A. Ponnusami.

The financial support by Science and Engineering Research Board (SERB) through Teachers Associateship for Research Excellence (TAR/2021/000016) is sincerely acknowledged by the authors Vishwas Mahesh and Dineshkumar Harursampath.

REFERENCES

[1] Lee, J., Boyd IV, J.G. and Lagoudas, D.C. 2005. Effective properties of three-phase electro-magneto-elastic composites. *International Journal of Engineering Science*, 43(10), pp. 790–825.

[2] Vinyas, M. and Kattimani, S.C. 2018. Finite element evaluation of free vibration characteristics of magneto-electro-elastic rectangular plates in hygrothermal environment using higher-order shear deformation theory. *Composite Structures*, 202, pp. 1339–1352.

[3] Vinyas, M. and Kattimani, S.C. 2017. Static analysis of stepped functionally graded magneto-electro-elastic plates in thermal environment: a finite element study. *Composite Structures*, 178, pp. 63–86.

[4] Vinyas, M. and Kattimani, S.C. 2017. Hygrothermal analysis of magneto-electro-elastic plate using 3D finite element analysis. *Composite Structures*, 180, pp. 617–637.

[5] Vinyas, M. 2021. Computational analysis of smart magneto-electro-elastic materials and structures: Review and classification. *Archives of Computational Methods in Engineering*, 28(3), pp. 1205–1248.

[6] Vinyas, M., Sunny, K.K., Harursampath, D., Nguyen-Thoi, T. and Loja, M.A.R. 2019. Influence of interphase on the multi-physics coupled frequency of three-phase smart magneto-electro-elastic composite plates. *Composite Structures*, 226, p. 111254.

[7] Vinyas, M., Nischith, G., Loja, M.A.R., Ebrahimi, F. and Duc, N.D. 2019. Numerical analysis of the vibration response of skew magneto-electro-elastic plates based on the higher-order shear deformation theory. *Composite Structures*, 214, pp. 132–142.

[8] Vinyas, M. and Harursampath, D. 2021. Computational evaluation of electro-magnetic circuits' effect on the coupled response of multifunctional magneto-electro-elastic composites plates exposed to hygrothermal fields. *Proceedings of the Institution of Mechanical Engineers, Part C: Journal of Mechanical Engineering Science*, 235(15), pp. 2832–2850.

[9] Udupa, G., Rao, S.S. and Gangadharan, K.V. 2014. Functionally graded composite materials: An overview. *Procedia Materials Science*, 5, pp. 1291–1299.

[10] Naebe, M. and Shirvanimoghaddam, K. 2016. Functionally graded materials: A review of fabrication and properties. *Applied Materials Today*, 5, pp. 223–245.

[11] Huang, D.J., Ding, H.J. and Chen, W.Q. 2007. Analytical solution for functionally graded magneto-electro-elastic plane beams. *International Journal of Engineering Science*, 45(2–8), pp. 467–485.

[12] Tang, Y., Ma, Z.S., Ding, Q. and Wang, T. 2021. Dynamic interaction between bi-directional functionally graded materials and magneto-electro-elastic fields: A nano-structure analysis. *Composite Structures*, 264, p. 113746.

[13] Kiran, M.C., Kattimani, S.C. and Vinyas, M. 2018. Porosity influence on structural behaviour of skew functionally graded magneto-electro-elastic plate. *Composite Structures*, 191, pp. 36–77.

[14] Kiran, M.C. and Kattimani, S.C. 2018. Assessment of porosity influence on vibration and static behaviour of functionally graded magneto-electro-elastic plate: A finite element study. *European Journal of Mechanics-A/Solids*, 71, pp. 258–277.

[15] Esayas, L.S. and Kattimani, S. 2021. Effect of porosity on active damping of geometrically nonlinear vibrations of a functionally graded magneto-electro-elastic plate. *Defence Technology*, 18, pp. 891–906.

[16] Vinyas, M. 2020. On frequency response of porous functionally graded magneto-electro-elastic circular and annular plates with different electro-magnetic conditions using HSDT. *Composite Structures*, 240, p. 112044.

[17] Mahesh, V. 2021. Porosity effect on the nonlinear deflection of functionally graded magneto-electro-elastic smart shells under combined loading. *Mechanics of Advanced Materials and Structures*, pp. 1–27.

[18] Heshmati, M. and Amini, Y. 2019. A comprehensive study on the functionally graded piezoelectric energy harvesting from vibrations of a graded beam under travelling multi-oscillators. *Applied Mathematical Modelling*, 66, pp. 344–361.

[19] Qi, L. 2019. Energy harvesting properties of the functionally graded flexoelectric microbeam energy harvesters. *Energy*, 171, pp. 721–730.

[20] Motlagh, P.L., Anamagh, M.R., Bediz, B. and Basdogan, I. 2021. Electromechanical analysis of functionally graded panels with surface-integrated piezo-patches for optimal energy harvesting. *Composite Structures*, 263, p. 113714.

[21] Gomez, M., Reid, R., Ohara, B. and Lee, H. 2013. Influence of electrical current variance and thermal resistances on optimum working conditions and geometry for thermoelectric energy harvesting. *Journal of Applied Physics*, 113(17), p. 174908.

[22] Wei, S., Hu, H. and He, S. 2013. Modeling and experimental investigation of an impact-driven piezoelectric energy harvester from human motion. *Smart Materials and Structures*, 22(10), p. 105020.

[23] Zhu, M., Worthington, E. and Njuguna, J. 2009. Analyses of power output of piezoelectric energy-harvesting devices directly connected to a load resistor using a coupled piezoelectric-circuit finite element method. *IEEE Transactions on Ultrasonics, Ferroelectrics, and Frequency Control*, 56(7), pp. 1309–1317.

[24] Zhang, Y.W., Chen, W.J., Ni, Z.Y., Zang, J. and Hou, S. 2020. Supersonic aerodynamic piezoelectric energy harvesting performance of functionally graded beams. *Composite Structures*, 233, p. 111537.

[25] Ma, M., Xia, S., Li, Z., Xu, Z. and Yao, X. 2014. Enhanced energy harvesting performance of the piezoelectric unimorph with perpendicular electrodes. *Applied Physics Letters*, 105(4), p. 043905.

[26] Shirbani, M.M., Shishesaz, M., Hajnayeb, A. and Sedighi, H.M. 2017. Coupled magneto-electro-mechanical lumped parameter model for a novel vibration-based magneto-electro-elastic energy harvesting systems. *Physica E: Low-Dimensional Systems and Nanostructures*, 90, pp. 158–169.

4 Mass Resonator Sensor and Its Inverse Problems

Yin Zhang

CONTENTS

4.1 THE WORKING AND DESIGN PRINCIPLES OF A MASS RESONATOR SENSOR

4.1.1 THE WORKING PRINCIPLE

Before discussing the working principle of a mass resonator sensor, it is informative for us to have a brief review on that of a conventional mass spectrometry. A charged analyte in a magnetic field is with the Lorentz force, which is described by the following expression:

$$\vec{F} = q\vec{v} \times \vec{B} = m\vec{a} \tag{4.1}$$

Here, $\vec{F}$ is the Lorentz force, q, $\vec{v}$, m and $\vec{a}$ are the analyte charge, velocity, mass and acceleration, respectively. $\vec{B}$ is the applied magnetic field. The ¯ accent indicates that the quantities are the vectors. Here q and m are the only two scalars. The working principle of a conventional mass spectrometry is based on Eqn. (4.1), from which the expression of mass is readily derived as $m = q\vec{v} \times \vec{B}/\vec{a}$. The analyte samples are firstly ionized and then injected into a chamber, in which a magnetic field is applied and the

DOI: 10.1201/9781003328032-4

charged analytes are deflected by the Lorentz force. At the end of the chamber there is an ion detector, which measures q. In a mass spectrometry, $\bar{v}$ and $\bar{B}$ are the control parameters, $\bar{a}$ can be deduced from the trajectory of a charged analyte. When these four quantities (q, $\bar{v}$, $\bar{B}$ and $\bar{a}$) are known, the mass of an analyte m is found. The discovery and development of biological macromolecules ionization methods, which were recognized in the 2002 Nobel Prize in chemistry [1], have significantly boosted the researches of proteomics [2, 3]. In comparison, the greatest advantage of a mass resonator sensor is that it can work with the neutral analytes, i.e., no ionization is required [4–6]. Actually, the lack of the proper ionization method is the very reason for the over 70 years delay of analyzing large macromolecules [1]. Furthermore, to ionize the thermally stable molecule is still chemically a challenge [5]. And ionization can cause the structure change or damage to heavy species such as a protein or a biological macromolecule, which degrades the samples [7].

As a mechanical device, the working principle of a mass resonator sensor is the shifts of resonant frequencies, which is induced by the adsorption of analytes and described by the following equation:

$$f' = \frac{1}{2\pi} \sqrt{\frac{k + \Delta k}{m + \Delta m}} \sqrt{1 - \frac{(c + \Delta c)^2}{4mK}} \tag{4.2}$$

In Eqn. (4.2), a simplified one degree of freedom (DOF) model is used for the resonator. Here, f' is the shifted resonant frequency after adsorption; k, m and c are the resonator effective stiffness, mass and damping coefficient, respectively; Δk, Δk and Δc are their corresponding changes due to the adsorption. In a mass resonator sensor, f', k, m and c are the measured/known quantities; and in practice, Δc can also be found by measuring the frequency response or phase angle of the system with adsorbates [8, 9]. Therefore, Δk and Δm are the only two unknowns to be determined. However, for a given f', there are infinite possible combinations of Δk and Δm which can result in the same f'. This is the mathematical origin of the inverse problems in the mass resonator sensor, which is also a major obstacle for its applications. To illustrate how to use the shift of resonant frequency to determine the (effective) mass, for the time being, we assume that the adsorption only induces the mass loading effect and there is no damping. Correspondingly, Eqn. (4.2) becomes the following:

$$f' = \frac{1}{2\pi} \sqrt{\frac{k}{m + \Delta m}} \approx f_o \left(1 - \frac{\Delta m}{2m}\right) \tag{4.3}$$

Here, $f_o = \sqrt{k/m}/2\pi$ is the system resonant frequency with no adsorption. From Eqn. (4.3), Δm can be easily derived as $\Delta m = 2m(f_o - f')/f_o = 2m\Delta f/f_o$. Here, $\Delta f = f_o - f'$ is the shift of resonant frequency. We emphasize here that Δm is the effective mass induced by the adsorption, which is dependent on both the adsorbate position and mass. This, again, gives rise to the inverse problem. And we will elaborate this inverse problem issue later in details. Equation (4.2) or (4.3) is the sensing mechanism of a mechanical resonator, which was firstly proposed by Thundat et al. in 1995 as the sensitive mass sensors [10–12]. However, the caution should be taken that Δk, Δk, Δm and m are all effective ones, which vary for different structures and for different modes of the same structure. Equation (4.2) or (4.3), which should only be used to qualitatively explain the working mechanism, can cause confusions or errors when analyzing a real mechanical resonator [13]. The continuum model as discussed later in details should be used to more accurately describe a mechanical resonator rather than the above one DOF model.

4.1.2 THE DESIGN PRINCIPLE

In the mass resonator sensor applications, there are two important figures of merit: The minimum detectable mass change and ring-down time [14]. The minimum detectable mass change (δm) is given as follows [14]:

$$\delta m = 2m \left(\frac{2\pi}{Q\omega_o \tau} \frac{k_B T}{U_o} \right)^{1/2} \tag{4.4}$$

Here m is the resonator effective mass, $\omega_o = \sqrt{k/m} = 2\pi f_o$ is the circular resonant frequency, τ is the average time employed for the measurement, Q is the quality factor, k_B is the Boltzman constant, T is the absolute temperature and U_o is the energy stored in the resonator.

Besides τ, the parameter of δm is determined by the responsitivity and the noise level. In Eqn. (4.4), an implicit assumption is that the measurement noise is dominated by thermomechanical fluctuations [14] and $k_B T$ is the corresponding thermal energy, the effect of which on the resonator is indicated by $\left[2\pi \, k_B T/(Q\omega_o \tau U_o) \right]^{1/2}$ [4]. The responsivity (R) is the slope of the sensor output as a function of the input parameter to be measured [15], which is defined for a mass resonator sensor as $R = \partial\omega_o/\partial m \approx -\omega_o/(2m)$ [15]. The conventional mass spectrometry measurements usually involve the measurement of $\sim 10^8$ molecules [16]. The demand on characterizing the proteome at the single-cell or single-molecule level is huge because it can accelerate the identification of protein, disease biomarkers and, thus, new drug development [17]. In comparison, the silicon resonator with the gigahertz (10^9 Hz) resonant frequency and $Q = 10^5$ can (theoretically) achieve the resolution of a δm much less than a Dalton in the room temperature of $T = 300K$ [18]. A Dalton (Da, $\approx 1.66 \times 10^{-27}$ kg), by definition, is 1/12 of the mass of an unbound neutral atom of carbon-12 in its nuclear and electronic ground state and at rest, which physically is the (approximate) average mass of a proton and a neutron. The ultimate goal of any detection method is to achieve the level at which individual quanta of a measured entity can be resolved [19]. For a mass resonator sensor, the holy grail of measurement is to obtain the resolution of one Dalton, which has been achieved in 2012 by the carbon nanotube mechanical resonator [20]. Clearly, Eqn. (4.4) indicates that the ultimate limit of the resonator mass is confined by the thermomechanical fluctuations, which can lead to the intrinsic noise exceeding the sought-after signal [19]. Besides the benefit of working with the neutral analytes, the second advantage of the mass resonator sensor is that the atomic resolution can be achieved. The third advantage is that the mass resonator sensor uses the shifts of resonant frequencies induced by the adsorption, which can significantly lower the rate of false positive and false negative measurements in some biological applications [21, 22].

The second figure of merit is the ring-down time [14], which is a measure of the time width of the step in the output signal due to a sudden mass change on the resonator and is also often referred to as the response time [15]. The ring-down time (t_{RD}) is given by the following equation [14]:

$$t_{RD} = \frac{Q}{\omega_o} \tag{4.5}$$

These two figures of merit as described by Eqns. (4.4) and (4.5) offer three major guidelines for the design of a mass resonator sensor: To increase U_o, ω_o and Q. Increasing U_o and ω_o results in both the smaller δm and t_{RD}. Although increasing Q yields a smaller δm, it causes the undesirable increase of t_{RD}, which physically slows down the resonator response to the mass change. In the experiments of a mass resonator sensor, it is the steady state which is measured [4]. For the system with a high Q, its transient state lasts longer and thus yields a larger t_{RD} [23]. However, as elaborated later, a high-quality factor is a much sought-after property in a resonator [24]. Besides achieving a smaller δm, a higher Q also increases the resonator sensitivity [14]. Sensitivity here is defined as the smallest detectable change of the input signal with a specified signal-to-noise ratio (SNR) [15]. Here, the trade-off is to achieve a higher Q rather than a smaller t_{RD}. Besides the previous three resonator design guidelines, an obvious approach of making δm smaller is to reduce the thermomechanical fluctuation by lowering the temperature. And this is the reason for that many mass resonator sensors are operated in a cryogenic environment [25–27]. Huge efforts have been infused to develop the various methods, such as photothermal, backaction, sideband and feedback [27], to cool down a resonator. The temperature as low as 1.4 μK has been reached [28]. In such low temperature, a macroscopic resonator can reach its ground state, in which its quantum mechanics characteristics of motion are demonstrated [28–30].

The first guideline of designing a mass resonator sensor is to enhance the energy stored in the resonator (U_o), which leads to a smaller δm. Two major approaches can be used to increase U_o. One is to exert tension on the resonator structures. From the viewpoint of continuum mechanics, many resonators are the one-dimensional (1D) structures of string, beam and arch, or the 2D structures of membrane, plate and shell. The potential energy of these structures consists of two parts: Bending and stretching energies. Tension enhances the energy of the stretching part, which also leads to the increase of the quality factor Q [31, 32]. Because the bending loss at the edge is the dominant source of damping, tension increases the stretching energy and thus decreases the proportion of bending energy in the structure vibration energy [15]. It is thus called the damping dilution for this damping reduction by tension mechanism [15]. With the presence of tension, the quality factor for a clamped-clamped beam is given as $Q = \left[\frac{(n\pi)^2}{12} \left(\frac{h}{L} \right)^2 \frac{E}{\sigma} + \frac{1}{\sqrt{3}} \frac{h}{L} \sqrt{\frac{E}{\sigma}} \right]^{-1}$ (h, L, E: The beam thickness, length and Young's modulus, respectively; σ: tensile stress) [15]. Here, we emphasize that the damping dilution mechanism works for the clamped-clamped beam but not for a cantilever beam [32]. Another benefit of tension is that its presence also increases the structure effective stiffness, which leads to the desirable increase of ω_o. Moreover, the presence of tension can delay the onset of the nonlinearity due to the large amplitude vibration [33], which as discussed later can significantly complicate the analysis of the inverse problems. The second approach is to increase the external driving force, which leads to a larger vibration amplitude. Here the caution should be taken: When the vibration amplitude is large, the linear governing equation can no longer be valid because of the geometrical nonlinearity. For the doubly clamped, doubly hinged and hinged-clamped non-slack beam structures, this geometrical nonlinearity is due to the mid-plane stretching [34], which results in an additional tensile axial force that stiffens the structure [35, 36]. Therefore, it is often referred to as the tension-dominant nonlinearity [37, 38]. The critical amplitude for the tension-dominant nonlinearity onset of a doubly clamped nanomechanical resonator is studied by Postma et al. [39]. They showed that the beam aspect ratio and the axial loading are the two major factors determining this critical amplitude [39]. For the cantilever beam, this geometrical nonlinearity results from the nonlinearity of curvature and thus referred to as the curvature-dominant nonlinearity [37, 38]. Mathematically, both the tension-dominant and curvature-dominant nonlinearities introduce a hardening cubic nonlinearity term and as a result, the linear governing equation becomes the nonlinear Duffing equation. The nonlinearity of the Duffing equation is evaluated by its *nonlinearity coefficient* [40], the variation of which as the function of the beam aspect ratio and the axial loading is systematically studied in [38]. When the electrostatic force is the driving/actuation force, it introduces the softening quadratic and cubic nonlinear terms, which results from the Taylor series expansion of the nonlinear electrostatic force [35]. The softening effect of the electrostatic/physical force competes with the hardening effect of the geometric nonlinearity, the response of the resonator can be either a hardening or a softening Duffing oscillator, or even close to a linear oscillator [41–43]. In general, the nonlinearity of the Duffing oscillator makes the system frequency response much more complicate, for example, the mode coupling and vibration localization [44]. Strictly speaking, the concept of resonant frequency is for a linear system. As a result, simple relation of the shifted resonant frequency versus adsorbated mass as given by Eqns. (4.2) and (4.3) can no longer hold for a nonlinear system. Other disadvantages are that the quality factor decreases in the nonlinear vibrations [40, 45] and antiresonance response due to the nonlinearity [46]. Both the hardening and softening effects due to the nonlinearity can result in the saddle-node instability [35, 38, 47], which leads to a sudden change in the vibration amplitude of a resonator. This instability can be effectively used to realize a binary state of a nanomechanical resonator for the applications of a memory element [48] or a switch [49]. Cautions should also be taken when the electrostatic force is the driving/actuation force. The electrostatic force can induce the pull-in instability [35, 47], at which the mechanical force can no longer balance the electrostatic one. As a result, the structure collapses and collides with the electrode, which often causes the failure of the system [50, 51]. Furthermore, the electrostatic force is inversely proportional to the square of the gap distance between the resonator and electrode [35, 47]. The Taylor series expansion of the electrostatic force contains the quadratic and cubic terms, which results in the two equilibria within the gap distance [35] and the symmetry-breaking of the potential (due to the quadratic term of the electrostatic force) [52]. Therefore, the presence of the nonlinear electrostatic force can significantly complicate the resonator frequency response [52], which may cause an enormous difficulty of solving the inverse problem for the mass-sensing applications.

The second design guideline is to increase the resonant frequency of ω_o or more generally, the resonant frequencies of the continuous system. The major disadvantage of a mechanical resonator as compared with an electrical or a magnetic one is its low speed of operation [53], or say, small ω_o. Achieving a higher fundamental resonant frequency is always a major driving force in the development of a mass resonator sensor. For a continuum structure, there are infinite resonant frequencies. Generally, increasing the fundamental resonant frequency will also increase the rest. There are five major approaches of increasing the resonant frequencies (ω_o): Boundary conditions, higher modes, scaling down, material selection and tension. For a rectangular beam structure without an axial force and damping, its ith angular resonant frequency (ω_i) of transverse bending is given as follows [54, 55]:

$$\omega_i = \sqrt{\frac{EI}{m_B}} \frac{\beta_i^2}{L^2} = \beta_i^2 \frac{h}{L^2} \sqrt{\frac{E}{12\rho}} \tag{4.6}$$

Here, E is the beam Young's modulus, $m_B = \rho bh$ (ρ: density, b: beam width and h: beam thickness) is the beam mass per unit length, L is the beam length, β_i is the dimensionless constant totally determined by the beam boundary conditions and is obtained by solving a transcendental equation [54, 55]. For the first modes of the clamped-clamped (C-C), clamped-hinged (C-H), hinged-hinged (H-H) and cantilever (clamped-free, C-F) beams, $\beta_1 = 4.73$ (C-C), $\beta_1 = 3.9266$ (C-H), $\beta_1 = 3.14159 = \pi$ (H-H) and $\beta_1 = 1.8751$ (C-F) [54, 55]. Clearly, the clamped-clamped beam is with the largest β_1 and cantilever beam is with the smallest one. One thing needs to be pointed out here is that the free-free beam shares the same β_is with the clamped-clamped beam if its "first mode", which is a rigid body motion, i.e., $\beta_0 = 0$, is excluded [54]. However, the free-free beam cannot be used in the resonator design. One of the breakthroughs in the development of the resonator is that its first fundamental resonant frequency ($\omega_1/2\pi$) surpasses that of a microwave, i.e., higher than the gigahertz, which was firstly achieved by a clamped-clamped silicon carbide nanobeam [56]. The fundamental resonant frequency of a resonator keeps rising: The clamped-clamped silicon beam with the paddles on its both sides reached 1.49 GHz [57]; the clamped-clamped carbon nanotube resonator achieved the 2 GHz [20] and 3.12 GHz [58]. Here we emphasize that when the Euler-Bernoulli beam model in Eqn. (4.6) is used, it means that the continuum model can still be adequately applied to the nanomechanical resonator [15, 59]. The validity of the continuum modeling is corroborated by the atomic simulations for the monolayer graphene [60, 61].

From the analysis of the boundary conditions, it seems that we should use the clamped-clamped beam and discard the cantilever. However, another factor of design, i.e., the quality factor, should also be considered. One of the major damping mechanisms is the clamping loss [15], i.e., the energy dissipates through the beam connection points, which is also referred to as the attachment loss [62, 63] or support loss [64]. There are two clampings for the clamped-clamped beam and only one for the cantilever beam. Therefore, the clamping loss of the cantilever beam is less, which yields a larger quality factor [62]. Moreover, due to its small stiffness, the free end of a cantilever is with a large displacement, which is convenient and suitable for the displacement sensing via an optical measurement [21, 22]. As a result, the clamped-clamped and cantilever beams are the two dominant beam structures used in the resonators.

The second approach is to excite the higher modes. For any type of the beam structure, β_i increases monotonically with the mode number i [54, 55]. For example, $\beta_1 = 4.73$, $\beta_2 = 7.8532$, $\beta_3 = 10.9956$ and so on for the clamped-clamped beam; $\beta_1 = 1.8751$, $\beta_2 = 4.6941$, $\beta_3 = 7.8547$ and so on for the cantilever beam. When the higher modes are excited, the corresponding resonant frequencies are larger. For a cantilever beam, it also brings an additional benefit of a larger quality factor [65, 66]. It is emphasized here that for other types of structure such as the clamped-clamped beam or a plate with four edges clamped, the higher-mode excitation can even decrease the quality factor [67]. Besides, two cautions should also be taken here. One is that as the higher modes are with the larger wave numbers, the energy/bending deformation of the beam is more evenly distributed along the beam span, which makes the response amplitude smaller and, therefore, lower the signal-to-noise ratio and poses a measurement difficulty. The other is that as the wavelengths of the higher modes are smaller, the Euler–Bernoulli beam model for a slender beam may not hold any more and the Timoshenko beam model for a thick beam should be adopted [68]. The switching of the models on the resonator requires a different computation and explanation on the resonant frequency shifts, which can cause a significant difficulty of solving the related inverse problem [68].

The third approach of the scaling down is indicated by the h/L^2 term in Eqn. (4.6), which is the size effect of the resonant frequency. When the dimensions of a resonator are scaled down, h/L^2 decreases and thus $\omega_i \propto L^{-1}$ (assuming that the ratio of h/L keeps constant). This is the major reason why the resonator becomes smaller and reaches the nanometer scale [56]. Furthermore, with a smaller size, the (effective) mass of a resonator (m) is also smaller [69], which yields the smaller minimum detectable mass change (δm) as indicated by Eqn. (4.4). However, the scaling down of a mechanical resonator also brings three major disadvantages in its applications. The first disadvantage of a very small resonator is to specify the adsorbate location [16, 20, 70–72], which becomes extremely difficult if not impossible. This is also the very physical reason for the inverse problems arising in the mass resonator sensor applications [72]. The second disadvantage is that with the diminishing dimensions, the output signal diminishes as well. A large array of coupled nanomechanical resonators may be needed to enhance the output signal via the synchronization mechanism [73]. The third disadvantage is that the mechanical properties of a micro/nano-resonator may not be well controlled during its fabrication. Because of its small size, the mechanical properties of a micro/nano-resonator are very sensitive to the fabrication imprecision and defects [74]. Each resonator may differ from one another and require its own calibration, which poses a tremendous difficulty.

The fourth approach of the material selection is indicated by the $\sqrt{E/(12\rho)}$ term in Eqn. (4.6). In essence, E is associated with the resonator stiffness; ρ is with the inertial mass. The materials with the high E/ρ ratios, such as carbon nanotube [20, 26], graphene [19, 67, 75, 76], silicon carbide [56, 74] and silicon nitride [77, 78], are frequently used in the mass resonator sensors. The fifth approach of enhancing tension is not indicated in Eqn. (4.6). As mentioned before on how to enhance U_o, the axial tension/compression stiffens/softens a structure, which increases the stiffness and thus the resonant frequencies. Modulating the resonant frequencies through the axial force or temperature, the so-called tunability is also a much sought-after property of a resonator [39, 76]. When the tension is very large, a beam behaves like a string [79] and plate like a membrane [80]. The experiment shows that the beam behaves like a string when the tensile strain reaches 0.35% or larger [81]. For a string, its resonant frequencies are proportional to the square root of the tension, i.e., $\omega_i \propto \sqrt{T/m_s}$ (T: tension, m_s: string mass per unit length) [8, 9]. The clamped-clamped beam is suitable for this tension-enhancement mechanism [31, 32, 77, 81, 82]. Physically, the exertion of tension on a micro/nano-cantilever beam is rather difficult, which is usually done by the residual stress approach during its fabrication process [32]. One caution here is on the material selection. Most of the macroscopic materials yield around the tensile strain of 0.2%. To have the significant improvement of the resonant frequency, the tensile strain usually is required to surpass this 0.2% value of strain. If the material cannot take the high tension, damage or plastic deformation can occur, which leads to the undesirable variations of the resonator mechanical properties. So far, graphene [76], carbon nanotube [20] and silicon nitride [31, 77] are the three major materials used for the tension-enhancement mechanism. For example, graphene failure strain is up to 12% [83]; the silicon nitride nanowire in the experiments is under the tensile stress of 830 MPa [31] and 1.4 GPa [84], respectively. Here we need to emphasize that although the above analysis is applied for a rectangular beam, the main conclusions also hold for the beams with the other types of cross-sections.

The third design guideline is to increase the quality factor of a resonator. The quality factor is defined as follows [15]:

$$Q = 2\pi \frac{U}{\Delta U} \tag{4.7}$$

Here, U is the maximum energy stored in a resonator and ΔU is the energy dissipated per cycle. The quality factor (Q) is a dimensionless quantity, which in essence evaluates the energy dissipation and phase angle in the steady state: A higher Q means less energy dissipation and smaller (lagging) phase angle. However, Eqn. (4.7) is not that straightforward to indicate the quality factor effect on the resonator performance and we should view it from another different angle. The equation of motion for a one DOF system under an external harmonic force driving is given as the following [8, 9]:

$$\ddot{x} + 2\xi\omega_o\dot{x} + \omega_o^2 x = \frac{F}{m}\cos(\omega t) = A\omega_o^2 \cos(\omega t) \tag{4.8}$$

where $\omega_o = \sqrt{k/m}$ (k: spring stiffness and m: mass) is the resonant/natural frequency of the undamped system; F and ω are the amplitude and angular frequency of the external driving force, respectively. $A = F/k$ is the static elongation of the spring under constant force F. The dimensionless viscous damping factor (ξ) is defined as $\xi = c/(2\omega_o m)$ and c is the coefficient of viscous damping [8]. The steady state of the forced vibration, which mathematically is the particular solution of Eqn. (4.8) and with the same oscillating of ω as the external force, is with the following relation [8]:

$$\frac{|x_s|}{A} = G(\omega) = \frac{1}{\sqrt{\left[1 - \left(\frac{\omega}{\omega_o}\right)^2\right]^2 + \left(2\xi\frac{\omega}{\omega_o}\right)^2}} \tag{4.9}$$

where $|x_s|$ is the amplitude of the steady state, which varies as the function of the driving frequency ω. Clearly, $G(\omega)$ is a measure of the system response to a harmonic excitation of frequency ω, which is often known as the frequency response [8]. The maximum of $G(\omega)$ is $G_{max} = 1/\left(2\xi\sqrt{1-\xi^2}\right) = Q$ when $\omega = \omega_o\sqrt{1-2\xi^2}$ [8], which physically corresponds to the resonance of the system. Clearly, another physical meaning of the quality factor Q is the (maximum) amplification factor of the system steady state at the resonance as compared with the static magnitude of A. When damping is very small, $Q \cong 1/2\xi$. Clearly, smaller damping corresponds to larger quality factor. The quality factor also determines the bandwidth of the system as $\Delta\omega \cong 1.09\omega_o/Q$ [85]. Here $\Delta\omega = \omega_2 - \omega_1$ is the bandwidth and ω_1, ω_2 are found as the half-power points, which are defined mathematically as $G(\omega_1) = G(\omega_2) = Q/\sqrt{2}$ [8]. In terms of the frequency response, a large Q means the large magnitude of the peak and narrower bandwidth. The presence of damping also generates another effect of phase angle, i.e., the steady state oscillates as $x_s = AG(\omega)\cos(\omega t - \varphi)$. Here the phase angle is given $\varphi = \tan^{-1}\left[\frac{2\xi\omega/\omega_o}{1-(\omega/\omega_o)^2}\right]$ [8]. Physically, the presence of a phase angle means that the system steady state response lags behind the excitation force. Experimentally, to obtain the frequency response as indicated by Eqn. (4.9) involves the acquisition of successive steady state amplitude-driving frequency ($|x_s| - \omega$) curves. It is also very time-consuming especially when the quality factor is large because it takes a long time for the transient response to die out [24]. Therefore, the frequency response measurement is not recommended in the resonator applications [15]. In comparison, measuring the phase angle is a much more efficient method, which only requires one measurement. In the above expression of the phase angle $\varphi = \tan^{-1}\left[\frac{2\xi\omega/\omega_o}{1-(\omega/\omega_o)^2}\right]$, ω is the input signal and ω_o is known; once φ is obtained experimentally, ξ and thus Q are found. The measurement of phase angle is often performed by a closed loop system called phase-locked loop (PLL) [15]. This same working principle has also been successfully developed into the phase-imaging method in the amplitude-modulation atomic force microscopy (AM-AFM) capable of performing the non-destructive subsurface imaging with high spatial resolution on the *salmonella typhimurium* cells covered by an extracellular polymeric capsule [86], in which the different phase angles are caused by different dampings induced by the inhomogeneity of the composite materials.

4.1.3 DAMPING MECHANISMS

Damping is a phenomenological model on the system energy dissipation. In Eqn. (4.8), damping is modeled as the linear viscous damping. Damping can be nonlinear [85, 87] and with other different models. For example, the model of the hysteretic damping is presented as follows [88]:

$$m\ddot{x} + \frac{h}{\omega}\dot{x} + kx = F\cos(\omega t) \tag{4.10}$$

where h is the coefficient of hysteretic damping and the quality factor of the system is given as $Q = h/k$ [88]. The hysteretic damping force is $\frac{h}{\omega}\dot{x}$, which is dependent on both the driving frequency ω and velocity $\dot{x}$. In comparison, the viscous damping force is $c\dot{x}$, which is not dependent on ω. The viscous damping is the dominant model used for the mass resonator sensors [38]. The hysteretic damping model is mainly used for the metal structures [87, 88] and many resonators are often coated with the metal layers for certain functionality. It is well possible that for quite some resonators, both the hysteretic and viscous dampings can be present. To accurately model a real damping force is always a challenge [88]. Generally speaking, both the hysteretic and viscous dampings are the simplified models of characterizing a real complex dissipation process, which is usually dependent on driving frequency and highly nonlinear [87]. Furthermore, we restrict our discussion on the linear damping. Various nonlinear dampings can arise in a resonator, for example, from the interaction of the mechanical resonator with a thermal bath of harmonic DOFs [85].

A major obstacle to the development of the sensitive mass resonator sensor is that the quality factor decreases significantly when the device dimensions are scaled down as shown in Figure 4.1 by Mohanty et al. [89]. This size-dependent property of the quality factor can be attributed to the large and ever-increasing surface contribution to the energy dissipation as the resonator dimensions become smaller, or say, the surface loss is the dominant mechanism of energy dissipation [90]. It is suggested that the surface dissipation is dependent on the surface defect density and independent of the mechanical properties of the bulk [91]. Here the concept of surface defect can be very broad, which includes the adsorption layer [92], dangling bonds [93] and chemical oxidation [94]. Surface roughness, which further increases the surface area, can make the surface loss even larger [90]. The effective methods of mitigating the surface dissipation are to reduce the surface defects via the chemical/surface treatment [93] or the heat treatment such as annealing [64], which removes oxidation or other surface termination layer. Although huge efforts have been infused into the study of the damping mechanisms and the solutions to improving the quality factor, a clear physical picture on the various damping mechanisms in a micro/nanostructure is still elusive [89]. Various damping mechanism are summarized by Schmid et al. [15]

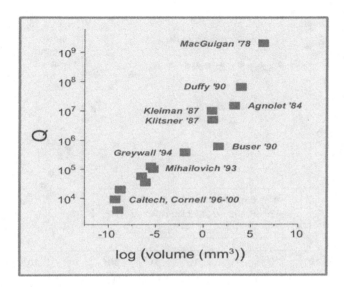

FIGURE 4.1 The overall trend of the size dependence of quality factor in mechanical resonators. (Reprinted figure with permission from P. Mohanty et al. [89]. Copyright 2002 by the American Physical Society. https://doi.org/10.1103/PhysRevB.66.085416)

as four general and broad categories: medium interaction, clamping, intrinsic and other losses. The fourth category of other loss covers all other damping mechanisms which are not covered by the first three mechanisms.

The mechanism of the medium interaction loss is rather straightforward: The resonator interacts with the ambient medium during the vibration and, therefore, the energy is transferred from the resonator to the medium. This loss mechanism is also referred to as the momentum-exchange damping [23]. When the ambient medium is air, this medium interaction loss mechanism can be further classified into two different damping mechanisms of the ballistic and fluid regions, which is determined by the Knudsen number. The dimensionless Knudsen number (K_n) is defined as the ratio of the mean free length of gas to the characteristic length scale of the resonator [15]. The ballistic region is with $K_n < 1$, in which gas cannot be treated as a continuum and the dissipation is due to the impacts of the non-interacting gas molecules. The fluidic region is with $K_n > 1$, in which gas can be treated as a viscous fluid. The medium interaction loss can be the predominant energy loss mechanism for a micro/nanomechanical resonator when the pressure is above 10^{-2} mbar [95]. Lavrik and Datskos [96] found that the ambient air damping at the atmospheric pressure is so large that the resonant frequencies of different resonators are almost only dependent on the beam width and independent on its thickness. The solution to the medium interaction loss is simple: To put the resonator in a lower pressure/vacuum chamber to reduce/eliminate the interaction [4, 20, 64, 95, 96]. It is interesting to notice that in the SU-8 microstring experiment, when the air pressure is around five Pascal, which is already in the ballistic region, further reduction of the air pressure has no use of improving the quality factor as shown in Figure 4.2 excerpted from the book by Schmid et al. [15].

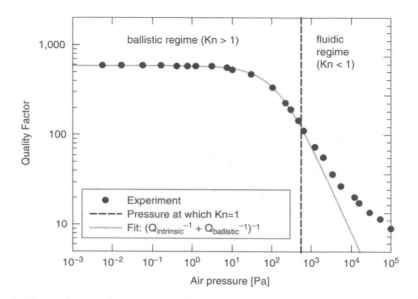

FIGURE 4.2 Measured quality factors plotted against air pressure, for a 14 μm wide SU-8 microstring. The dashed line shows the pressure $p = 550$ Pa at which $Kn = 1$, which represents the transition pressure between the fluidic and the ballistic regime. (Reprinted figure with permission from S. Schmid, L.G. Villanueva and M.L. Roukes [15]. Copyright 2016 by the American Physical Society. https://doi.org/10.1007/978-3-319-28691-4)

The clamping/attachment/support loss [62, 63] is caused by that of the energy of a resonator transmits into the base/support, which can be a primary dissipation of a chunky microcantilever beam [64]. The support and resonator are with different geometric dimension or, more generally, they can also be two different materials, which results in the so-called *acoustic mismatch* [15]. This *acoustic mismatch* or, say, the discontinuity at the junction cross-section, is responsible for the clamping loss. According to the D'Alembert solution, the vibration of a resonator is a standing wave, which is superposed by the two traveling waves. The *acoustic mismatch* is characterized by the impedance, which is a concept borrowed from the electric circuit theory. The impedance together with the area difference between the resonator and support cross-sections determines the transmission and reflection of an incidence wave [97]. The transmitted waves are the source of the resonator energy dissipation. The clamping loss of a cantilever beam with the Poisson ratio of 0.3 is characterized in Figure 4.3 by Photiadis et al. [63]. Clearly, the damping of the clamping loss (Q^{-1}) in Figure 4.3 is divided into two regions, the expression of which is given by the following two equations:

$$Q^{-1} \approx 0.95 \frac{w}{l} \frac{h^2}{h_b^2} \tag{4.11}$$

$$Q^{-1} \approx 0.31 \frac{w}{l} \left(\frac{h}{l}\right)^4 \tag{4.12}$$

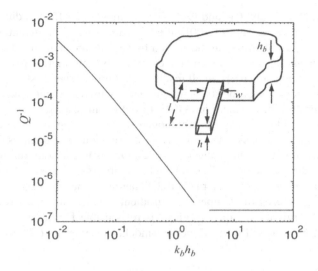

FIGURE 4.3 Predicted contribution of radiation to the loss factor for the fundamental mode of a cantilever beam attached to a substrate of thickness h_b. Here, k_b is the wave number of the transmitted shear wave inside the substrate. The beam is taken to have dimensions $l/h = 20$ and $l/w = 10$ (l, h, w: the beam length, thickness and width). (Reprinted from D.M. Photiadis and J.A. Judge [63], with permission of AIP Publishing.)

Here, l, w and h are the beam length, width and thickness of a resonator, respectively; h_b is the substrate thickness. As seen in Figure 4.3, the two regions are demarcated by the parameter of $k_b h_b$. Here, $k_b = 1/\lambda_b$ is the wave number of the propagating wave transmitted into the substrate and is its wavelength. Eqn. (4.11) applies when $k_b h_b < 1$ and Eqn. (4.12) applies when $k_b h_b > 1$ [63]. Clearly, to reduce the damping, an effective method is to geometrically increase the substrate thickness of h_b and to reduce the resonator aspect ratio of w/l. Similarly to that of Figure 4.2, the damping of Q^{-1} saturates once the $k_b h_b$ value reaches 1, which is the limit of reducing damping by increasing the substrate thickness. An implicit assumption used in Eqns. (4.11) and (4.12) is that the materials of the resonator and substrate are the same. Therefore, besides the geometric approaches of modulating h_b and w/l, another effective approach of reducing damping is the material selection, which is to reduce the ratio of E/E_b (E and E_b: Young's moduli of the resonator and substrate, respectively) because $Q^{-1} \propto E/E_b$ [63]. From the viewpoint of impedance, a larger E/E_b ratio means a larger impedance mismatch between the resonator and substrate, which results in that more incidence wave from the resonator is reflected back rather than propagating/transmitting through the substrate [97]. More reflection and less transmission mean less energy dissipation and more energy stored in the resonator system.

The intrinsic damping by definition is those loss mechanisms taking place either on the surface or in the bulk of a resonators. Here we only review three major types: thermoelastic, anelastic and Akhiezer dampings. For other intrinsic dampings, the reader should refer to the topic review by Imboden and Mohanty [90], and the books by Schmid et al. [15] and Cleland [98]. The thermoelastic effect is due to the inhomogeneities of stress. Zener [99, 100] called this damping effect the internal friction. The inhomogeneity of stress can be caused by various mechanisms. For example, a cavity, the orientation or anisotropy of a crystal, vibrating motion pattern [100] and material inhomogeneity [101] can all result in an inhomogeneous stress field inside a resonator. The presence of the stress inhomogeneity increases entropy, which is then associated with the heat generation [101]. In the beam resonator case, the bending motion is responsible for the stress inhomogeneity: According to the Euler–Bernoulli beam theory, the stress inside a beam varies linearly along its thickness direction and the beam neutral plane demarcates two zones: One is under compressive stress and the other is under tensile stress. The zone with compression becomes warmer and the zone with tension becomes cooler [15, 102]. Therefore, the stress gradient due to the beam bending induces the temperature gradient. As a result, the irreversible heat flow driven by this temperature gradient is responsible for the energy dissipation [100, 103]. More specifically, the Zener thermoelastic damping (Q_Z^{-1}) is characterized by the following equation [103]:

$$Q_Z^{-1} = \Delta_E \frac{\omega \tau_Z}{1+(\omega \tau_Z)^2} \tag{4.13}$$

where τ_Z is the thermal relaxation time and $\tau_Z = h^2/(\pi^2 \chi)$ (h: beam thickness and χ: the thermal diffusivity of the beam material); ω is the beam-vibrating (circular) frequency. Here, Δ_E is a dimensionless quantity called the *relaxation strength of the modulus* defined as follows [100, 103]:

$$\Delta_E = \frac{E_a - E_i}{E_i} = \frac{E_i \alpha^2 T_o}{C_p} \tag{4.14}$$

where E_a is the adiabatic value of Young's modulus, which corresponds to the unrelaxed state; E_i is the isothermal value of Young's modulus, which corresponds to the relaxed state. Here α, T_o and C_p are the coefficient of thermal expansion, the ambient temperature and the specific heat at constant pressure.

In Figure 4.4, Q_Z^{-1}/Δ_E, or say, $\omega \tau_Z/[1+(\omega \tau_Z)^2]$, as the function of ξ, is plotted. Here $\xi = h\sqrt{\omega_o/(2\chi)}$ and ω_o is the isothermal value of ω and χ is the thermal diffusivity of the solid material of the beam [103]. Physically, ξ is a dimensionless parameter indicating approximately the square root of the ratio of the relaxation time (τ_Z) to the resonator period. As clearly seen in Figure 4.4, Zener's Eqn. (4.13) captures the thermoelastic damping rather accurately. In Figure 4.4, the peak value of Q_Z^{-1}/Δ_E is around 0.494 at $\xi \approx 2.225$. Here Eqns. (4.13) and (4.14) in conjunction with Figure 4.3 provide two major approaches of reducing the thermoelastic damping. The first is to reduce Q_Z^{-1}/Δ_E by modulating the driving frequency of the resonator. Therefore, either a smaller or a larger ξ value than 2.225 can both result in a smaller Q_Z^{-1}/Δ_E. When ξ is very small, the relaxation time is very small compared to the resonator period, which means that the heat has no time to relax and thus the adiabatic state is kept. Therefore, very little energy is dissipated. In comparison, when ξ is large,

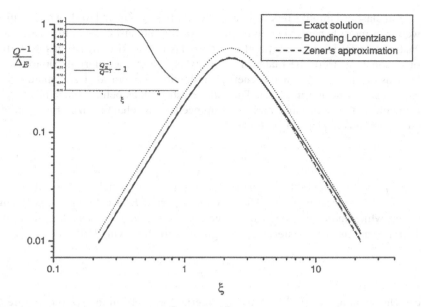

FIGURE 4.4 Q_Z^{-1}/Δ_E, or say, $\omega\tau_Z/\left[1+(\omega\tau_Z)^2\right]$, as the function of ξ, is plotted for the small flexural vibrations of a thin beam. The solid line is the result derived by Lifshitz and Roukes, the dotted line is the bounding Lorentzian and the dashed lined is Zener's derivation of Eqn. (4.13). (Reprinted figure with permission from R. Lifshitz and M.L. Roukes (2000) [103]. Copyright 2000 by the American Physical Society. https://doi.org/10.1103/PhysRevB.61.5600).

the relaxation time is very large compared to the resonator period, which means that the resonator stays more or less in an equilibrium state, i.e., the isothermal state. And again, very little energy is dissipated. In summary, the first approach is to avoid certain driving frequency range of ξ around 2.225. The second approach is to reduce Δ_E. For the single metal crystal with the hexagonal symmetry (for example, zinc), its Δ_E value depends on the angle (θ) between the principal axis of the crystal and axis of the beam (length direction). As given in Table 3 of the Zener's classical paper [100], the difference of the Δ_E value can be as large as 165 times as θ changes from 0° to 90° for zinc at 100 K. Zener's Tables 3 and 4 in [100] also show that for different metal at different temperatures, the Δ_E value in general is in a decreasing trend as θ changes from 0° to 90° and $\theta = 90°$ always results in the smallest Δ_E. Therefore, aligning the resonator length axis perpendicular to the principal axis of the crystal is the most effective approach when the single metal crystal is used as the building material of the resonator. For the polycrystalline metal and anisotropic metal cases, this Δ_E value varies in a large range for different metals as seen in Tables 1 and 6 in [100] and tungsten is with the smallest Δ_E. Once the resonator material is selected, the material properties such as E_i, α and C_p in Eqn. (4.14) are fixed (in certain range). Reducing the ambient temperature of T_o is also an effective method in case that the value of $E_i\alpha^2 T_o/C_p$ does not increase significantly.

The anelastic damping model is also developed by Zener [104], which can be viewed as a more general version of the above thermoelastic damping model. The anelastic material is special type of the viscoelastic material, which has no lasting deformation [15]. Here the material viscosity is due to the internal friction when it deforms, which results in the energy dissipation. For an anelastic material, its stress and strain are not in phase: The induced strain lags behind the applied stress and, therefore, the stress-strain relationship needs to be characterized by a complex Young's modulus (E^*) given as follows:

$$E^* = E_1 + iE_2 \tag{4.15}$$

The real part of E_1 is called the storage Young's modulus, which indicates the elastic effect of a viscoleastic material. The imaginary part of E_2 is called the loss Young's modulus, which indicates the viscous effect, or say, the energy dissipation. The quality factor of an anelastic material (Q_a) is given as the following [15]:

$$Q_a = \frac{E_1}{E_2} \tag{4.16}$$

The anelastic dampings (Q_a^{-1}) of various materials are well summarized in Figure 4.5 by Ashby [105]. It is noticed at the lower right corner of Figure 4.5, the silicon carbide and silicon nitride are the two materials with the smallest Q_a^{-1}, which is a major reason for them to be frequently used for the building material of a resonator [32, 56, 74, 77, 78, 106]. Moreover, this anelastic damping model is also used for the surface friction, which can be the dominant energy dissipation in a micro/nanomechanical resonator because of its high surface-to-volume ratio [95, 106]. Similarly, based on the Zener's anelastic damping model [104], the dissipative surface stress model is also developed [107].

The Akhiezer damping (Q_{Ak}^{-1}), which is also often referred to as the phonon-phonon interaction loss [15, 98], is with the following expression similar to the Zener thermoelastic damping of Eqn. (4.13) [98, 108]:

$$Q_{Ak}^{-1} = \frac{C_V T_o \lambda_{avg}^2}{\rho v^2} \frac{\omega \tau_{ph-ph}}{1 + \left(\omega \tau_{ph-ph}\right)^2} \tag{4.17}$$

where C_V and T_o are the ambient temperature and the specific heat at constant volume, respectively; ρ is the resonator density and $v = \sqrt{E/\rho}$ (E: resonator Young's modulus) is the sound velocity; τ_{ph-ph} is the phonon relaxation time and ω is the resonator (circular) vibration frequency; λ_{avg} is the mean value of the Grüneisen parameter, which is a dimensionless quantity characterizing the volume-changing effect of a crystal lattice has on its vibrational properties. Very similar to the mechanism of the thermoelastic damping, the Akhiezer damping results from the coupling of the strain field and phonon. The temperature gradient arises due to the strain variations: Temperature decreases for positive dilation (stretching) and increases for negative dilation (compression). As a result, the diffusive motions between the phonons with different temperatures remove the energy from the system strain energy [98]. Again,

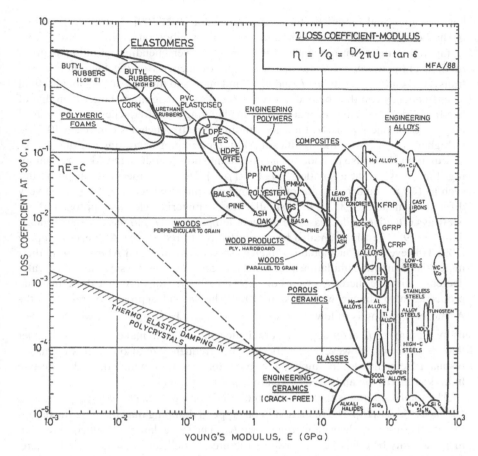

FIGURE 4.5 The anelastic dampings of various materials. Here the abscissa of Young's modulus E corresponds to our E_1 in Eqn. (4.16), the ordinate of loss coefficient η is our Q_a^{-1}. (Reprinted from Acta Metallur, 37(5), M.F. Ashby (1989) [105], with permission from Elsevier.)

the irreversible heat flow is responsible for the Akhiezer damping. In essence, these three intrinsic loss mechanisms of thermoelastic, anelastic and Akhiezer dampings can all be attributed to one thing: relaxation. Phonon, which is the lattice vibration, is with very high frequency. In Eqn. (4.17), τ_{ph-ph} is with the order of picosecond (10^{-12} s) and ω is usually with the order of gigahertz (10^9 Hz) or megahertz (10^6 Hz) for some nanomechanical resonators. The typical value of $\omega\tau_{ph-ph}$ is in the order of 10^{-2} or less [108]. Small value of $\omega\tau_{ph-ph}$ means the small Akhiezer damping as indicated by Eqn. (4.17). The Akhiezer damping becomes significant when the resonator vibrating frequency (ω) is 10 GHz or higher. Though attainable, it is still a huge technical challenge to fabricate a nanobeam resonator with the transverse-bending frequency higher than 1 GHz [56]. The Akhiezer damping of the bending mode of a resonator is very small. The Akhiezer damping mainly results from the beam longitudinal vibrations for a nano-resonator [108, 109]. The eigenfrequency of the beam longitudinal vibration, or say, the bar vibration, is $\propto L^{-1}\sqrt{E/\rho}$ [8, 9]. Compared to the eigenfrequency of the beam transversebending vibration as given in Eqn. (4.6), $\omega_l/\omega_b \approx L/h$ (ω_l and ω_b: eigenfrequencies of the longitudinal and transversebending vibrations; L and h: beam length and thickness). The longitudinal eigenfrequency can thus be one or even two orders of magnitude larger than that of bending. For those nano-resonators with the bending eigenfrequency of hundreds of megahertz or one gigahertz, their longitudinal eigenfrequencies can reach 10 GHz or higher, which causes the significant Akhiezer damping according to Eqn. (4.17).

Equation (4.17) also provides us an insight into how to reduce the Akhiezer damping. Firstly, we want a smaller $C_V T_o \lambda_{avg}^2 / (\rho v^2)$. However, the material properties such as C_V, λ_{avg} and τ_{ph-ph} are rather complexly interconnected and it will be very difficult to make $C_V T_o \lambda_{avg}^2 / (\rho v^2)$ smaller from a material selection approach. But it is always effective to decrease the ambient temperature (T_o) to reduce the Akhiezer damping [98]. Tension can also be very effective to reduce the Akhiezer damping [109]. However, caution should be taken as shown by Kunal and Aluru [109] in their molecular dynamic simulation that the Akhiezer damping does not monotonically decrease with the increase of tension. The possible reasons are that those material properties also change with tension [109]. The second approach of reducing the Akhiezer damping is to lower the value of $\omega\tau_{ph-ph} / \left[1 + \left(\omega\tau_{ph-ph}\right)^2\right]$, which reaches its maximum of 1/2 at $\omega\tau_{ph-ph} = 1$. A realistic approach is to make $\omega\tau_{ph-ph} \ll 1$ rather than $\omega\tau_{ph-ph} \gg 1$. However, a smaller driving frequency ω also requires a smaller resonant frequency ω_o to amplify the resonator vibration, which degrades the two figures of merit, δm and t_{RD}, as given in Eqns. (4.4) and (4.5).

The total damping is the summation of all dampings, i.e., $Q^{-1} = Q_{medium}^{-1} + Q_{clamping}^{-1} + Q_{intrinsic}^{-1} + Q_{other}^{-1}$ [15]. The reduction in one type of damping may increase the damping of other type, or deteriorate δm and/or t_{RD}. The same damping reduction mechanism can only be effective for certain structures. A delicate balance is needed here. For example, the damping dilution mechanism of the damping reduction by tension only works for the clamped-clamped beam and is totally ineffective for the cantilever beam [32]. For this damping dilution mechanism to work, it depends on the interactions of thermoelastic, anelastic and clamping dampings [67, 102]. The other example is that the damping reduction mechanism by the higher mode excitation only works for the cantilever beam but not the clamped-clamped beam [65–67]. A specific analysis is required for each individual analysis. In the above two examples, the key analysis is the shape of the structure deformation. The beam-bending energy, which is responsible for the energy dissipation, depends on the curvature. In contrast, the stretching energy only depends on the slope. The large curvature is around two places: the antinodes, which are peaks/valleys of the beam vibration, and edges. In comparison, the nodes, which are the immovable points/lines of the vibration, are with rather small curvatures. As seen in Figure 4.6 excerpted from [102] by Schimd et al., the first mode shape of the clamped-clamped beam experiences a dramatic shape change with the increase in the axial tension. The beam shape under large tension is like a string, which significantly decreases the curvatures at both the antinode (center) and edges. As a result, the energy dissipation due to bending decreases. In contrast,

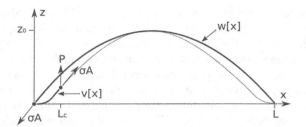

FIGURE 4.6 Deformation shapes of the clamped-clamped beam with no (dashed line) and large (solid line) tension. The beam shape with a large tension is close to that of string. (Reprinted figure with permission from S. Schmid, K.D. Jensen, K.H. Nielsen and A. Boisen (2011) [102]. Copyright 2011 by the American Physical Society. https://doi.org/10.1103/PhysRevB.84.165307)

as seen in Figure 4.7 [35], the axial tension has almost no alleviation on the curvature at the clamped end of a cantilever and the overall curvature reduction of the whole beam is very limited, which should be the mechanism responsible for the failure of the damping dilution mechanism on the cantilever as shown in the experiments [32]. However, when higher modes are excited, both the number and curvature of the antinodes increase. Because of the increasing contribution of the antinodal bending

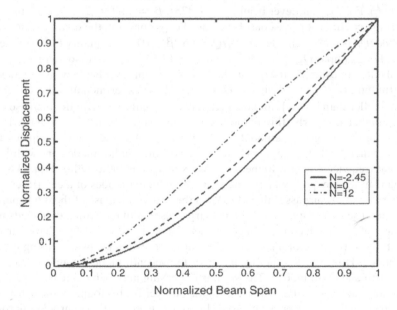

FIGURE 4.7 Deformation shapes of the cantilever beam with different axial tension. Here the negative/positive value of the normalized axial force N physically corresponds to the compress/tension. Here, $N = -2.45$ is the value very close to the buckling load of a cantilever beam. (Reprinted from Sens. Actuators A, 127, Y. Zhang and Y.P. Zhao (2006) [35], with permission from Elsevier.)

energy, the anelastic energy dissipation energy increases, which should lead to a smaller quality factor. The experiments show that the decrease of quality factor is rather surprisingly small. For the clamped-clamped structure, Yu's explanation [67] is that under a large tension, the high-curvature zone of higher modes is still local-ized around the edges and, therefore, the energy dissipation induced by the effect of the edge curvature is dominant. Instead of a small decrease, the excitation of higher modes actually enhances the quality factor as mentioned before. Here we offer an explanation in conjunction with Eqn. (4.8). The quality factor for the one DOF sys-tem is with the following expression:

$$Q \cong \frac{1}{2\xi} = \frac{\omega_o m}{c} = \frac{\sqrt{km}}{c} \qquad (4.18)$$

where $\xi = c/(2\omega_o m)$, c is the coefficient of viscous damping counting for the anelas-tic effect of the cantilever beam, $\omega_o = \sqrt{k/m}$ is the resonant frequency of a mode of an undamped cantilever beam. Here k and m are the effective spring stiffness and mass of the beam vibrating in the corresponding mode. In the Galerkin method, when the normalized mode shape is used to discretize the cantilever beam [35, 51], the effective mass m has no change for different mode. However, effective spring stiffness k for different mode varies and is proportional to β_i^4. Here β_i (i: the mode number) is the constant determined by the beam boundary conditions as discussed in Eqn. (4.6). For the cantilever beam, $\beta_1 = 1.8751$, $\beta_2 = 4.6941$, $\beta_3 = 7.8547$ and so on [54]. If the coefficient of viscous damping c keeps constant, the quality factor of the ith mode (Q_i) is with such relation: $Q_i/Q_1 = (\beta_i/\beta_1)^2$ (Q_1 : the quality factor of the first mode). For example, $Q_2/Q_1 \approx 6.26$ and $Q_3/Q_1 \approx 17.54$. In comparison with the experi-mental data, for example, the quality factors of the cantilever beam with the thickness of 1 μm [66], $Q_2/Q_1 \approx 3.84$ and $Q_3/Q_1 \approx 7.35$. Those experimental ratios are consis-tently smaller than $(\beta_i/\beta_1)^2$ and the difference also enlarges with the increase of the mode number i. Clearly, this is the result of the assumption of constant c. Higher modes are with higher antinodal bending energy and, therefore, dissipates more energy, which leads to a higher c. At the same time, higher modes are with higher resonant frequency, or say, higher effective stiffness as indicated in Eqn. (4.8). The mechanism for the quality factor increase of the higher modes of a cantilever is that the increase of ω_o surpasses that of c. With the further increase of the mode number, c can increase more rapidly and finally surpasses that of ω_o, which then leads to the decrease of quality factor. As seen in their Table 1 presented by Ghatkesar et al. [66], the quality factor of the 1 μm-thick cantilever beam keeps increasing until the sixth mode. However, the quality factor of the seventh mode is smaller than those of the fifth and sixth modes. The similar increasing and then decreasing pattern of the quality factor with the increase of mode number is also found in the 4 μm-thick and 7 μm-thick cantilever beams [66]. In comparison, a silicon carbide micro-disk resonator, the central area of which is supported by an underneath silicon oxide ped-estal, shows complex patterns of the quality factor changes as a function of the mode number [74]. The micro-disk resonator is driven by the Brownian thermomechanical fluctuations. The smaller resonator as shown by Wang et al. [74] in Figure 4.2 shows the monotonical decrease of the quality factor with the increase of mode number.

In contrast, the quality factor change of the larger resonator as shown in Figure 4.4 is rather erratic [74].

Various damping mechanisms are rather complex and interconnected, which can make the job of damping reduction very difficult. While the three general strategies of enhancing the quality factor as summarized by Moser et al. [26] can offer a very useful guideline of reducing damping. The first is to improve the quality of resonator-building material, which includes exerting tension, optimizing the fabrication process to reduce contamination and thus decrease the material surface and internal frictions. The second is to isolate the resonator from its surrounding environment, or say, to reduce the medium interaction loss. The third is to operate resonators at cryogenic temperatures. Their carbon nanotube resonator is an ultra-clean one grown by chemical vapor deposition in the last step of the fabrication process, which makes it free of fabrication residues and thus reduces the surface friction [26]. The device is cooled down to 30 mK, the quality factor as high as five million is then achieved [26]. Similarly, the clamped-clamped silicon nitride resonator achieves the quality factor of 1 million via exerting large tensile force and operating in vacuum [77]. Finally, we would like to address that in the scenario in which a resonator is operated close to the intrinsic limit of thermomechanical noise, a large damping can be used to suppress the thermal fluctuations and, therefore, a better signal-to-noise ratio can be achieved [110], which is a more or less surprising result. Furthermore, a large damping can also achieve a better frequency stability at the same time and in that particular scenario, a large damping is helpful for the improvement of nanomechanical resonator [110].

4.2 INVERSE PROBLEMS OF MASS RESONATOR SENSORS

4.2.1 ORIGINS OF THE INVERSE PROBLEM IN MASSZ RESONATOR SENSOR

In Eqn. (4.2), the origin of the inverse problem is briefly discussed. In this section, we focus on the beam model of continuum theory rather than the one DOF theory for a detailed discussion. The adsorption can induce the changes of the beam mass, stiffness and damping. The corresponding governing equation of the beam-free vibration is as follows [111–113]:

$$\left[m + \Delta m + \sum_{i=1}^{K} M_i \delta \left(x - x_i \right) \right] \frac{\partial^2 w}{\partial t^2} + \left(c + \Delta c \right) \frac{\partial w}{\partial t} - \left(N + \Delta N \right) \frac{\partial^2 w}{\partial x^2} + \left(EI + \Delta EI \right) \frac{\partial^4 w}{\partial x^4} = 0 \quad (4.19)$$

where m is the beam mass per unit length with the unit of $kg \cdot m^{-1}$, c is the beam viscous damping, N is the beam axial force and EI is the beam-bending stiffness (E: Young's modulus and I: moment of inertia). Here Δm, Δc, ΔN and ΔEI are their corresponding changes after the adsorption. For induced mass effect, Δm is due to the distributed mass as seen in Figure 4.8 and with the same unit of $kg \cdot m^{-1}$ as m. Sometimes, the size of the adsorbates is so small and, therefore, the adsorbate is modeled as the concentrated mass [12, 15, 113]. Here M_i is the cocentrated mass with unit of kg, δ is the Dirac delta function [113], x_i is the adsorbate location and K is the number of adsorbate. The induced mass effect is due to both the distributed mass of Δm and

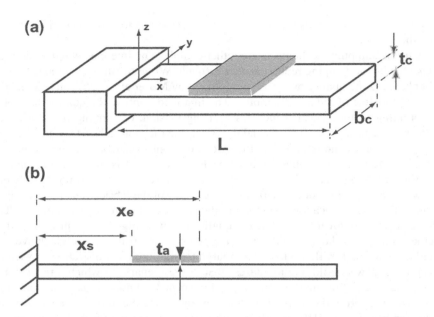

FIGURE 4.8 (a) Schematic diagram of a cantilever resonator with molecules adsorbed on its surface and the coordinate system. (b) The adsorbed layer is (assumed) uniformly distributed from x_s to x_e with a thickness of t_a and the same width as that of the cantilever. (Reprinted from Sens. Actuators B: Chem, 202, Y. Zhang (2014) [112], with permission from Elsevier.)

the concentrated mass of $\sum_{i=1}^{K} M_i \delta(x - x_i)$. Here Δm and M_i are the distributed and concentrated masses of adsorbates. However, their influences on the resonator resonant frequencies/eigenfrequencies also depend in their locations. The location effect in the concentrated mass is embodied in the Dirac delta function and x_i. When a concentrated mass is on the nodes of an excited mode, the mass effect is zero and, therefore, the corresponding resonant frequency is unchanged. In comparison, when a concentrated mass is on the antinodes, the mass effect is the maximum resulting in the largest eigenfrequency reduction in the corresponding mode. The nodes and antinodes are different for the different modes of a given structure. For different structures or the same structure with different boundary conditions, their mode shapes are completely different, the same Δm and M_i will result in different eigenfrequency shifts. The effect of the distributed mass also depends on its location, which is given as x_s (starting point) and x_e (ending point) as seen in Figure 4.8. Equation (4.19) is a generalized version of the governing equation indicating the adsorption effect, which can vary depending on x_s and x_e [112]. In summary, the mass effect on the resonator eigenfrequency depends on four factors: (1) magnitude of Δm and M_i, (2) their locations (x_s and x_e for the distributed mass, and x_i for the concentrated mass), (3) the excited mode and (4) the beam boundary conditions.

Both ΔN and ΔEI result in the change of the beam stiffness. Physically, the added mass effect is quite straightforward. Here we give a brief discussion on how an adsorption induces the changes of the axial force and bending stiffness. One of the stiffness change mechanisms is also very straightforward: The adsorption forms a new layer on the resonator and the resonator is like a composite structure. The elastic properties and geometry of this new layer together with those of the resonator determine the stiffness of the composite structure. Moreover, the adsorption can also induce the formation of alloys or chemical bondings. For example, the adsorption of hydrogen can form the palladium-hydrogen hydride and the adsorption of mercury can form the mercury-gold or the mercury-silver alloy [114]. For the adsorption with chemical bonding, the van der Waals (vdW) interaction for neutral adsorption and the Coulomb/electrostatic interaction for the charged adsorption can exist, which leads to the changes of both surface stress and surface elasticity [115]. The coexistence of different interactions and their competition determine the final state of surface stress and surface elasticity. For example, the surface stress induced by the self-assembly of alkanethiols on gold is determined by the attractive vdW and repulsive electrostatic forces [116], which are dependent on the alkanethiol coverage and chain length. For the adsorption of double-stranded deoxyribonucleic acid (DNA), Eom et al. [117] showed that the change of the beam-bending stiffness is due to the change of the intermolecular interaction, which consists of three major parts: hydration repulsion, electrostatic repulsion and configurational entropic effect. Dareing and Thundat [118] presented a model showing that the arrangement of the adsorbates on a resonator surface changes the Lennard-Jones (L-J) potential, which induces the in-plane elastic stress. The above various interactions can be classified into two types: adsorbate-surface and adsorbate-adsorbate interactions [119]. Furthermore, Hu et al. [120] showed that besides the changes of the L-J potential and in-plane stress, adsorption also changes the structure boundary condition, which effectively impacts the overall structure stiffness. In the gas adsorption experiment, the small vapor gas molecules can interpenetrate the polymer chains of polymethyl methacrylate (PMMA) films coated on a resonator surface [121], which changes the Young's modulus of PMMA and thus the stiffness of the resonator system. The adsorbed gases with nonzero dipole moments interact more strongly with the PMMA and stiffen the PMMA layer more significantly [121]. An adsorbate can induce the electronic and mechanical distortions of a graphene [122, 123], which imparts a tension to a suspended graphene resonator and thus leads to the increase of its resonant frequency [33]. In a broader sense, an adsorbate changes the Gibbs free energy of the system [124], which may induce surface stress or other effects. The charged density waves (CDWs), which result from the electron-phonon coupling, change the electron distribution inside a crystal and thus distort the lattice [125]. As a result, the large stress due to the lattice distortion can change the effective stiffness of a $NbSe_2$ resonator and, therefore, its resonant frequency [125]. All of these alter the overall system elastic properties and thus the stiffness of the composite structures.

There are several mechanisms for an adsorption resulting in the change of a resonator axial force (ΔN). For example, when neutral hydrogen or helium is adsorbed, the surface dissociation of hydrogen/helium subsequently causes the formation of interstitial hydrogen in a metal alloy lattice, which distorts the lattice [126]. Therefore, this lattice distortion relieves the built-in tensile stress of a resonator, which is responsible for the decrease of resonant frequency [126]. The ion adsorption can result in the surface stress, which leads to the change of ΔN and thus the resonator eigenfrequency [127]. In the adsorption of water ions on a silicon resonator, the oxidized surface (SiO2) will be hydroxylated by a reaction with OH and H radials generated from water ions [127]. This hydroxylation can also change the resonator stiffness. Moreover, if the resonator driving force is an electrostatic one, the adsorbed ions form a gradient field of charges, which interacts with an electrode. As mentioned before, the electrostatic force softens the system, which effectively reduces the resonator stiffness [127]. Sometimes the characterization of this softening effect can become very complicated: Besides the gate voltage, the quantized excess charges on a resonator surface can also have an impact on the electrostatic force [27, 128]. The adsorbate layer absorbing the light, which induces thermal expansion and thus thermal stress due to the photothermal effect, can also relieve the build-in stress [77, 129]. Because of its small dimensions, the micro/nanomechanical resonator is with a small heat capacity and this photothermal effect can be very significant. For example, in the experiment of a 530 μm × 530 μm × 50 nm silicon nitride drum resonator, two 10-nm Au adsorbate particles reduce the drum built-in stress from 30 MPa to 0.8 MPa through this photothermal effect [129], which then causes a significant shift of resonant frequency. Similarly, for the resonators driven by the laser heating, the thermal stress can significantly alter the in-plane tension [130]. Generally, the built-in stress, which often results from the top-down process, is unknown and treated as a contamination [131]. Furthermore, the compressive force due to the different thermal expansion at different ambient temperature can also change the build-in stress of a clamped-clamped structure [76].

The mass change of Δm and M_i, which mathematically can only be positive, physically can only result in the decrease or at most no change of the resonator eigenfrequency as indicated in Eqns. (2), (3) and (19). In contrast, ΔN and ΔEI, which can be either positive or negative, physically can either increase or decrease the resonator eigenfrequency. When Δm, M_i, ΔN and ΔEI are the known quantities and suppose that the corresponding m, N and EI are also known, Eqn. (4.19) presents a forward or direct problem. The eigenfrequency of a given mode is uniquely determined by Eqn. (4.19), which is a straightforward cause-effect relation, i.e., the Δm, M_i, ΔN and ΔEI (plus the adsorbate positions) are the causes which result in the effect of the resonator eigenfrequency shifts. However, in the applications of the mass resonator sensor, the (shifts of) eigenfrequencies, which are the effect, are the measured quantities and we need to find the causes. This gives rise to the inverse problem and we need to think in a different paradigm of the effect-cause. Mathematically, for a given/measured eigenfrequency, there are infinite possible combinations of Δm, M_i, ΔN and ΔEI [111–113]. Here the resonator damping c and Δc can be measured before and after an adsorption [12, 65], which helps to reduce the complexity of the inverse problem. The simplest version of the inverse problem is to assume that the adsorption is with the mass loading effect only. For a concentrated mass, there are only two unknowns: adsorbate mass and its location. The assumption of the mass effect may not hold in some scenarios. The vivid examples are that the adsorption of the Escherichia coli (*E. coli*) bacteria [132] and organic molecules of alkanethiol [133] on a silicon resonator causes the increase of resonant frequency. The mass effect alone cannot explain the increase of resonant frequency as the adsorbate mass can only reduce the resonant frequency. Another example is in the gas adsorption experiment [121]. As seen in McCaig's Table 1 [121], the resonant frequency can either increase or decrease depending on what kind of gas is adsorbed in the PMMA coating surface. For example, the adsorption of hexane reduces the resonant frequency, while that of ethyl acetate enhances it [121]. If the stiffness effect is not considered, the adsorbed mass determined from the eigenfrequency shifts will be wrongly evaluated. Of course, the consideration of the stiffness effect can greatly complicate the inverse problem [72]. Furthermore, the counteractive effects of the mass and stiffness can have significant impact on the sensitivity of the resonator sensor [134].

An implicit assumption that the adsorbates don't move on the resonator surface is also used in Eqn. (4.19). When the physiosorbed atoms are with relatively large kinetic energy and small adhesion energy, surface diffusion can occur [135–137]. The resonator eigenfrequency can also be shifted due to the change of adsorbate position driven by the surface diffusion. In the severe scenario, the resonator vibration can even have a qualitative change referred to as the diffusion-induced biostability [136, 137], in which the resonator suddenly switches to another equilibrium configuration. Besides the position change, the moving adsorbates on the resonator surface can also result in an effective axial compression due to the (generalized) Coriolis effect [138] and thus reduce the stiffness. In the mass sensing of the proteins or biomolecules which are transported to the resonator by a fluid via the microfluidic channels [139–142], the effect of the moving fluid mass may need to be considered for an accurate evaluation for the mass when the moving velocity is large in the scenario of the high throughput application. When the adsorbates are charged, the coupling between the resonator mechanical motion and electron transport can further complicate the problem [143]. Besides the problem of the moving adsorbates, the adsorption-desorption processes constantly occur on the resonator surface [59, 135, 144] and these processes can even be influenced by the resonator motions [145], which leads to a very strong coupling [143]. This further complicates the analysis.

4.2.2 CALIBRATION OF A MASS RESONATOR SENSOR

4.2.2.1 Surface Effect

In the above inverse problem, the mechanical properties of m, N and EI are assumed the known properties and the unknowns are Δm, M_i, ΔN and ΔEI. However, as the dimensions of the resonator become small, those mechanical properties may not be accurately characterized and a calibration is thus needed. For example, the bending stiffness $EI = Ebh^3/12$ for a rectangular cross section beam (b, h: the beam width and thickness, E: the beam Young's modulus). The manufacturing tolerances and material variations cause the bending stiffness of one resonator to differ slightly from another, which introduces the deviations from the design values [146, 147]. The manufacturing errors mainly impact the resonator geometry such as the width, thickness and length. As the resonator is scaled down, the ratio of the surface-to-volume increases, the surface effects stand out. Here the surface effects on the resonator mechanical property are mainly two types: surface elasticity and surface stress. Usually, Young's modulus E is measured from a macroscopic material. For a microscopic material, Young's modulus is an important evaluation parameter on the

FIGURE 4.9   The core–shell model of a circular nanowire. The core corresponds to the bulk and the shell corresponds to the surface. (Reprinted from J. He and C.M. Lilley (2008), [153], with permission of AIP Publishing.)

material elasticity and it is very sensitive to the defect and surface layer, which is a result of the interatomic forces between the constituent atoms [148]. The presence of defect or doping directly impacts the interatomic forces, which then changes the material elastic properties [148]. The surface effect has a very significant influence on a silicon nitride microcantilever, which makes its effective Young's modulus to deviate dramatically from that of a macroscopic one [149]. The surface effect is size-dependent, which stands out in various micro/nanostructures [150, 151].

The application of the ansatz that nanostructure = bulk + surface [152] in continuum mechanics leads to the so-called core–shell model as seen in Figure 4.9. The formation of a surface layer makes a nano/micro-structure a *de-facto* composite structure [36, 153–155]. The core (bulk) and shell (surface) are with two different elastic properties. Surface reconstruction [156] and surface relaxation [157, 158] are the two major mechanisms responsible for the formation of a surface layer. The presence of a surface layer changes the effective Young's modulus of a nano/micro-structure [36, 153, 154]. Besides surface elasticity, the formation of a surface layer also induces surface stress, which changes the structure axial force together with the residual stress due to fabrication.

Here we need to have a brief discussion on these three different quantities: surface elasticity, surface stress and surface energy. Although these three quantities physically are very different, they all share the same unit of $N \cdot m^{-1}$. In a surface layer, the total surface stress (τ) is given as follows [36, 153–155]:

$$\tau = \sigma + C_s \varepsilon \qquad (4.20)$$

where ε is the dimensionless strain and C_s is the surface modulus. Surprisingly, C_s can be either positive or negative [152, 159]. Physically, τ is the result of charge redistribution as the electrons respond to the effects of terminating a solid at a surface [156, 160]. In Eqn. (4.20), τ consists of two parts: σ and $C_s \varepsilon$. Here σ, which is strain-independent, is often referred to as (constant) surface stress [161–164]; $C_s \varepsilon$, which is strain dependent, is often referred to as surface elasticity [162–164]. The presence of surface elasticity (C_s) essentially changes the effective Young's modulus and thus the overall stiffness of a micro/nanostructure [165]. Here we emphasize that τ, σ and C_s are all with the same unit of $N \cdot m^{-1}$. The strain ε is dimensionless and ε is not the surface strain. The surface strain, which is with the unit of meter, is defined as the difference between the ideal and actual positions of a surface plane [166]. From the viewpoint of the finite phases, the total surface stress τ is an excess of the bulk stress tensor [166]. Surface is a special type of interface, which is the solid-vacuum

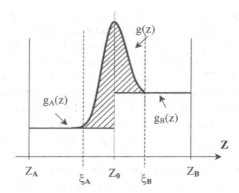

FIGURE 4.10 The schematic on an interfacial excess quantity, which is the shaded area in the figure. (Reprinted from Surface Science Reports, 54, P. Müller and A. Saúl (2004), with permission from Elsevier.)

interface. The definition of an interfacial excess quantity, which is also demonstrated in Figure 4.10, is given as follows [166]:

$$g^{interf} = \int_{Z_A}^{Z_B} g(z)\,dz - g_A\left(Z_A - Z_0\right) - g_B\left(Z_B - Z_0\right) \tag{4.21}$$

where g_A and g_B are the bulk densities in different phases; g^{interf} is the interfacial excess quantity per unit interface area. In Figure 4.10 [166], Z is the axis perpendicular to the interface, ξ_A and ξ_B define the interface thickness. The definition of g^{interf} is independent of Z_A and Z_B as far as $Z_A < \xi_A$ and $Z_B > \xi_B$. Clearly, g^{interf}, as indicated in Eqn. (4.21), is the accumulated quantity evaluating the change of density $g(z)$ across an interface. When $g(z)$ is the bulk stress, the corresponding g^{interf} is the surface/interfacial stress. Because of the integration as given in Eqn. (4.21), bulk stress and surface stress are with different units of $N \cdot m^{-2}$ and $N \cdot m^{-1}$, respectively.

From the thermodynamics viewpoint, the total surface stress τ is a tensor associated with the reversible work to elastically stretch a pre-existing surface [167]. In comparison, surface energy (γ) is thermodynamically defined as the reversible work per unit area involved in forming a surface, which exposes new atoms [168]. Therefore, the concept of surface energy arises in fracture mechanics, in which new surfaces are created, or in contact mechanics, in which two surfaces are combined into one interface [169, 170]. Again, from the viewpoint of the finite phase, surface energy is the excess of the elastic energy density of the bulk [166]. The surface stress and surface energy are related by the following Shuttleworth equation [166]:

$$\tau_{ij} = \gamma\delta_{ij} + \frac{\partial\gamma}{\partial\varepsilon_{ij}^{\parallel}} \tag{4.22}$$

where δ_{ij} is the Kronecker delta and $\varepsilon_{ij}^{\parallel}$ is the bulk strain parallel to the interface/surface. Mathematically, surface energy (γ) is a scalar and its value is always positive.

In comparison, surface stress τ_{ij} is a second-order tensor, its value can be either positive or negative. In general, the values of γ and τ_{ij} are significantly different in many solids [166]. Another thing needs to be emphasized here is that surface stress can only arise in a solid due to the kinematic constraint that the bulk atoms are unable to move out to a surface [152]. In contrast, for a liquid whose bulk atoms can move to a surface, there is only surface energy and no surface stress in a liquid [152].

When an adsorption occurs and the deposition layer thickness is very small compared to that of the resonator, only the surface stress effect needs to be considered [171], for example, in the receptor-ligand type of adsorption [172, 173] or the self-assembled monolayer [114]. When the thickness of a deposition layer is relatively large, the stiffness of the adsorption material itself can be very significant [132, 133]. In this scenario, the interface of a resonator-adsorption layer is formed. The thickness, Young's modulus and Poisson's ratio of the interface layer together with the dimensions and elastic properties of the resonator and adsorption layer determine a complex stress distribution inside this composite structure [174, 175], which can result in the different deflections and different (effective) stiffness.

There is a debate on whether a surface stress can change the stiffness of a cantilever beam/plate structure and thus its eigenfrequency. This debate can be traced back to the experiment by Lagowski et al. [161] in 1975, in which they found the eigenfrequency of a GaAs cantilever increases with h/L^2 (h, L: beam thickness and length). Here h/L^2 is a characteristic parameter indicating the resonator dimension and the related discussion is also presented in Eqn. (6). Lagowski et al. [161] ascribed the mechanism of the increasing eigenfrequency to surface stress, which is the strain-independent σ as indicated in Eqn. (20). Gurtin et al. [162] rebutted the Lagowski's conclusion [161] and they concluded that the eigenfrequencies of a cantilever is independent of σ, and the increase of the eigenfrequency is due to the surface elasticity, which is the strain-dependent $C_s\varepsilon$ in Eqn. (20). Lachut and Sader [163, 164] further studied the problem and they agreed with Gurtin et al. [162] that surface stress has almost no effect on the eigenfrequency of a cantilever beam/plate. Their reason is that the surface stress contribution to the axial load is zero and thus no stiffness change because of the free end of the cantilevered beam/plate [163]. The change of the axial load of a cantilever beam/plate due to surface stress is *in direct violation of Newton's third law* [163]. According to Lachut and Sader [163], surface stress can have very little effect on the stiffness/eigenfrequency of a cantilever structure and this little effect is due to the Poisson's effect, which can only be revealed by the three-dimensional modeling. Lachut and Sader's conclusion [163, 164] is corroborated in the experiment by Karabalin et al. [176] that eigenfrequencies of various cantilever beams have almost no changes as the surface stress varies; in comparison, those of the clamped-clamped beams change dramatically. However, the experiment by Finot et al. [114] on this particular issue shows that the surface stress, due to the charge repulsion in ionic solutions or electrochemical conditions, can change the stiffness and eigenfrequencies of a cantilever, which is consistent with the adsorption experiment of ions [127]. Ono and Eshashi's explanation [127] for the eigenfrequency shifts is that surface stress changes the axial force of a cantilever beam as indicated in their Eqn. (4.4), which is in direct contradiction with the Lachut and Sader's conclusion [163].

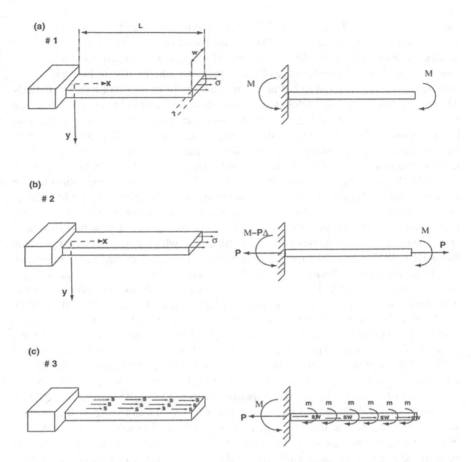

FIGURE 4.11 Three different models of surface stress. (a) #1 model is the Stoney model, in which the surface stress effect is modeled as a concentrated bending moment applied at the free end. (b) #2 model is to model the surface effect as a concentrated bending moment and a concentrated axial force applied at the free end. (c) #3 model is to model the surface stress as the distributed bending moment and stress along the cantilever length. (Source: Y. Zhang, Q. Ren and Y.P. Zhao (2004) [171]).

Three different models of the surface stress effect on a cantilever beam are presented in Figure 4.11 excerpted from [171]. In Figure 4.11(a), the #1 model is to account for the surface stress effect as a concentrated bending moment applied at the free end. This model leads to the famous Stoney formula as the following [177–180]:

$$\kappa = \frac{6\sigma}{E^* h^2} \tag{4.23}$$

where κ is the uniform curvature of the cantilever, σ is the surface stress defined in Eqn. (4.20) with the unit of $N \cdot m^{-1}$, $E^* = E/(1-v)$ is the biaxial modulus (E, v: beam Young's modulus and Poisson's ratio) [178], h is the beam thickness. In Eqn. (4.23),

Stoney originally used E instead of E^* [171, 177]. Although the Stoney formula serves as a cornerstone for the analysis of curvature-based measurements, its accuracy is always an issue [171, 178–180]. In the modeling aspect, there are four major factors responsible for the inaccuracy of the Stoney formula. Firstly, in Eqn. (4.23), the curvature of κ is (assumed to be) uniform, which is in general not true [180]. Only when the surface stress is modeled as a concentrated moment applied at the free end and only for the cantilever beam, can the assumption of a uniform curvature hold [171]. Secondly, for many bilayer structures, for example, the film/substrate, when the film thickness is relatively large, the film stiffness needs to be accounted, which leads to the inaccuracy of the Stoney formula because Eqn. (4.23) does not account for the stiffness of the deposition layer or film [178, 179]. The Stoney formula can be viewed as the limit case of the film/substrate bilayer structure as the film thickness approaches zero [179, 180]. Thirdly, σ is assumed as a constant across the film/deposition layer thickness. In general, this is not true either. The actual (residual) stress inside a thin film varies across the thickness, which is the so-called gradient stress [181, 182]. The model for the gradient stress is rather complex, which is also dependent on the distribution of gradient stress, and simple relation of Eqn. (4.23) is not valid any more [183, 184]. Fourthly, the Stoney formula does not account for the axial load effect of surface stress. In Figure 4.11(b) and (c), #2 and #3 models account for the axial load as a concentrated load and a distributed load, respectively. A direct result of incorporating the axial load effect is that the curvature is no longer uniform along the beam length [171, 180]. The error of the Stoney formula leads to an error of 533% is observed in a cantilever beam experiment [185]. The reason for the #2 and #3 models to incorporate the axial load is due to the beam model. In a beam model (both the Euler–Bernoulli and Timoshenko beam models), the governing equation is to describe the deflection of the neutral surface. As seen in Figure 4.11(c), surface stress is in-plane and acting on the beam upper surface. To use the beam model, we have to move the surface stress from the surface to the neutral plane, which results in both the bending moment and axial load [171]. And the in-plane axial load leads to the $(N + \Delta N)\dfrac{\partial^2 w}{\partial x^2}$ term in Eqn. (4.19), which physically is a distributed transverse load given by the Young–Laplace formula [186, 187] and causes the structure stiffness change. Here, there is an implicit assumption of the in-plane surface stress in the Young–Laplace formula and in Figure 4.11. A real surface is a rough one, both the in-plane and out-of-plane stresses can be induced due to the roughness, which can result in the surface stress effect significantly different from those predicted by the Young–Laplace formula [188].

Because it is still a heated debated issue, it is worth spending some time discussing whether surface stress results in the axial load effect. As mentioned earlier, Lachut and Sader concluded that surface stress cannot change the axial load of a cantilever beam, otherwise it will be *in direct violation of Newton's third law* [163]. We start our discussion on this Newton's third law issue, which is traced back to the bimetallic thermostat model developed by Timoshenko in 1925 [189]. Timoshenko's model is for a beam/plate consisting of two metals with two different coefficients of thermal expansion [189]. When the temperature varies, the two metals will expand differently if they are not constrained. However, they are constrained by their interface and they must expand as a whole. The following two rules apply for the analysis of the bimetallic thermostat [190–192]:

> *Rule 1: The compression of one layer + the extension of the other layer = difference in "free" lengths.*
> *Rule 2: The tensile force applied to the short layer by the long layer is equal in magnitude to the compressive force applied to the long layer by the short layer.*

Here Rule 1 is a geometrical constraint to require the two layers of a bimetallic thermostat to expand as a whole. The detailed information and its physical meaning can be clearly seen in Figs. 1 and 2 of [191]. Rule 2 is the Newton's third law. For a clamped-clamped bimetallic thermostat, a compressive axial load is generated due to the thermal expansion by the two clamped ends. For a clamped-clamped beam, the total strain is zero because the whole beam length does not vary during the thermal expansion. The total strain consists of two parts: mechanical and thermal strains. The thermal strain due to the thermal expansion is positive (elongation), the mechanical strain must then be negative to cancel the thermal strain. As a result, the compressive axial load is generated because stress can only result from the mechanical strain. However, for a cantilever beam, because its free end has no constraint and thus no external load can be exerted. Therefore, the net axial load is zero, or say, the magnitude of the tension in one layer must be equal to that of the compression in the other layer. Therefore, a cantilevered bimetallic thermostat can generate bending moment but no axial load, which is in agreement with Lachut and Sader's conclusion as discussed previously [163]. Here we need to emphasize that for the above two rules to apply, the following two assumptions must also be satisfied: ideal interface and no initial/residual stress. Firstly, an ideal interface means that its thickness is zero and the strain/displacement is continuous across the interface. But the stress can have a sudden change/discontinuity across an ideal interface due to the continuity of strain and discontinuity of elastic property across the interface. The bimetallic thermostat model given by Timoshenko [189] assumes such ideal interface. However, a real interface deviates more or less from an ideal one due to various defects, such as dislocation [174, 193], twin and dangling bonds [194], accumulated around the interface or its interface–void–interface formation patterns [195], which is typical in a micro-composite structure. All of these reduce the composite interface adhesion and thus its mechanical strength [194]. The continuity of the strain/displacement across a non-ideal interface does not hold, which is demonstrated in Figure 4.5 of [175], because of the interfacial slip and finite thickness of the interface layer [193]. For example, the Cu/Si interface layer thickness is around a nanometer [196]. Instead of the above two simple rules, a rather complex analysis is needed to analyze such non-ideal interface [174, 175, 193]. There are two major types of models on the non-ideal interface: shear lag model [197] and lap shear model [198], which are with a deformable/compliant interface of finite thickness and strength. Of course, Newton's third law is always valid for the non-ideal interface case. But things now become much more complicated here: As for the interface layer, there are three layers in the non-ideal interface model rather than the two layers in an ideal interface model. The interface layer plays a vital role in the force transfer between the two layers

[174, 175, 193] and the experimentally measured stresses in the two layers of the micron size are significantly different from those predicated by an ideal interface model [199]. One major reason resulting in different surface stress measurements is that the surface stress inside a film is only partially transferred to the substrate/structure and an adhesion layer is usually needed to enhance the transfer for an accurate measurement [200]. The slippage of a film/coating is a major mechanism for the loss of stress transfer [201]. The second assumption is that there is no initial/residual stress. The pre-existing initial/residual stress alters the atomic arrangement, or say, distorts the lattice, which thus alters the interatomic forces. The elasticity of the structure, for example, (effective) Young's modulus will be correspondingly altered due to the change of interatomic force [148]. Park and Klein used the surface Cauchy–Born (SCB) model to simulate the surface stress effect and their conclusion is that the eigenfrequency of a cantilever beam is indeed changed by surface stress [202, 203]. Surface stress is due to the surface relaxation and/or surface reconstruction, which both distort the lattice or change the atomic arrangement and thus change the structure stiffness. A continuum mechanics approach shows that there is a difference between the original and relaxed states of a nanobeam, which results in the residual stress for both the cantilever and clamped-clamped beams [204]. When a gold film is deposited on a silicon microcantilever, surface/residual stress is generated due to the formation of grain boundary during the coalescence of individually nucleated clusters [160]. More specifically, the voids and channels between the grains determine the magnitude of a surface stress [200]. This surface/residual stress is dependent on the two control parameters of the deposition speed and layer thickness [160]. The surface stress can even have the tensile-compressive transition as the two control parameters are varied [160]. In a broader sense, the above two assumptions are the same. The effect of the initial/residual stress creates a defected interface/surface. For example, the bonding of two wafers can generate significant dislocations around their interface when the wafer roughness is large. Those dislocations induce a permanently deformed strain field, which thus induces the residual stress [174, 193]. Again, Newton's third law still holds for the bonded wafers with a damaged interface. For the bonded structure to deform as a whole, the two forces acting on the two wafers due to the deformation must obey the Newton third law [193]. But the pre-existing residual stresses/strains are independent parameters determined totally by the atomic arrangements after bonding, which can be clearly seen in Eqns. (10) and (12) of [193]. Besides the wafer bonding, there are various processes/mechanisms causing the residual stress and its gradients during a fabrication, for example, interfacial atomic diffusion [182], epitaxial growth [167, 179], etching [205], low-pressure chemical vapor deposition (LPCVD) [205] and sputtering [199, 206]. In the above processes, large deformation and thus defect are often involved. For example, the lattice mismatch between the film and substrate crystals during epitaxial growth induces very large strain around the interface [179] and the misfit dislocations are generated as a result of the glide of threading dislocations [167].

Moreover, a real surface/interface can behave as either an ideal or non-ideal one depending on the applied load and its mechanical strength. If the surface/interface experiences the permanent deformation (as induced by dislocations [167, 193]), which results in residual stress, the elastic property of the structure as a whole can be altered [159, 207]. In those scenarios, surface stress needs to be treated as residual stress [120, 207–209]. This residual stress scenario is like a localized defect induced by an adsorption and a localized defect can have an impact on the overall mechanical properties of a specimen [210, 211]. When the adsorption-induced surface stress is rather small and the interface of the adsorption layer-cantilever beam behaves as an ideal one, there is no net axial force due to Newton's third law. If the bending stiffness of the adsorption layer is zero or rather small, the stiffness and thus the eigenfrequency of a cantilever do not change. The experiment by Finot et al. [114] actually shows that both the scenarios can occur. In their experiment, the surface stress due to the charge repulsion in ionic solutions or electrochemical conditions changes the cantilever eigenfrequency; in contrast, the surface stress due to the swelling of a film bends the cantilever but has no effect on its eigenfrequency [114].

4.2.2.2 Multilayer Resonator

In general, the resonator is a composite structure as seen in Figure 4.12 excerpted from [176]. Various coating layers are deposited on the resonator for various purposes. One major purpose is to enhance the sensitivity [121, 145, 176, 212, 213]. For example, certain coating layer can increase the interaction with adsorbate, resulting in a larger surface stress [145, 213]. The second purpose is to enhance the interface strength. For example, the chromium/gold/chromium (Cr/Au/Cr) film are sputtered and patterned on a silicon resonator; the chromium layer is an adhesion layer to enhance the bonding between the gold and silicon [214]. The third one is to enhance the reflectivity for the measurement purpose, for example, a gold layer on a silicon cantilever beam [12]. This multilayer composite structure significantly increases the difficulty in both modeling/characterization of the micro/nanostructures and solving the related inverse problems. Firstly, it is difficult to monitor/measure the geometry of the deposited layer because of their small size. For example, the uncertainty of

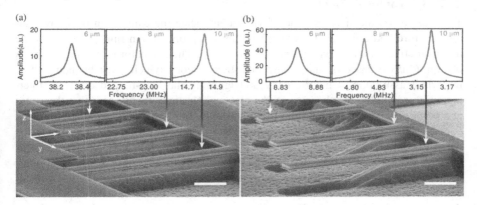

FIGURE 4.12 (a) The clamped-clamped beams and (b) cantilever beams. Both the clamped-clamped and cantilever beams are the composite beam of the thickness of 320 nm, which consist of four layers: 20 nm aluminum nitride (AlN), 100 nm molybdenum (Mo), 100 nm AlN and 100 nm Mo. (Reprinted figure with permission from R.B. Karabalin, L.G. Villanueva, M.H. Matheny, J.E. Sader, and M.L. Roukes, Stress-induced variations in the stiffness of micro- and nanocantilever beams. Phys. Rev. Lett., 108, 236101 [2012], Ref. [176]. Copyright 2012 by the American Physical Society. https://doi.org/10.1103/PhysRevLett.108.236101)

the deposition layer thickness poses a major challenge on the accuracy in the mass resonator application [12]. Secondly, during its fabrication, the mechanical properties of the resonator are sensitive to the defects, impurity and fabrication error [215]. Different fabrication procedures can result in different interface conditions and residual stress [160], which causes the resonator mechanical properties to differ [174, 175, 193]. The worst scenario is that the mechanical properties of the resonators can vary from one to another; even they are manufactured by the same procedures and at the same time [146, 147, 216]. The poor reproducibility is a serious problem which is frequently encountered in fabrication of a small mechanical device [160].

Mahmoud [13] performed the theoretical analysis and finite element (FE) computation on the three-layer microcantilever and compared the results with the experimental ones [217]. It is noticed that the resonant frequencies and flexural rigidity predicted by the theory and FE are consistently higher than the experimental ones. It is observed that Mahmoud's theory [13] and FE computation are based on the ideal interface model. A possible mechanism is the Poisson's ratio mismatch between different layers, which results in a constraint force in the beam width direction and thus changes the beam flexural rigidity. However, this Poisson's ratio mismatch mechanism was ruled out by Ilic et al. [218]. Another possible mechanism proposed here is that different layers form non-ideal interfaces, which yields smaller flexural rigidity as compared with that of the ideal interfaces. Furthermore, when the residual interfacial/surface stress acts in both the width and length direction, the effective Young's modulus is the biaxial one as given in Eqn. (4.23), which also helps to reduce the flexural rigidity. Similarly, compared with the experimentally measured resonant frequencies of a seven-layer microcantilever [219], Mahmoud's theoretical results [13] and Zurn's FE computational results [219] are, again, shown to be consistently higher.

4.2.2.3 Nonlocal Effect

There are also other factors which further complicate the calibration of the resonator structure and thus the analysis of the inverse problem. An important one is the nonlocal effect of a micro/nanostructure [68, 220]. The resonator governing equation of Eqn. (4.19) is based on the classical elasticity theory, which is established on two important assumptions: homogeneity and ignoring the long-ranged effect of interatomic forces [221]. In other words, the classical elasticity theory says that the stress at a point depends on the strain at that point only. In an atomistic simulation, this is equivalent to saying that the forces acting on an atom of a solid are only due to the nearest neighbor atoms [222]. This is the reason why the classical elasticity theory is also the local theory. In general, the local theory is not true because the range of interatomic force is not confined in the nearest neighbors [223]. When the dimensions of a structure become small, the prediction of the local theory and the experimental data are shown to have a significant difference or even be wrongful, for example, on the dispersion relation of a wave propagating in a nanostructure [224]. The nonlocal theory of continuum mechanics [223–229] was developed in essence to incorporate the long-range effect of interatomic force, which mainly results from the discreteness and inhomogeneity of material micro-structure [230, 231]. The discreteness and inhomogeneity of a polycrystal nanobeam due to its different grains is vividly demonstrated in Figure 4.13 excerpted from [232].

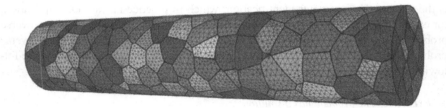

FIGURE 4.13 The 3D schematic of polycrystalline micro-structure, the inhomogeneity of which is mainly due to its different grains. (Source: Y. Wei, Y. Li, L. Zhu, Y. Liu, X. Lei, G. Wang, Y. Wu, Z. Mi, J. Liu, H. Wang and H. Gao (2014) [232]).

A typical version of the nonlocal theory is presented as the following [233]:

$$\left(1 - l_1^2 \nabla^2\right) t_{ij}(x) = C_{ijkl} \left(1 - l_2^2 \nabla^2\right) \epsilon_{ij}(x) \tag{4.24}$$

where $t_{ij}(x)$ is the nonlocal stress at the reference point x and $\epsilon_{ij}(x)$ is the classical (dimensionless) strain; $\nabla^2 = \partial^2 / \partial x_1^2 + \partial^2 / \partial x_2^2$ is the Laplacian operator and C_{ijkl} is the elastic modulus tensor. Here l_1 and l_2 are two characteristic lengths. When $l_1 = 0$ and $l_2 \neq 0$, i.e., that the stress gradient terms are gone, Eqn. (4.24) is reduced to the strain gradient elasticity model. In comparison, when $l_2 = 0$ and $l_1 \neq 0$, i.e., that the strain gradient terms are gone, Eqn. (4.24) is reduced to the nonlocal elasticity model. The strain gradient effect stiffens a structure and in contrast, the stress gradient effect softens a structure. If this nonlocal effect is introduced, the governing equation of a resonator vibration can be rather complex as compared to Eqn. (4.19) [68, 220]. However, the introduction of these two lengths of l_1 and l_2 is purely based on *mathematical structure* [234], or say, that the dimensional analysis requires $l_1^2 \nabla^2$ and $l_2^2 \nabla^2$ to be dimensionless in Eqn. (4.24). In other words, there is no clear physical picture on the two lengths of l_1 and l_2 though they are more or less the parameters characterizing the range of the interatomic forces [223]. According to Yang and Lakes [235, 236], the appearance of l_1 and l_2 in the nonlocal theory is purely phenomenological. The full utility of the nonlocal theories hinges on one's ability to determine these characteristic lengths [237]. The work on the evaluation of the characteristic lengths is quite few and certainly no clear consensus exists [230]. To find these two characteristic lengths forms an inverse problem, which can be solved by measuring the eigenfrequency shifts of a nanostructure [68, 220].

4.2.2.4   Property of an Inverse Problem

In Eqn. (4.19), when all the parameters such as mass, axial force and bending stiffness are given together with the boundary conditions to find the system's eigenfrequencies, it formulates a *direct problem*. In the applications of a mass resonator sensor, the (shifts of) eigenfrequencies are the measured quantities; we need to use these measured information to find the adsorption-induced parameters of Δm (distributed mass), M_i (concentrated mass together with its position x_i), ΔN and ΔEI (variations of the axial force and bending stiffness). Finding those coefficients in a differential equation through some of its solutions formulates an *inverse problem* [238]. The above inverse problem is the type of the parameter identification, which in practice has a wide application, for example, the structural damage identification [239]. In general, solving an inverse problem is much more difficult than solving a direct one because most of the inverse problems are *ill-posed* [239–241]. A *well-posed* problem is required to satisfy the following three conditions introduced firstly in 1902 by Jacques Hadamard [239]:

> **Existence:** *The problem has a solution.*
> **Uniqueness:** *The problem has only one solution.*
> **Continuity:** *The solution is a continuous function of the data.*

An ill-posed or improperly posed problem [240] is the one which fails to satisfy one or more of the above three Hadamard's criteria. The third criterion of continuity is also often referred to as the stability of a solution [238, 240]. Most of the inverse problems are ill-posed [240], which often violate the second or the third criterion or both. Every inverse problem requires a particular approach and a uniform theory of solving an inverse problem cannot be expected [240].

Compared with the approaches at the early stage using the brute force to solve the inverse problems, there are some progresses in the study of the inverse problem. However, even today, it is still appropriate to quote the following Gladwell's words to comment on the current status of the inverse problem study: *The study of inverse problems is at an earlier stage of evolution than that of direct problems* [239]. Actually, most of the inverse problems are unsolvable because of the lack of information, which cannot be remedied by any mathematical trickery [240]. Luckily, the lack of information is not an issue in the inverse problems of a mass resonator sensor. There are infinite eigenvalues in the vibration problem of a continuous system. Physically, those eigenvalues correspond to the resonant frequencies, which can be measured in an experiment.

Using the shifts of resonant frequencies to determine the parameters in Eqn. (4.19) formulates a specific type of inverse problem: inverse Sturm–Liouville problems. The eigenvalue problem of Eqn. (4.19) is a Sturm–Liouville problem [238], which is with a self-adjoint operator. The function of a self-adjoint operator in the eigenvalue problem of a continuous system is like a symmetric matrix for the discrete system [9], which ensures some properties of its eigenvalue and eigenvectors, for example, the orthogonal property of the eigenvectors. Here we need to emphasize that the Sturm–Liouville eigenvalue problem is of second order only. Here Eqn. (4.19) is the fourth order differential equation due to its bending term. However, once the separation of variables method is used, the eigenvalue problem of Eqn. (4.19) is of second order [8, 9]. The specific difficulty of solving the inverse problem of Eqn. (4.19) resides in the property of the ill-posed problem. In 1966, Prof. Kaz proposed a famous inverse problem of *"Can one hear the shape of a drum?"* [242]. Here the drum is a two-dimensional structure of membrane, whose restoring force is provided by tension. More specifically, Kaz's question is: Can we use the resonance frequencies of a drum to determine its shape? Kaz's study shows that the resonance frequencies can tell us some very specific properties of a drum. For example, we can "hear" the area and circumference of a drum, and some connectivity properties (i.e., whether there is a hole and the number of holes) [242]. However, in the end, Kaz was not sure whether we can hear the shape of a drum [242]. The answer to Kaz' question is, *"You can't hear the shape of a drum"* [243]. In 1992, Gordon, Webb and Wolpert constructed two drums with different shapes but the same eigenfrequencies [244]. In this inverse problem, the property of uniqueness breaks down, which is also the scenario of the inverse problem of Eqn. (4.19).

4.2.2.5   Solving the Inverse Problems

A simple version of the inverse problem of the mass resonator sensor is to assume that only the effect of the concentrated mass is present in a resonator and as a result, Eqn. (4.19) becomes the following:

$$\left[m + \sum_{i=1}^{K} M_i \delta(x - x_i) \right] \frac{\partial^2 w}{\partial t^2} + c \frac{\partial w}{\partial t} - N \frac{\partial^2 w}{\partial x^2} + EI \frac{\partial^4 w}{\partial x^4} = 0 \qquad (4.25)$$

As seen in Figure 4.14, the adsorption effect of atoms on a nanomechanical resonator is modeled as concentrated masses [70–72]. The simplest case is presented in Figure 4.15, in which there is only one adsorbate [113, 245], i.e., $K = 1$ in Eqn. (4.25). As one of the most marvelous achievements of a nanomechanical resonator, it is so sensitive that the decreasing shift of the eigenfrequency induced by an atom adsorption can be detected [20, 70, 82], in which the step-wise resonant frequency shift is observed. The step-wise resonant frequency shift indicates the discrete nature of the adsorbates arriving on the surfaces of a nanomechanical resonator one by one, which is also the hallmark of sensing an individual adsorption event [16]. However, there are two parameters to be determined by the shifts of the resonator eigenfrequencies: concentrated mass (M_1) and its position (x_1). In practice, the evaporated atoms are sprayed onto the resonator surface and their landing positions are extremely difficult to be controlled or measured [16, 20, 71]. In the application of a mass resonator sensor, finding the concentrated mass is the desired goal and its position is an undesired one. However, the concentrated mass and its position are the two inseparable variables determining the shifts of the resonator eigenfrequencies. Although some instruments such as optical microscopy [139] and scanning electron microscopy (SEM) [246] can be used to determine the adsorbate location, there are limitations. For example, there is not enough contrast between a cell and solution for an optical imaging [139] and SEM cannot be applied to the non-metallic materials. As commented by Knobel [72], before a practical mass spectrometer can be made, the most important problem to be solved is to determine the atom landing positions. Of course, for some special applications, the spatially and chemically discriminant binding sites can be designed, which specifies the adsorbate location. For example, a gold nanodot is deposited around the free end of a silicon nitride cantilever as biomolecular-tethering site, with which a double-stranded DNA can bind [247].

When the damping term is ignored, the eigenvalue problem of Eqn. (4.19) is formulated as the following:

$$\det \left[\mathbb{K} - \lambda \mathbb{M} \right] = 0 \qquad (4.26)$$

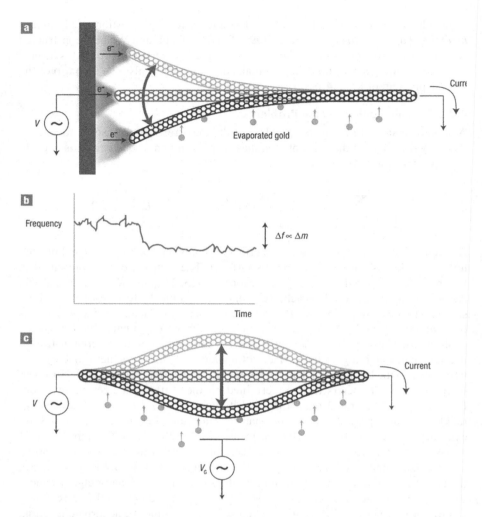

FIGURE 4.14   (a) The adsorption of atoms on the surface of the carbon nanotube resonator of a cantilever beam. (b) The decreasing shifts of the resonant frequency due to the mass effect of the adsorbed atoms. (c) The adsorption of atoms on the surface of the carbon nanotube resonator of a clamped-clamped beam. (Reprinted by permission from Springer Nature Customer Service Centre GmbH: Springer Nature, Nat. Nanotech, R.G. Knobel, Copyright 2008.)

Where det[] means the determinant of a matrix; $\mathbb{K}$ and $\mathbb{M}$ are the stiffness and mass matrices, respectively; λ is the eigenvalue and more specifically, λ is the square of eigenfrequency, i.e., $\lambda = \omega^2$. The stiffness matrix $\mathbb{K}$ is a $n \times n$ matrix depending on the numerical discretization method, the beam dimensions and the parameters of $N + \Delta N$ and $EI + \Delta EI$ in Eqn. (4.19). Similarly, the mass matrix $\mathbb{M}$ is a $n \times n$ matrix depending on the numerical discretization method, the beam dimensions and the parameters of Δm, M_i and x_i in Eqn. (4.19). The presence of damping makes the

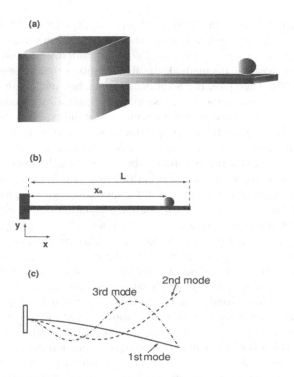

FIGURE 4.15 (a) Schematic diagram of a cantilever resonator with an accreted particle; (b) the coordinate system, the particle position and the beam length; (c) the first three modes of a uniform cantilever. (Source: Y. Zhang and Y. Liu (2014) [113]).

eigenvalue problem rather complex, which is the so-called damped nongyroscopic system type of eigenvalue problem and is formulated as the following [248]:

$$\det[\mathbb{K}^* + \lambda^* \mathbb{M}^*] = 0 \tag{4.27}$$

Now $\mathbb{K}^* = \begin{pmatrix} \mathbb{M} & 0 \\ 0 & -\mathbb{K} \end{pmatrix}$ and $\mathbb{M}^* = \begin{pmatrix} \mathbb{C} & \mathbb{K} \\ \mathbb{K} & 0 \end{pmatrix}$ are the $2n \times 2n$ matrices and $\mathbb{C}$ here is the $n \times n$ damping matrix; λ^* now is the eigenfrequency rather than its square. Mathematically, the eigenvalue problem of Eqn. (4.27) is much more difficult because λ^* is now a complex number due to the presence of damping [8, 9, 248]. Some examples on the eigenvalue computation of Eqn. (4.27) are presented in [113, 249–251].

For simplicity, we focus only on the eigenvalue problem of Eqn. (4.26). In conjunction with Eqn. (4.25) which assumes only the concentrated mass effect, $\mathbb{K}$ keeps constant and $\mathbb{M}$ changes according to M_i and x_i. Even for the simplest case of one adsorbate, there are two unknowns and multimode resonance experiment/model is required to extract these two unknowns [16, 113, 214, 245, 251]. A method of solving the above inverse problem is the least square method. Here the least square method is applied to find M_i and x_i, which minimize the difference between the predicted resonant frequencies (by Eqns. [25] and [26]) and experimentally measured ones [12]. For the least square method to work smoothly, there are some requirements on the condition of the experimental data [239]. Quite often when solving an inverse problem via the least square method, the ill-condition problem of matrix is often encountered [239]. Therefore, the additional approaches such as the Morozov and Tikhonov regularization methods are needed in the minimization computation [239, 241], which in essence is to enhance the numerical stability. Hanay et al. [252] presented a method called inertial imaging, which can determine multiple variables of the mass, position, molecular size and shape of individual adsorbates. The inertial imaging method solves the inverse problem by formulating the spatial moments from the zeroth to the third, which are also optimized by the least square method. An alternative inverse problem formulation of the inertial imaging is provided by Sader et al. [253], which involves solving the nonlinear equation via the Newton-Rhapson method. Besides the least square method, there are other inverse problem-solving methods such as the joint probability density function method [7, 16], fitting procedure method [245], energy method [254] and the direct comparison method [113, 251]. All these methods in essence are the brute force methods and their working principles are the same, which depends on the multimode resonance experiment/ model. The basic idea behind these methods is to go through all the possible combinations of the concentrated mass and position for each mode. For example, the joint probability density function method is to maximize the joint probability density function, which is equivalent to minimizing a function indicating the resonant frequency difference of the predicted and the measured [7]. As mentioned before, for a given/measured resonant frequency shift of a mode, the possible combinations of the concentrated mass and position are infinite. However, this combination set is different for different mode. Mathematically, the intersection of these different combination sets is the unique solution of the inverse problem [113, 251]. To explain why this method works, it is appropriate to quote the following words from Pascal's Pensées, which are also quoted in chapter 5 *(Inverse Problems for Some More General Systems)* of Gladwell's book [239]:

Words differently arranged have a different meaning, and meanings differently arranged have different effects.

Specifically, for the inverse problem of the concentrated mass and its position to be determined by the shifts of resonant frequencies, the working mechanisms are the following two [79, 251]:

1. For a specific mode, the concentrated mass (M_i) and its position (x_i) have different impact on its resonant frequency.
2. Different resonant frequency responds differently to a given set of M_i and x_i.

We give the following example to demonstrate how the above two working mechanisms can be used to find the two convolving and coupled parameters M_1 and x_1 for the single adsorbate on a clamped-clamped beam case [251]. In Figure 4.16, the dimensionless first two resonant frequencies (ω_1 and ω_2) are presented as the

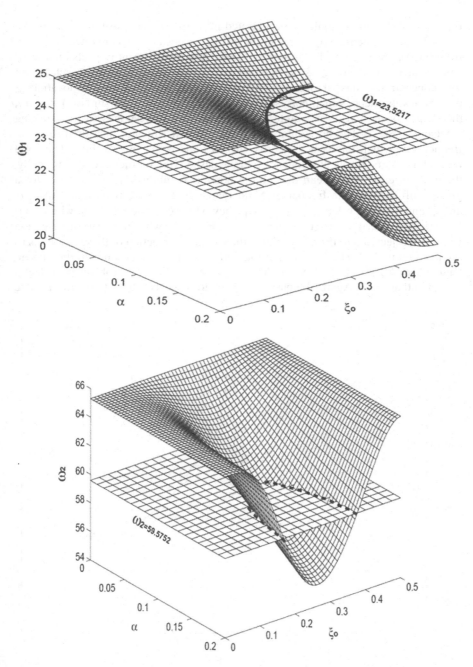

FIGURE 4.16   The left is the variation of the dimensionless first resonant frequency (ω_1) as the function of the dimensionless concentrated mass (α) and its position (ξ_o). The right is the dimensionless second resonant frequency (ω_2) as the function of α and ξ_o. (Source: Y. Zhang and Y.P. Zha (2015) [251]).

functions of the dimensionless concentrated mass (α) and its position (ξ_o), respectively. For a clamped-clamped beam, the adsorbate position (ξ_o) only needs to vary from 0 to 0.5 because of the symmetry [251]. Here the different impacts on a resonant frequency of α and ξ_o as the two independent variables are seen in each figure: The concentrated mass α determines the variation magnitude and its location ξ_o determines its shape, which corresponds to the working mechanism No. 1. Cleary, the shapes of ω_1 and ω_2 are quite different and this corresponds to the working mechanism No. 2. The horizontal planes are the experimental measured resonant frequencies, which intersect the curvy surfaces and result in two lines. The solid line is for the first resonant frequency and the dashed line for the second. Physically, these intersection lines mean the infinite possible combinations of α and ξ_o for a given (shift of) resonant frequency. Mathematically, there are two unknowns of α and ξ_o and the shift of one resonant frequency only gives one equation, which cannot uniquely determine these two parameters. However, when these two intersection lines are projected into the α - ξ_o plane, there is an intersection of these two lines as seen in Figure 4.17. The intersection marked as a circle corresponds to the present value of $(\alpha, \xi_o) = (0.1, 0.3)$ exactly and, therefore, the inverse problem is perfectly solved in that sense. Again, the physical reason for these two lines to intersect is due

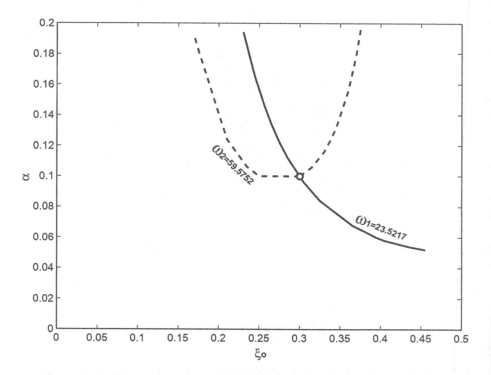

FIGURE 4.17 The projections of the two intersection curves obtained in Figure 4.16 into the α-ξ_o plane. The intersection of the two curves is marked with a circle, which corresponds to $(\alpha, \xi_o) = (0.1, 0.3)$ exactly. (Source: Y. Zhang and Y.P. Zhao (2015) [251]).

to the above two mechanics. When the number of unknowns are more than two, the above graphic solution method cannot presented. The general procedure for formulation and solution to this type of the inverse problem can be found in [79, 251], in which the Newton–Rhapson method is needed to solve the problem. For the Newton–Rhapson method to start, an initial guess on the unknown parameters is needed. The good news is that in solving such inverse problems formulated in [79, 251], the Newton–Rhapson method is not sensitive to the initial guess. Here, this direct comparison method of solving the inverse problem requires a huge amount of computations of the eigenvalue problem of Eqn. (4.26) or (4.27), due to the huge combinations of the various concentrated mass and position. The eigenvalue problem here is to solve a transcendental equation, which can result in a large computation and can thus be very time-consuming [249]. Various algorithms are developed to perform an efficient computation on the eigenvalue problems of the concentrated mass and spring [249]. The increase of one more number of the unknowns can increase several orders of magnitude of the eigenvalue problem computation. Some efficient algorithms particularly designed for solving such inverse problem can be of significant help [113, 255]. For example, the approximate method based on the Galerkin method [113] and the modified Dunkerley method [255] provide rather more efficient approaches on the eigenfrequency computations of the beam with the concentrated adsorbates.

There is an issue on the overspecified data [240]. Here the two unknowns are determined by the two (shifts of) eigenfrequencies. While, more eigenfrequencies can be measured by the experiment, i.e., more data can be provided. For example, when three eigenfrequencies are measured, you can pick either two of the three, i.e., ω_1 and ω_2, ω_1 and ω_3, ω_2 and ω_3, to determine the two unknowns. And either of these three combinations results in the same result. In contrast, different combinations in the Belardinelli's method [256], which in essence is the least square method of minimizing, yield different results. In the least square formulation of the problem, there is no such issue and in general more data can make the error smaller. For example, Ma et al. [257] used the shifts of five resonant frequencies to determine two unknowns. Depending on the problem formulation, sometimes solving the inverse problem of two unknowns may require three measured eigenfrequencies [113]. Another issue is that when solving the inverse problem as presented in Figures 4.16 and 4.17, the two "measured" eigenfrequencies are the exact ones, which is the major reason for yielding the exact values of the concentrated mass and its position. In reality, there are errors in the measured eigenfrequencies. When the errors of the measured eigenfrequencies are considered, the errors of the concentrated mass and its position are magnified significantly and disproportionally [79]. This disproportional increase of the errors in an inverse problem is a vivid embodiment of its ill-posed problem nature, which violates the third criterion of Hadamard [239]: continuity. This violation of continuity is also quite often encountered in other type of the inverse problems, for example, the indentation problems of a film/substrate structure [258, 259]. In one word, the above direct comparison method has a very low tolerance for the errors of the measured eigenfrequencies. However, with the development of the state-of-art technology, the current sensitive nanomechanical resonator is capable of detecting the shifts of eigenfrequencies caused by the adsorption event of a protein or a molecule or an atom [16, 20, 70, 71, 82]. Moreover, the current read-out technology for a nano-mechanical resonator has made the frequency resolution at the level of parts-per-billion (ppb) [260], which can technically ensure a rather accurate measurement of the eigenfrequencies.

There is one more issue of the inverse problem of a single adsorbate on a two-dimensional (2D) resonator such as a membrane or plate [75]. Compared with the one-dimensional structure such as a string, beam and arch, there is an additional trouble called the degenerate mode in a two-dimensional resonator. For a 2D structure, there is a mathematical scenario that an eigenvalue can have two linearly independent eigenvectors [9, 261]. Physically this means that an eigenfrequency/resonant frequency can have two mode shapes and a linear combination of these two modes is also a mode for the eigenfrequencies, which is the scenario of degenerate mode [9, 261]. In a 2D drum resonator [262, 263], these degenerate modes are difficult to be controlled and monitored. Here the drum resonator is a circular membrane structure and its degenerate modes correspond to the non-axi-symmetric, or say, the azimuth angle-dependent modes [9, 261]. Mathematically, the presence of the degenerate modes of a 2D structure also poses a significant challenge of solving the inverse problem. The solution is to avoid the resonant frequencies of the degenerate modes and only excite those non-degenerate modes by tuning the excitation frequency in an experiment [75]. In a circular membrane, the non-degenerate modes are the azimuth angle-independent and thus axisym-metric [9, 261].

This direct comparison method has also been applied to solve other types of inverse problem via the shifts of resonant frequencies, for example, surface elastic-ity and surface stress [36], the nonlocal parameters [68, 220], mass and induced in-plane axial force [79, 111], mass and stiffness [112]. In essence, this direct com-parison method of using different shifts of resonant frequencies can be applied to various inverse problems. If a resonator can be excited to vibrate in different dimensions, for example, a resonator vibrates transversely in both the thickness and width directions [17], or say, the out-of-plane and in-plane directions [142], more information can be provided to solve the inverse problem. Gil-Santos et al. [17] have used the shifts of the two first resonant frequencies in these two directions to deter-mine the mass and stiffness of the adsorbates. Similarly, the resonator torsional modes together with its bending can also serve to offer more information [257, 264]. Moreover, a torsional mode is with a smaller damping and a better sensitivity than those of a flexural bending mode [257]. When a uniform adsorption layer is formed on a cantilever, the variations of the resonant frequency and quality factor can be used to extract the mass and stiffness of the adsorption layer [265]. Similar to the direct method of solving an inverse problem via the shifts of resonant frequencies, the information extracted from a microcantilever deflection can be used to deter-mine its residual stress gradients [183]; the information extracted from the inden-tation curves can be used to determine the Young's moduli of film and substrate, and the thickness of the film layer [258, 259]; the information extracted from the deflection shape of an adhered beam can be used to determine its adhesion energy and residual stress [266].

4.3 SUMMARY

Compared with a conventional mass spectrometry which utilizes the Lorentz force as the sensing mechanism and, therefore, requires the ionization of an analyte, the greatest advantage of a mass resonator sensor is the capability of working with neutral analytes. Besides avoiding the degrading or damage of an analyte due to ionization, there is also a huge demand for sensing the mass of a neutral sample. As a mechanical device, the sensing mechanism of a mass resonator sensor is the shifts of the resonant frequencies due to an adsorption. With the scaling-down of its dimension and the proper material selection, the current state-of-art mass resonator sensor is with the fundamental resonant frequency of the gigahertz and capability of detecting the shift of the resonant frequencies induced by the mass of an adsorbate much less than a Dalton, which satisfies the requirements of almost all the practical applications. A major obstacle in the fabrication of a mass resonator sensor is the significant decrease of its quality factor (Q-factor) with its dimensions decreasing to the micro/nanometer scale. Enormous efforts have been infused to study the mechanisms of Q-factor and various effective approaches for the enhancement of Q-factor have also been proposed. With the significant progresses in the fabrications of micro/nanomechanical devices and the related measurement tools in recent years, the capability of a mass resonator sensor of accurately and swiftly detecting the resonant frequency shifts of a tiny mass has been dramatically improved. However, there is still a long way to go for the practical mass-sensing application of a mass resonator sensor. The reason is that there is still a huge gap between detecting the resonant frequency shifts and determining the mass of an adsorbate. This gap is mainly caused by the slow/little progress and the huge challenges in solving the inverse problem, which is often an ill-posed problem. Besides the adsorbate mass, its position, the surface stress/elasticity, stiffness and nonlocality can all induce the resonant frequency shifts of a mass resonator sensor. The calibrations of the mechanical properties of a mass resonator sensor become a challenging inverse problem too. The main mathematical challenges are the following two: properly modeling those effects and accurately solving the inverse problem. As commented in this chapter, there are still arguments and differences in modeling those effects, which stand out and even become dominant with the scaling-down of the device dimensions. The other challenge is to solve the inverse problem of determining the adsorbate mass from the shifts of resonant frequencies. An outstanding problem is the numerical stability due to the property of such ill-posed problem: A tiny error is the (measured) data of resonant frequencies can cause a huge error in the final result of the adsorbate mass or no solution with a given (high) accuracy requirement. Various methods have been taken to tackle the problem and a satisfying general solution approach has not arrived yet though some special techniques may work fine but only for some specific problems. The holy grail of the mass resonator sensor problems is to solve the inverse problem. Compared with fast development of the micro/nanomechanical device fabrication, the progresses of solving the inverse problems are insanely slow. Besides reviewing and commenting on the bottle-neck inverse problems of a mass resonator sensor, another goal of this chapter is an academic call for more researchers to enter the field and intensify the related studies.

ACKNOWLEDGMENT

This work was supported by the National Natural Science Foundation of China (NSFC No. 11772335).

REFERENCES

[1] The Royal Swedish Academy of Sciences, Mass spectrometry (MS) and nuclear magnetic resonance (NMR) applied to biological macromolecules. Advanced information on the Nobel Prize in Chemistry 2002, 9 October 2002.

[2] R. Aebersold and M. Mann. Mass spectrometry-based proteomics. *Nature*, 422, 198 (2003).

[3] B. Domon and R. Aebersold. Mass spectrometry and protein analysis. *Science*, 312, 212 (2006).

[4] K.L. Ekinci, X. Huang and M.L. Roukes. Ultrasensitive nanoelectromechanical mass detection. *Appl. Phys. Lett.*, 84, 4469 (2004).

[5] A. Boisen. Nanomechanical systems: Mass spec goes nanomechanical. *Nat. Nanotech.*, 4, 404 (2009).

[6] W. Hiebert. Devices reach single-proton limit. *Nat. Nanotech.*, 7, 278 (2012).

[7] O. Malvar, J.J. Ruz, P.M. Kosaka, C.M. Domínguez, E. Gil-Santos, M. Calleja and J. Tamayo. Mass and stiffness spectrometry of nanoparticles and whole intact bacteria by multimode nanomechanical resonators. *Nat. Commun.*, 7, 13452 (2016).

[8] L. Meirovitch. *Fundamentals of Vibrations*. Boston, MA: McGraw-Hill Company, 2001.

[9] L. Meirovitch. *Analytical Methods in Vibrations*. New York, NY: Macmillan, 1967.

[10] G.Y. Chen, T. Thundat, E.A. Wachter and R.J. Warmack. Adsorption-induced surface stress and its effects on resonance frequency of mocrocantilevers. *J. Appl. Phys.*, 77, 3618 (1995).

[11] T. Thundat, E.A. Wachter, S.L. Sharp and R.J. Warmack. Detection of mercury vapor using resonating microcantilevers. *Appl. Phys. Lett.*, 66, 1695 (1995).

[12] S. Dohn, S. Schmid, F. Amiot and A. Boisen. Position and mass determination of multiple particles using cantilever based mass sensors. *Appl. Phys. Lett.*, 97, 044103 (2010).

[13] M.A. Mahmoud. Validity and accuracy of resonance shift prediction formulas for microcantilevers: A review and comparative study. *Crit. Rev. Solid State Mater. Sci.*, 41, 386 (2016).

[14] E. Buks and B. Yurke. Mass detection with a nonlinear nanomechanical resonator. *Phys. Rev. E.*, 74, 046619 (2006).

[15] S. Schmid, L.G. Villanueva and M.L. Roukes. *Fundamentals of Nanomechanical Resonators*. Switzerland: Springer International Publishing AG, 2016.

[16] M.S. Hanay, S. Kelber, A.K. Naik, D. Chi, S. Hentz, E.C. Bullard, E. Colinet, L. Duraffourg and M.L. Roukes. Single-protein nanomechanical mass spectrometry in real time. *Nat. Nanotech.*, 7, 602 (2012).

[17] E. Gil-Santos, D. Ramos, J. Martinez, M. Fernandez-Regulez, R. Garcia, A. San Paulo, M. Calleja and J. Tamayo. Nanomechanical mass sensing and stiffness spectrometry based on two-dimensional vibrations of resonant nanowires with yoctogram resolution. *Nat. Nanotech.*, 5, 641 (2010).

[18] K.L. Ekinci, Y.T. Yang and M.L. Roukes. Ultimate limits to inertial mass sensing based upon nanoelectromechanical systems. *J. Appl. Phys.*, 95, 2682 (2004).

[19] F. Schedin, A.K. Geim, S.V. Morozov, E.W. Hill, P. Blake, M.I. Katsnelson and K.S. Novoselov. Detection of individual gas molecules adsorbed on graphene. *Nat. Mater.*, 6, 652 (2007).

[20] J. Chaste, A. Eichler, J. Moser, G. Ceballos, R. Rurali and A. Bachtold. A nanome-chanical mass sensor with yoctogram resolution. *Nat. Nanotech.*, 7, 301 (2012).

[21] P.M. Kosaka, V. Pini, J.J. Ruz, R.A. da Silva, M.U. González, D. Ramos, M. Calleja and J. Tamayo. Detection of cancer biomarkers in serum using a hybrid mechanical and optoplasmonic nanosensor. *Nat. Nanotech.*, 9, 1047 (2014).

[22] G. Longo, L. Alonso-Sarduy, L. Marques Rio, A. Bizzini, A. Trampuz, J. Notz, G. Dietler and S. Kasas. Rapid detection of bacterial resistance to antibiotics using AFM cantilevers as nanomechanical sensors. *Nat. Nanotech.*, 8, 522 (2013).

[23] N. Liu, F. Giesen, M. Belov, J. Losby, J. Moroz, A.E. Fraser, G. McKinnon, T.J. Clement, V. Sauer, W.K. Hiebert and M.R. Freeman. Time-domain control of ultrahigh frequency nanomechanical systems. *Nat. Nanotech.*, 3, 715 (2008).

[24] R. Lifshitz and M.T. Cross. Nonlinear dynamics of nanomechanical and microme-chanical resonators. In: H.G. Schuster (ed.), *Reviews of Nonlinear Dynamics and Complexity*. Weinbeim, Germany: Wiley-VCH Verlag GmbH & Co. KGaA, 2009.

[25] B. Witkamp, M. Poot and H.S.J. van der Zant. Bending-mode vibration of a suspended nanotube resonator. *Nano Lett.*, 6, 2904 (2006).

[26] J. Moser, A. Eichler, J. Güttinger, M.I. Dykman and A. Bachtold. Nanotube mechanical resonators with quality factors of up to 5 million. *Nat. Nanotech.*, 9, 1007 (2013).

[27] M. Poot and H.S.J. van der Zant. Mechanical systems in the quantum regime. *Phys. Rep.*, 511, 273 (2012).

[28] B. Abbott et al. Observation of a kilogram-scale oscillator near its quantum ground state. *New J. Phys.*, 11, 073032 (2009).

[29] A. Hopkins, K. Jacobs, S. Habib and K. Schwab. Feedback cooling of a nanomechani-cal resonator. *Phys. Rev. B*, 68, 235328 (2003).

[30] T.J. Kippenberg and K.J. Vahala, Cavity optomechanics: Back-action at the mesoscale. *Science*, 321, 1172 (2008).

[31] Q.P. Unterreithmeier, T. Faust and J. P. Kotthaus. Damping of nanomechanical resona-tors. *Phys. Rev. Lett.*, 105, 027205 (2010).

[32] S.S. Verbridge, J.M. Parpia, R.B. Reichenbach, L.M. Bellan and H.G. Craighead. High quality factor resonance at room temperature with nanostrings under high tensile stress. *J. Appl. Phys.*, 99, 124304 (2006).

[33] C. Chen, S. Rosenblatt, K.I. Bolotin, W. Kalb, P. Kim, I. Kymissis, H.L. Stormer, T.F. Heinz and J. Hone. Performance of monolayer graphene nanomechanical resonators with electrical readout. *Nat. Nanotechnol.*, 4, 861 (2009).

[34] P.H. McDonald Jr. and N.C. Raleigh. Nonlinear dynamic coupling in a beam vibration. *J. Appl. Mech.*, 22, 573 (1955).

[35] Y. Zhang and Y.P. Zhao. Numerical and analytical study on the pull-in instability of micro-structure under electrostatic loading. *Sens. Actuators A*, 127, 366 (2006).

[36] Y. Zhang, L. Zuo and H. Zhao. Determining the effects of surface elasticity and surface stress by measuring the shifts of resonant frequencies. *P. Roy. Soc. A-Math. Phy.*, 469, 20130449 (2013).

[37] A.H. Nayfeh and D.T. Mook. *Nonlinear Oscillation*. New York: John Wiley & Sons, Inc., 1979.

[38] Y. Zhang, Y. Liu and K.D. Murphy. Nonlinear dynamic response of beam and its appli-cation in nanomechanical resonator. *Acta Mech. Sin.*, 28, 190 (2012).

[39] P. Postma, I. Kozinsky, A. Husain and M.L. Roukes. Dynamic range of nanotube- and nanowire-based electromechanical systems. *Appl. Phys. Lett.*, 86, 223105 (2005).

[40] V. Peano and M. Thorwart. Macroscopic quantum effects in a strongly driven nanome-chanical resonator. *Phys. Rev. B*, 70, 235401 (2004).

[41] M. Agarwal, S. Chandorkar, R.N. Candler, B. Kim, M.A. Hopcroft, R. Melamud, C.M. Jha, T.W. Kenny and B. Murmann. Optimal drive condition for nonlinearity reduction in electrostatic microresonators. *Appl. Phys. Lett.*, 89, 214105 (2006).

[42] L.C. Shao, M. Palaniapan and W.W. Tan. The nonlinearity cancellation phenomenon in micromechanical resonators. *J. Micromech. Microeng.*, 18, 065014 (2008).

[43] L.C. Shao, M. Palaniapan, W.W. Tan and L. Khine. Nonlinearity in micromechanical free–free beam resonators: Modeling and experimental verification. *J. Micromech. Microeng.*, 18, 025017 (2008).

[44] A. Eichler, M. del Álamo Ruiz, J.A. Plaza and A. Bachtold. Strong coupling between mechanical modes in a nanotube resonator. *Phys. Rev. Lett.*, 109, 025503 (2012).

[45] V. Sazonova, Y. Yaish, H. Üstünel, D. Roundy, T.A. Arias and P. L. McEuen. A tunable carbon nanotube electromechanical oscillator. *Nature*, 431, 284 (2004).

[46] M.I. Dykman and M.V. Fistul. Multiphoton antiresonance. *Phys. Rev. B*, 71, 140508 (R) (2005).

[47] Y. Zhang, Y. Wang, Z. Li, Y. Huang and D. Li. Snap-through and pull-in instabilities of an arch-shaped beam under electrostatic loading. *J. Microelectromech. Syst.*, 16, 684 (2007).

[48] R.L. Badzey, G. Zolfagharkhani, A. Gaidarzhy and P. Mohanty. A controllable nanomechanical memory element. *Appl. Phys. Lett.*, 85, 3587 (2004).

[49] R.L. Badzey, G. Zolfagharkhani, A. Gaidarzhy and P. Mohanty. Temperature dependence of a nanomechanical switch. *Appl. Phys. Lett.*, 86, 023106 (2005).

[50] Y. Zhang, Y. Wang and Z. Li. Analytical method of predicating the instabilities of a micro arch-shaped beam under electrostatic loading. *Microsyst. Technol.*, 16, 909 (2010).

[51] Y. Zhang and Y.P. Zhao. A precise model for the shape of an adhered microcantilever. *Sens. Actuators A: Phys.*, 171, 381 (2011).

[52] A. Eichler, J. Moser, M.I. Dykman and A. Bachtold. Symmetry breaking in a mechanical resonator made from a carbon nanotube. *Nat. Commun.*, 4, 2843 (2013).

[53] A. Erbe, H. Krömmer, A. Kraus, R.H. Blick, G. Corso and K. Richter. Mechanical mixing in nonlinear nanomechanical resonators. *Appl. Phys. Lett.*, 77, 3102 (2000).

[54] T.C. Chang and R.R. Craig Jr. Normal modes of uniform beams. *J. Eng. Mech.*, 95, 1027 (1969).

[55] S.S. Rao. *Vibration of Continuous Systems*, 2nd edition. Hoboken, NJ: John Wiley & Sons, Inc., 2019.

[56] X.M.H. Huang, C.A. Zorman, M. Mehregany and M.L. Roukes. Nanodevice motion at microwave frequencies. *Nature*, 421, 496 (2003).

[57] A. Gaidarzhy, G. Zolfagharkhani, R.L. Badzey and P. Mohanty. Evidence for quantized displacement in macroscopic nanomechanical oscillators. *Phys. Rev. Lett.*, 94, 030402 (2005).

[58] D. Garcia-Sanchez, A.S. Paulo, M.J. Esplandiu, F. Perez-Murano, L. Forró, A. Aguasca and A. Bachtold. Mechanical detection of carbon nanotube resonator vibrations. *Phys. Rev. Lett.*, 99, 085501 (2007).

[59] K.L. Ekinci and M.L. Roukes. Nannomechanical systems. *Rev. Sci. Instrum.*, 76, 061101 (2005).

[60] Y. Wei, B. Wang, J. Wu, R. Yang and M.L. Dunn. Bending rigidity and Gaussian stiffness of single-layered graphene. *Nano Lett.*, 13, 26 (2013).

[61] J. Atalaya, A. Isacsson and J.M. Kinaret. Continuum elastic modeling of graphene resonators. *Nano Lett.*, 8, 4196 (2008).

[62] J.A. Judge, D.M. Photiadis, J.F. Vignola, B.H. Houston and J. Jarzynski. Attachment loss of micromechanical and nanomechanical resonators in the limits of thick and thin support structures. *J. Appl. Phys.*, 101, 013521 (2007).

[63] D.M. Photiadis and J.A. Judge. Attachment losses of high Q oscillators. *Appl. Phys. Lett.*, 85, 482 (2004).

[64] J. Yang, T. Ono and M. Esashi. Surface effects and high quality factors in ultrathin single-crystal silicon cantilevers. *Appl. Phys. Lett.*, 77, 3860 (2000).

[65] S. Dohn, R. Sandberg, W. Svendsen and A. Boisen. Enhanced functionality of cantilever based mass sensors using higher modes. *Appl. Phys. Lett.*, 86, 233501 (2005).

[66] M.K. Ghatkesar, V. Barwich, T. Braun, J. Ramseyer, Ch. Gerber, M. Hegner, H.P. Lang, U. Drechsler and M. Despont. Higher modes of vibration increase mass sensitivity in nanomechanical microcantilevers. *Nanotechnology*, 18, 445502 (2007).

[67] P.L. Yu, T.P. Purdy and C.A. Regal. Control of material damping in high-Q membrane microresonators. *Phys. Rev. Lett.*, 108, 083603 (2012).

[68] Y. Zhang. Frequency spectra of nonlocal Timoshenko beams and an effective method of determining nonlocal effect. *Int. J. Mech. Sci.*, 128–129, 572 (2017).

[69] B. Lassagne, D. Garcia-Sanchez, A. Aguasca and A. Bachtold. Ultrasensitive mass sensing with a nanotube electromechanical resonator. *Nano Lett.*, 8, 3735 (2008).

[70] K. Jensen, K. Kim and A. Zettl. An atomic-resolution nanomechanical mass sensor. *Nat. Nanotech.*, 3, 533 (2008).

[71] A.K. Naik, M.S. Hanay, W.K. Hiebert, X.L. Feng and M.L. Roukes. Towards single-molecule nanomechanical mass spectrometry. *Nat. Nanotech.*, 4, 445 (2009).

[72] R.G. Knobel. Weighing single atoms with a nanotube. *Nat. Nanotech.*, 3, 525 (2008).

[73] M.C. Cross, A. Zumdieck, R. Lifshitz and J.L. Rogers. Synchronization by nonlinear frequency pulling. *Phys. Rev. Lett.*, 93, 224101 (2004).

[74] Z. Wang, J. Lee and P. Feng. Spatial mapping of multimode Brownian motions in high-frequency silicon carbide microdisk resonators. *Nat. Commun.*, 5, 5158 (2014).

[75] Y. Zhang and Y.P. Zhao. Detecting the mass and position of an adsorbate on a drum resonator. *P. Roy. Soc. A-Math. Phy.*, 470, 20140418 (2014).

[76] V. Singh, S. Sengupta, H. S. Solanki, R. Dhall, A. Allain, S. Dhara, P. Pant and M.M. Deshmukh. Probing thermal expansion of graphene and modal dispersion at low-temperature using graphene nanoelectromechanical systems resonators. *Nanotechnology*, 21, 165204 (2010).

[77] T. Larsen, S. Schmid, L.G. Villanueva and A. Boisen. Photothermal analysis of individual nanoparticulate samples using micromechanical resonators. *ACS Nano*, 7, 6188 (2013).

[78] V. Pini, J. Tamayo, E. Gil-Santos, D. Ramos, P. Kosaka, H. Tong, C. van Rijn and M. Calleja. Shedding light on axial stress effect on resonance frequencies of nanocantilevers. *ACS Nano*, 5, 4269 (2011).

[79] Y. Zhang and Y.P. Zhao. Mass and force sensing of an adsorbate on a string resonator sensor. *Sens. Actuators B: Chem.*, 221, 305 (2015).

[80] Y. Zhang. Large Deflection of clamped circular plate and accuracy of its approximate analytical solution. *Sci. China Phys. Mech.*, 59, 624602 (2016).

[81] Z.Y. Ning, T.W. Shi, M.Q. Fu, Y. Guo, X.L. Wei, S. Gao and Q. Chen. Transversally and axially tunable carbon nanotube resonators in situ fabricated and studied inside a scanning electron microscope. *Nano Lett.*, 14, 1221 (2014).

[82] H. Chiu, P. Hung, H. Postma and M. Bockrat, Atomic-scale mass sensing using carbon nanotube resonators. *Nano Lett.*, 8, 4342 (2008).

[83] C. Lee, X. Wei, J.W. Kysar and J. Hone. Measurement of the elastic properties and intrinsic strength of monolayer graphene. *Science*, 321, 385 (2008).

[84] S. Verbridge, D. Shapiro, H.G. Craighead and J.M. Parpia. Macroscopic tuning of nanomechanics: Substrate bending for reversible control of frequency and quality factor of nanostring resonators. *Nano Lett.*, 7, 1728 (2007).

[85] A. Eichler, J. Moser, J. Chaste, M. Zdrojek, I. Wilson-Rae and A. Bachtold. Nonlinear damping in mechanical resonators made from carbon nanotubes and graphene. *Nat. Nanotechnol.*, 6, 339 (2011).

[86] R. Garcia, R. Magerle and R. Perez. Nanoscale compositional mapping with gentle forces. *Nat. Mater.*, 6, 405 (2007).

[87] S. Biswas, P. Jana and A. Chatterjee. Hysteretic damping in an elastic body with frictional microcracks. *Int. J. Mech. Sci.*, 108-109, 61 (2016).

[88] R.E.D. Bishop and D.C. Johnson. *The Mechanics of Vibration.* Cambridge: Cambridge University Press, 1960.

[89] P. Mohanty, D.A. Harrington, K.L. Ekinci, Y.T. Yang, M.J. Murphy and M.L. Roukes. Intrinsic dissipation in high-frequency micromechanical resonators. *Phys. Rev. B*, 66, 085416 (2002).

[90] M. Imboden and P. Mohanty. Dissipation in nanoelectromechanical systems. *Phys. Rep.*, 534, 89 (2014).

[91] Y. Wang, J. Henry, D. Sengupta and M. Hines. Methyl monolayers suppress mechanical energy dissipation in micromechanical silicon resonators. *Appl. Phys. Lett.*, 85, 5736 (2004).

[92] J. Henry, Y. Wang, D. Sengupta and M. Hines. Understanding the effects of surface chemistry on Q: mechanical energy dissipation in alkyl-terminated (C_1–C_{18}) micromechanical silicon resonators. *J. Phys. Chem. B*, 111, 88 (2007).

[93] J. Yang, T. Ono and M. Esashi. Investigating surface stress: Surface loss in ultrathin single-crystal silicon cantilevers. *J. Vac. Sci. Technol. B*, 19, 551 (2001).

[94] R. Mihailovich and N. MacDonald. Dissipation measurements of vacuum-operated single crystal silicon microresonators. *Sens. Actuators A*, 50, 199 (1995).

[95] J. Yang, T. Ono and M. Esashi. Energy dissipation in submicrometer thick single-crystal cantilevers. *J. Microelectromech. Syst.*, 11, 775 (2002).

[96] N.V. Lavrik and P.G. Datskos. Femtogram mass detection using photothermally actuated nanomechanical resonators. *Appl. Phys. Lett.*, 82, 2697 (2003).

[97] K. Graff. *Wave Motion in Elastic Solids.* Oxford: Clarendon Press, 1975.

[98] A.N. Cleland. *Foundations of Nanomechanics.* Berlin: Springer, 2003.

[99] C. Zener. Internal friction in solids I. Theory of internal friction in reeds. *Phys. Rev.*, 52, 230 (1937).

[100] C. Zener. Internal friction in solids II. General theory of thermoelastic internal friction. *Phys. Rev.*, 53, 90 (1938).

[101] S. Prabhakar and S. Vengallatore. Thermoelastic damping in bilayered micromechanical beam resonators. *J. Micromech. Microeng.*, 17, 532 (2007).

[102] S. Schmid, K.D. Jensen, K.H. Nielsen and A. Boisen. Damping mechanisms in high-Q micro and nanomechanical string resonators. *Phys. Rev. B*, 84, 165307 (2011).

[103] R. Lifshitz and M.L. Roukes. Thermoelastic damping in micro- and nanomechanical systems. *Phys. Rev. B*, 61, 5600 (2000).

[104] C. Zener. *Elasticity and Anelasticity of Metals.* Chicago: University of Chicago Press, 1948.

[105] M.F. Ashby. Overview No. 80: On the engineering properties of materials. *Acta Metallur.*, 37, 1273 (1989).

[106] K.Y. Yasumura, T.D. Stowe, E.M. Chow, T. Pfafman, T.W. Kenny, B.C. Stipe and D. Rugar. Quality factors in micron- and submicron-thick cantilevers. *J. Microelectromech. Syst.*, 9, 117 (2000).

[107] C.Q. Ru. Size effect of dissipative surface stress on quality factor of microbeams. *Appl. Phys. Lett.*, 94, 051905 (2009).

[108] K. Kunal and N.R. Aluru. Akhiezer damping in nanostructures. *Phys. Rev. B*, 84, 245450 (2011).

[109] K. Kunal and N.R. Aluru. Intrinsic dissipation in a nano-mechanical resonator. *J. Appl. Phys.*, 116, 094304 (2014).

[110] S.K. Roy, V.T.K. Sauer, J.N. Westwood-Bachman, A. Venkatasubramanian and W.K. Hiebert. Improving mechanical sensor performance through larger damping. *Science*, 360, 1203 (2018).

[111] Y. Zhang. Determining the adsorption-induced surface stress and mass by measuring the shifts of resonant frequencies. *Sens. Actuators A: Phys.*, 194, 169 (2013).

[112] Y. Zhang. Detecting the stiffness and mass of biochemical adsorbates by a resonator sensor. *Sens. Actuators B: Chem.*, 202, 286 (2014).

[113] Y. Zhang and Y. Liu. Detecting both the mass and position of an accreted particle by a micro/nano-mechanical resonator sensor. *Sensors*, 14, 16296 (2014).

[114] E. Finot, A. Fabre, A. Passian and T. Thundat. Dynamic and static manifestation of molecular absorption in thin films probed by a microcantilever. *Phys. Rev. Appl.*, 1, 024001 (2014).

[115] X. Yi and H.L. Duan. Surface stress induced by interactions of adsorbates and its effect on deformation and frequency of microcantilever sensors. *J. Mech. Phys. Solids*, 57, 1254 (2009).

[116] R. Berger, E. Delamarche, H. Lang, Ch. Gerber, J.K. Gimzewski, E. Meyer and H. Güntherodt. Surface stress in the self-assembly of alkanethiols on gold. *Science*, 276, 2021 (1997).

[117] K. Eom, T.Y. Kwon, D.S. Yoon, H.L. Lee and T.S. Kim. Dynamical response of nano-mechanical resonators to biomolecular interactions. *Phys. Rev. B*, 76, 113408 (2007).

[118] D.W. Dareing and T. Thundat. Simulation of adsorption-induced stress of a microcantilever sensor. *J. Appl. Phys.*, 97, 043526 (2005).

[119] N. Backmann, C. Zahnd, F. Huber, A. Bietsch, A. Plückthun, H. Lang, H. Güntherodt, M. Hegner and C. Gerber. A label-free immunosensor array using single-chain antibody fragments. *P. Natl. Acad. Sci. USA*, 102, 14587 (2005).

[120] K. Hu, W. Zhang, X. Shi, H. Yan, Z. Peng and G. Meng. Adsorption-induced surface effects on the dynamical characteristics of micromechanical resonant sensors for in situ real-time detection. *J. Appl. Mech.*, 83, 081009 (2016).

[121] H.C. McCaig, E. Myers, N.S. Lewis and M.L. Roukes. Vapor sensing characteristics of nanoelectromechanical chemical sensors functionalized using surface-initiated polymerization. *Nano Lett.*, 14, 3728 (2014).

[122] M. Chi and Y.P. Zhao. Adsorption of formaldehyde molecule on the intrinsic and Al-doped graphene: A first principle study. *Comput. Mater. Sci.*, 46, 1085 (2009).

[123] M. Chi and Y.P. Zhao. First principle study of the interaction and charge transfer between graphene and organic molecules. *Comput. Mater. Sci.*, 56, 79 (2012).

[124] N.V. Lavrik, M.J. Sepaniak and P.G. Datskos. Cantilever transducers as a platform for chemical and biological sensors. *Rev. Sci. Instrum.*, 75, 2229 (2004).

[125] S. Sengupta, H.S. Solanki, V. Singh, S. Dhara and M.M. Deshmukh. Electromechanical resonators as probes of the charge density wave transition at the nanoscale in $NbSe_2$. *Phys. Rev. B.*, 82, 155432 (2010).

[126] X. Huang, M. Manolidis, S. Jun and J. Hone. Nanomechanical hydrogen sensing. *Appl. Phys. Lett.*, 86, 143104 (2005).

[127] T. Ono and M. Esashi. Stress-induced mass detection with a micromechanical/nanomechanical silicon resonator. *Rev. Sci. Instrum.*, 76, 0930107 (2005).

[128] S. Sapmaz, Y.M. Blanter, L. Gurevich and H.S.J. van der Zant. Carbon nanotubes as nanoelectromechanical systems. *Phys. Rev. B.*, 67, 235414 (2003).

[129] M.H. Chien, M. Brameshuber, B.K. Rossboth, G.J. Schütz and S. Schmid. Single-molecule optical absorption imaging by nanomechanical photothermal sensing. *Proc. Natl. Acad. Sci. USA*, 115, 11150 (2018).

[130] A. Jöckel, M.R. Rakher, M. Korppi, S. Camerer, D. Hunger, M. Mader and P. Treutlein. Spectroscopy of mechanical dissipation in micro-mechanical membranes. *Appl. Phys. Lett.*, 99, 143109 (2011).

[131] S. Manzeli, D. Dumcenco, G.M. Marega and A. Kis. Self-sensing, tunable monolayer MoS_2 nanoelectromechanical resonators. *Nat. Commun.*, 10, 1 (2019).

[132] D. Ramos, J. Tamayo, J. Mertens, M. Calleja and A. Zaballos. Origin of the response of nanomechanical resonators to bacteria adsorption. *J. Appl. Phys.*, 100, 106105 (2006).

[133] J. Tamayo, D. Ramos, J. Mertens and M. Calleja. Effect of the adsorbate stiffness on the resonance of microcantilever sensors. *Appl. Phys. Lett.*, 89, 224104 (2006).

[134] W. Xiang, Y. Tian and X. Liu. Theoretical analysis of detection sensitivity in nano-resonator-based sensors for elasticity and density measurement. *Int. J. Mech. Sci.*, 197, 106309 (2021).

[135] Y. Yang, C. Callegari, X.L. Feng and M.L. Roukes. Surface adsorbate fluctuations and noise in nanoelectromechanical systems. *Nano Lett.*, 11, 1753 (2011).

[136] J. Atalaya, A. Isacsson and M.I. Dykman. Diffusion-induced biostability of driven nanomechanical resonators. *Phys. Rev. Lett.*, 106, 227202 (2011).

[137] J. Atalaya, A. Isacsson and M.I. Dykman. Diffusion-induced dephasing in nanomechanical resonators. *Phys. Rev. B*, 83, 045419 (2011).

[138] Y. Zhang. Steady state response of an infinite beam on a viscoelastic foundation with moving distributed mass and load. *Sci. China Phys. Mech.*, 63, 284611 (2020).

[139] T.P. Burg, M. Godin, S.M. Knudsen, W. Shen, G. Carlson, J.S. Foster, K. Babcock and S.R. Manalis. Weighing of biomolecules, single cells and single nanoparticles in fluid. *Nature*, 446, 1066 (2007).

[140] Y. Weng, F.F. Delgado, S. Son, T.P. Burg, S.C. Wasserman and S.R. Manalis. Mass sensors with mechanical traps for weighing single cells in different fluids. *Lab Chip*, 11, 4174 (2011).

[141] W.H. Grover, A.K. Bryan, M. Diez-Silva, S. Suresh, J.M. Higgins and S.R. Manalis. Measuring single-cell density. *Proc. Natl. Acad. Sci. USA*, 108, 10992 (2011).

[142] A. Martín-Pérez, D. Ramos, J. Tamayo and M. Calleja. Coherent optical transduction of suspended microcapillary resonators for multi-parameter sensing applications. *Sensors*, 19, 5069 (2019).

[143] B. Lassagne, Y. Tarakanov, J. Kinaret, D. Garcia-Sanchez and A. Bachtold. Coupling mechanics to charge transport in carbon nanotube mechanical resonators. *Science*, 325, 1107 (2009).

[144] R. Merkel, P. Nassoy, A. Leung, K. Ritchie and E. Evans. Energy landscapes of receptor–ligand bonds explored with dynamic force spectroscopy. *Nature*, 397, 50 (1999).

[145] F.M. Battiston, J.P. Ramseyer, H.P. Lang, M.K. Baller, Ch. Gerber, J.K. Gimzewski, E. Meyer and H.J. Güntherodt. A chemical sensor based on a microfabricated cantilever array with simultaneous resonance-frequency and bending readout. *Sens. Actuators B: Chem.*, 77, 122 (2001).

[146] M. Spletzer, A. Raman, A.Q. Wu and X. Xu. Ultrasensitive mass sensing using mode localization in coupled microcantilevers. *Appl. Phys. Lett.*, 88, 254102 (2006).

[147] M. Spletzer, A. Raman, H. Sumali and J.P. Sullivan. Highly sensitive mass detection and identification using vibration localization in coupled microcantilever arrays. *Appl. Phys. Lett.*, 92, 114102 (2008).

[148] M.A. Hopcroft, W.D. Nix, and T.W. Kenny. What is the Young's modulus of silicon? *J. Microelectromech. Syst.*, 19, 229 (2010).

[149] K.B. Gavan, H. Westra, E. van der Drift, W. Venstra and H. van der Zant. Size-dependent effective Young's modulus of silicon nitride cantilevers. *Appl. Phy. Lett.*, 94, 233108 (2009).

[150] M. Godin, V. Tabard-Cossa, P. Grütter and P. Williams. Quantitative surface stress measurements using a microcantilever. *Appl. Phy. Lett.*, 79, 551 (2001).

[151] Y. Zhang. Extracting nanobelt mechanical properties from nanoindentation. *J. Appl. Phys.*, 107, 123518 (2010).

[152] V.B. Shenoy. Atomic calculations of elastic properties of metallic fcc crystal surfaces. *Phys. Rev. B*, 71, 094104 (2005).

[153] J. He and C.M. Lilley. Surface stress effect on bending resonance of nanowires with different boundary conditions. *Appl. Phys. Lett.*, 93, 263108 (2008).

[154] J. He and C.M. Lilley. Surface effect on the elastic behavior of static bending nanowires. *Nano Lett.*, 8, 1798 (2008).

[155] P. Lu, H. P. Lee, C. Lu and S.J. O'Shea. Surface stress effects on the resonance properties of cantilever sensors. *Phys. Rev. B*, 72, 085405 (2005).

[156] H. Ibach. The role of surface stress in reconstruction, epitaxial growth and stabilization of mesoscopic structures. *Surf. Sci. Rep.*, 29, 195 (1999).

[157] J. Guo and Y.P. Zhao. The size-dependent elastic properties of nanofilms with surface effects. *J. Appl. Phys.*, 98, 074306 (2005).

[158] J. Guo and Y.P. Zhao. The size-dependent bending elastic properties of nanobeams with surface effects. *Nanotechnology*, 18, 295701 (2007).

[159] X. Li, H. Zhang and K.Y. Lee. Dependence of Young's modulus of nanowires on surface effect. *Int. J. Mech. Sci.*, 81, 120 (2014).

[160] J. Mertens, M. Calleja, D. Ramos, A. Tarýn and J. Tamayo. Role of the gold film nanostructure on the nanomechanical response of microcantilever sensors. *J. Appl. Phys.*, 101, 034904 (2007).

[161] J. Lagowski, H.C. Gatos and E.S. Sproles Jr. Surface stress and normal mode of vibration of thin crystals: GaAs. *Appl. Phys. Lett.*, 26, 493 (1975).

[162] M.E. Gurtin, X. Markenscoff and R.N. Thurston. Effect of surface stress on the natural frequency of thin crystals. *Appl. Phys. Lett.*, 29, 529 (1976).

[163] M.L. Lachut and J.E. Sader. Effect of surface stress on the stiffness of cantilever plates. *Phys. Rev. Lett.*, 99, 206102 (2007).

[164] M.L. Lachut and J.E. Sader. Effects of surface stress on thin elastic plates and beams. *Phys. Rev. B*, 85, 085440 (2012).

[165] F.Q. Yang. Effect of interfacial stresses on the elastic behavior of nanocomposite materials. *J. Appl. Phys.*, 99, 054306 (2006).

[166] P. Müller and A. Saúl. Elastic effects on surface physics. *Surf. Sci. Rep.*, 54, 157 (2004).

[167] R.C. Cammarata, K. Sieradzki and F. Spaepen. Simple model for interface stresses with application to misfit dislocation generation in epitaxial thin film. *J. Appl. Phys.*, 87, 1227 (2000).

[168] R.C. Cammarata and K. Sieradzki. Surface and interface stresses. *Annu. Rev. Mater. Sci.*, 24, 215 (1994).

[169] Y. Zhang. Transitions between different contact models. *J. Adhes. Sci. Technol.*, 22, 699 (2008).

[170] Y. Zhang. Adhesion map of spheres: Effects of curved contact interface and surface interaction outside contact region. *J. Adhes. Sci. Technol.*, 25, 1435 (2011).

[171] Y. Zhang, Q. Ren and Y.P. Zhao. Modelling analysis of surface stress on a rectangular cantilever beam. *J. Phys. D: Appl. Phys.*, 37, 2140 (2004).

[172] J. Fritz, M.K. Baller, H.P. Lang, T. Strunz, E. Meyer, H.J. Guntherodt, C.H. Gerber and J.K. Gimzewski. Stress at the solid-liquid interface of self-assembled monolayers on gold investigated with a nanomechanical sensor. *Langmuir*, 16, 9694 (2000).

[173] J. Fritz, M.K. Baller, H.P. Lang, H. Rothuizen, P. Vettiger, E. Meyer, H.J. Guntherodt, C.H. Gerber and J.K. Gimzewski. Translating biomolecular recognition into nanomechanics. *Science*, 288, 316 (2000).

[174] Y. Zhang. Interface layer effect on the stress distribution of a wafer-bonded bilayer structure. *J. Mater. Sci.*, 43, 88 (2008).

[175] Y. Zhang. Extended Stoney's formula for a film-substrate bilayer with the effect of interfacial slip. *J. Appl. Mech.*, 75, 011008 (2008).

[176] R.B. Karabalin, L.G. Villanueva, M.H. Matheny, J.E. Sader and M.L. Roukes. Stress-induced variations in the stiffness of micro- and nanocantilever beams. *Phys. Rev. Lett.*, 108, 236101 (2012).

[177] G.G. Stoney. The tension of metallic films deposited by electrolysis. *P. Roy. Soc. A-Math. Phy.*, 82, 172 (1909).

[178] L.B. Freund, J.A. Floro and E. Chason. Extensions of the Stoney formula for substrate curvature to configurations with thin substrates or large deformations. *Appl. Phys. Lett.*, 74, 1987 (1999).

[179] L.B. Freund and S. Suresh. *Thin Film Materials: Stress, Defect Formation and Surface Evolution.* Cambridge, UK: Cambridge University Press, 2003, chaps. 2 and 6.

[180] Y. Zhang and Y.P. Zhao. Applicability range of Stoney's formula and modified formulas for a film/substrate bilayer. *J. Appl. Phys.*, 99, 053513 (2006).

[181] W. Fang and J.A. Wickert. Comments on measuring thin-film stresses using bi-layer micromachined beams. *J. Micromech. Microeng.*, 5, 276 (1995).

[182] W. Fang and J.A. Wickert. Determining mean and gradient residual stresses in thin films using micromachined cantilevers. *J. Micromech. Microeng.*, 6, 301 (1996).

[183] Y. Zhang and Y.P. Zhao. An effective method of determining the residual stress gradients in a micro-cantilever. *Microsyst. Technol.*, 12, 357 (2006).

[184] Y. Zhang. Deflections and curvatures of film-substrate structure with the presence of gradient stress in MEMS application. *J. Micromech. Microeng.*, 17, 753 (2007).

[185] D.R. Evans and V.S.J. Craig. Sensing cantilever beam bending by the optical lever technique and its application to surface stress. *J. Phys. Chem. B*, 110, 5450 (2006).

[186] Z. Suo. Wrinkling of the oxide scale on an aluminum-containing alloy at high temperatures. *J. Mech. Phys. Solids*, 43, 829 (1995).

[187] P. Olsson and H.S. Park. On the importance of surface elastic contributions to the flexural rigidity of nanowires. *J. Mech. Phys. Solids*, 60, 2064 (2012).

[188] J. Weissmüller and H. Duan. Cantilever bending with rough surfaces. *Phys. Rev. Lett.*, 101, 146102 (2008).

[189] S. Timoshenko. Analysis of bi-metal thermostats. *J. Opt. Soc. Am.*, 11, 233 (1925).

[190] E.J. Hearn. *Mechanics of Materials*, 2nd edition. Oxford: Pergamon Press, 1985.

[191] Y. Zhang and Y.P. Zhao. A discussion on modeling shape memory alloy embedded in a composite laminate as axial force and elastic foundation. *Mater. Design*, 28, 1016 (2007).

[192] Y. Zhang and Y.P. Zhao. A study of composite beam with shape memory alloy arbitrarily embedded under thermal and mechanical loadings. *Mater. Design*, 28, 1096 (2007).

[193] Y. Zhang. Analysis of dislocation-induced strain field in an idealized wafer-bonded microstructure. *J. Phys. D: Appl. Phys.*, 40, 1118 (2007).

[194] F. Shi, S. MacLaren, C. Xu, K.Y. Cheng and K.C. Hsieh. Hybrid-integrated GaAs/GaAs and InP/GaAs semiconductors through wafer bonding technology: Interface adhesion and mechanical strength. *J. Appl. Phys.*, 93, 5750 (2003).

[195] K. Kendall. *Molecular Adhesion and Its Application: The Sticky Universe.* New York: Kluwer Academic/Plenum Publishers, 2001.

[196] C.E. Murray and I.C. Noyan. Finite-size effects in thin-film composites. *Philos. Mag. A*, 82, 3087 (2002).

[197] W.T. Chen and C.W. Nelson. Thermal stress in bonded joints. *IBM J. Res. Dev.*, 23, 179 (1979).

[198] E. Suhir. Stresses in bi-metal thermostats. *J. Appl. Mech.*, 53, 657 (1986).

[199] I.C. Noyan, C.E. Murray, J.S. Chey and C.C. Goldsmith. Finite size effects in stress analysis of interconnect structures. *Appl. Phys. Lett.*, 85, 724 (2004).

[200] V. Tabard-Cossa, M. Godin, I.J. Burgess, T. Monga, R.B. Lennox and P. Grutter. Microcantilever-based sensors: Effect of morphology, adhesion, and cleanliness of the sensing surface on Surface stress. *Anal. Chem.*, 79, 8136 (2007).

[201] J.J. Headrick, M.J. Sepaniak, N.V. Lavrik and P.G. Datskos. Enhancing chemi-mechanical transduction in microcantilever chemical sensing by surface modification. *Ultramicroscopy*, 97, 417 (2003).

[202] H.S. Park and P.A. Klein. Surface stress effects on the resonant properties of metal nanowires: The importance of finite deformation kinematics and the impact of the residual surface stress. *J. Mech. Phys. Solids*, 56, 3144 (2008).

[203] H.S. Park and P.A. Klein. Surface Cauchy-Born analysis of surface stress effects on metallic nanowires. *Phys. Rev. B*, 75, 085408 (2007).

[204] F. Song, G.L. Huang, H.S. Park and X.N. Liu. A continuum model for the mechanical behavior of nanowires including surface and surface-induced initial stresses. *Int. J. Solids Struct.*, 48, 2154 (2011).

[205] L. Elbrecht, U. Storm, R. Catanescu and J. Binder. Comparison of stress measurement techniques in surface micromaching. *J. Micromech. Microeng.*, 7, 151 (1997).

[206] X. Zhang, A. Misra, H. Wang, A. Lima, M. Hundley and R. Hoagland. Effects of deposition parameters on residual stresses, hardness and electrical resistivity of nanoscale twinned 330 stainless steel thin film. *J. Appl. Phys.*, 97, 094302 (2005).

[207] G.F. Wang and X.Q. Feng. Effects of surface elasticity and residual surface tension on the natural frequency of microbeams. *Appl. Phys. Lett.*, 90, 231904 (2007).

[208] Y. Li, J. Song, B. Fang and J. Zhang. Surface effects on the postbuckling of nanowires. *J. Phys. D: Appl. Phys.*, 44, 425304 (2011).

[209] Y. Li, B. Fang, J. Zhang and J. Song. Surface effects on the wrinkling of a stiff film bonded to a compliant substrate. *Thin Solid Films*, 520, 2077 (2012).

[210] K.D. Murphy and Y. Zhang. Vibration and stability of a cracked translating beam. *J. Sound Vibr.*, 237, 319 (2000).

[211] J. Wu and Y. Wei. Grain misorientation and grain-boundary rotation dependent mechanical properties in polycrystalline graphene. *J. Mech. Phys. Solids*, 61,1421 (2013).

[212] J.R. Barnes, R.J. Stephenson, M.E. Welland, Ch. Gerber and J.K. Gimzewski. Photothermla spectroscopy with femtojoule sensitivity using a micromechanical device. *Nature*, 372, 79 (1994).

[213] T. Thundat and L. Maya. Monitoring chemical and physical changes on subnanogram quantities of platinum dioxide. *Surf. Sci.*, 430, L546 (1999).

[214] H. Yu and X. Li. Bianalyte mass detection with a single resonant microcantilever. *Appl. Phys. Lett.*, 94, 011901 (2009).

[215] R. Lefevre, M. F. Goffman, V. Derycke, C. Miko, L. Forro, J. P. Bourgoin and P. Hesto. Scaling law in carbon nanotube electromechanical devices. *Phys. Rev. Lett.*, 95, 185504 (2005).

[216] Y. Zhang, Y. Petrov and Y.P. Zhao. Eigenfrequency loci crossings, veerings and mode splittings of two cantilevers coupled by an overhang. *J. Phys. Commun.*, 4, 085010 (2020).

[217] R. Sandberg, W. Svendsen, K. Molhave and A. Boisen. Temperature and pressure dependence of resonance in multi-layer microcantilevers. *J. Micromech. Microeng.*, 15, 1454 (2005).

[218] B. Ilic, S. Krylov and H.G. Craighead. Young's modulus and density measurements of thin atomic layer deposited films using resonant nanomechanics. *J. Appl. Phys.*, 108, 044317 (2010).

[219] S. Zurn, M. Hsieh, G. Smith, D. Markus, M. Zang, G. Hughes, Y. Nam, M. Arik and D. Polla. Fabrication and structural characterization of a resonant frequency PZT microcantilever. *Smart Mater. Struct.*, 10, 252 (2001).

[220] Y. Zhang and Y.P. Zhao. Measuring the nonlocal effects of a micro/nanobeam by the shifts of resonant frequencies. *Int. J. Solids Strut.*, 102–103, 259 (2016).

[221] A.E.H. Love. *The Mathematical Theory of Elasticity*. New York, NY: Dover Publications, 1927.

[222] A.W. McFarland and J. Colton. Role of material microstructure in plate stiffness with relevance to microcantilever sensors. *J. Micromech. Microeng.*, 15, 1060 (2005).

[223] E. Kröner. On the physical reality of torque stresses in continuum mechanics. *Int. J. Eng. Sci.*, 1, 261 (1963).

[224] A.C. Eringen. On differential equations of nonlocal elasticity and solutions of screw dislocation and surface waves. *J. Appl. Phys.*, 54, 4703 (1983).

[225] E. Cosserat and F. Cosserat. *Theorie des Corps Deformables*. Paris: A. Hermann et Fils, 1909.

[226] W.T. Koiter. Couple-stresses in the theory of elasticity I and II. *Koninkl. Ned. Akad. Wet. Ser. B*, 67, 17 (1964).

[227] R.D. Mindlin. Influence of couple-stresses on stress concentrations. *Exp. Mech.*, 3, 1 (1963).

[228] R.D. Mindlin. Micro-structure in linear elasticity. *Arch. Rat. Mech. Anal.*, 16, 51 (1964).

[229] R.D. Mindlin and H.F. Tiersten. Effects of couple-stresses in linear elasticity. *Arch. Rat. Mech. Anal.*, 11, 415 (1962).

[230] R. Maranganti and P. Sharma. Length scales at which classical elasticity breaks down for various materials. *Phys. Rev. Lett.*, 98, 195504 (2007).

[231] N.A. Fleck and J.R. Willis. Strain gradient plasticity: energetic or dissipative? *Acta Mech. Sin.*, 31, 465 (2015).

[232] Y. Wei, Y. Li, L. Zhu, Y. Liu, X. Lei, G. Wang, Y. Wu, Z. Mi, J. Liu, H. Wang and H. Gao. Evading the strength–ductility trade-off dilemma in steel through gradient hierarchical nanotwins. *Nat. Commun.*, 5, 3580 (2014).

[233] E.C. Aifantis. Update on a class of gradient theories. *Mech. Mater.*, 35, 259 (2003).

[234] E.C. Aifantis and J.R. Willis. The role of interfaces in enhancing the yield strength of composites and polycrystals. *J. Mech. Phys. Solids*, 53, 1047 (2005).

[235] J. Yang and R.S. Lakes. Transient study of couple stress effects in compact bone: Torsion. *J. Biomech. Eng.*, 103, 275 (1981).

[236] J. Yang and R.S. Lakes. Experimental study of micropolar and couple stress elasticity in compact bone in bending. *J. Biomech.*, 15, 91 (1982).

[237] R.K. Abu-Al-Rub and G.Z. Voyiadjis. Analytical and experimental determination of the material intrinsic length scale of strain gradient plasticity theory from micro- and nano-indentation experiments. *Int. J. Plast.*, 20, 1139 (2004).

[238] V.G. Romanov. *Inverse Problems of Mathematical Physics*. Utrecht, The Netherlands: Vnu Science Press BV, 1987.

[239] G.M.L. Gladwell. *Inverse Problems in Vibration*, 2nd edition. New York, NY: Kluwer Academic Publishers, 2004.

[240] G. Angler. *Inverse Problems in Differential Equations*. New York: Plenum Press, 1990.

[241] A. Kirsch. *An Introduction to the Mathematical Theory of Inverse Problems*, 3rd edition. Switzerland: Springer Nature, 2021.

[242] M. Kaz. Can one hear the shape of a drum? *Amer. Math. J.*, 73, 1 (1966).

[243] C. Gordon and D. Webb. You can't hear the shape of a drum. *American Scientist*, 84, 46 (1996).

[244] C. Gordon, D. Webb and S. Wolpert. Isospectral plane domains and surfaces via Riemannian orbifolds. *Invent. Math.*, 110, 1 (1992).

[245] S. Dohn, W. Svendsen, A. Boisen and O. Hansen. Mass and position determination of attached particles on cantilever based mass sensors. *Rev. Sci. Instrum.*, 78, 103303 (2007).

[246] B. Ilic, H.G. Craighead, S. Krylov, W. Senaratne, C. Ober and P. Neuzil. Attogram detection using nanoelectromechanical oscillators. *J. Appl. Phys.*, 95, 3694 (2004).

[247] B. Ilic, Y. Yang, K. Aubin, R. Reichenbach, S. Krylov and H.G. Craighead. Enumeration of DNA molecules bound to a nanomechanical oscillator. *Nano Lett.*, 5, 925 (2005).

[248] L. Meirovitch. *Computational Methods in Structural Dynamics*. Rockville, MD: Sijthoff & Noordhoff Inc., 1980.

[249] Y. Zhang. Eigenfrequency computation of beam/plate carrying concentrated mass/ spring. *J. Vibr. Acoust.*, 133, 021006 (2011).

[250] Y. Zhang and K D. Murphy. Multi-modal analysis on the intermittent contact dynamics of atomic force microscope. *J. Sound Vibr.*, 330, 5569 (2011).

[251] Y. Zhang and Y.P. Zhao. Mass and force sensing of an adsorbate on a beam resonator sensor. *Sensors*, 15, 14871 (2015).

[252] M.S. Hanay, S.I. Kelber, C.D. O'Connell, P. Mulvaney, J.E. Sader and M.L. Roukes. Inertial imaging with nanomechanical systems. *Nat. Nanotechnol.*, 10, 339 (2015).

[253] J.E. Sader, M.S. Hanay, A.P. Neumann and M.L. Roukes. Mass spectrometry using nanomechanical systems: beyond the point-mass approximation. *Nano Lett.*, 18, 1608 (2018).

[254] S. Schmid, S. Dohn and A. Boisen. Real-time particle mass spectrometry based on resonant micro strings. *Sensors*, 10, 8092 (2010).

[255] M.A. Mahmoud. Mass sensing of multiple particles adsorbed to microcantilever resonators. *Microsyst. Technol.*, 23, 711 (2017).

[256] P. Belardinelli, L. Hauzer, M. Šiškins, M.K. Ghatkesar and F. Alijani. Modal analysis for density and anisotropic elasticity identification of adsorbates on microcantilevers. *Appl. Phys. Lett.*, 113, 143102 (2018).

[257] S. Ma, H. Bai, S. Wang, L. Zhao, K. Yang, R. Fang and X. Zhou. Detecting the mass and position of a particle by the vibration of a cantilevered micro-plate. *Int. J. Mech. Sci.*, 172, 105413 (2020).

[258] Y. Zhang, Y. Zhao and Z. Cheng. Determining the layers' Young's moduli and thickness from the indentation of a bilayer structure. *J. Phys. D: Appl. Phys.*, 51, 065305 (2018).

[259] Y. Zhang, F. Gao, Z. Zheng and Z. Cheng. An inverse problem in film/substrate indentation: Extracting both the Young's modulus and thickness of film. *Acta Mech. Sin.*, 34, 1061 (2018).

[260] M. Modena, Y. Wang, D. Riedel and T.P. Burg. Resolution enhancement of suspended microchannel resonators for weighing of biomolecular complexes in solution. *Lab Chip*, 14, 342 (2014).

[261] R. Courant and D. Hilbert. *Methods of Mathematical Physics*, vol. 1. New York, NY: Interscience, 1961.

[262] V.P. Adiga, B. Ilic, R.A. Barton, I. Wilson-Rae, H.G. Craighead and J.M. Parpia. Approaching intrinsic performance in ultra-thin silicon nitride drum resonators. *J. Appl. Phys.*, 112, 064323 (2012).

[263] R.A. Barton, B. Ilic, A.M. van der Zande, W.S. Whitney, P.L. McEuen, J.M. Parpia and H.G. Craighead. High, size-dependent quality factor in an array of graphene mechanical resonators. *Nano Lett.*, 11, 1232 (2011).

[264] D. Ramos, J. Tamayo, J. Mertens, M. Calleja, L.G. Villanueva and A. Zaballos. Detection of bacteria based on the thermomechanical noise of a nanomechanical resonator: origin of the response and detection limits. *Nanotechnology*, 19, 035503 (2008).

[265] R.R. Grüter, Z. Khan, R. Paxman, J.W. Ndieyira, B. Dueck, B.A. Bircher, J.L. Yang, U. Drechsler, M. Depont, R.A. McKendry and B.W. Hoogenboom. Disentangling mechanical and mass effects on nanomechanical resonators. *Appl. Phys. Lett.*, 96, 023113 (2010).

[266] Y. Zhang and Y.P. Zhao. Determining both adhesion energy and residual stress by measuring the stiction shape of a microbeam. *Microsyst. Technol.*, 21, 919 (2015).

5 Axial-Wave Propagation of Carbon Nanorod-Conveying Fluid with Elastic Support Using Nonlocal Continuum Elasticity

V. Senthilkumar

CONTENTS

5.1 INTRODUCTION

The nanomechanic researchers extensively discussed the longitudinal-wave propagation [1] of carbon nanotube-conveying fluid phenomenon. The governing equations predicted the behavior of nanotubes filled with fluid using the continuum modeling approach. The critical velocity of the fluid plays a significant role in influencing the mechanical properties [2–5] of such structures. The wave propagation helps in understanding the mechanical properties in any system. So it is evident that the wave propagation study is essential before investigating the mechanical properties of the nanotube. The wave propagation study includes dispersion curve, phase velocity, and group velocity. Eringen developed the nonlocal continuum elasticity with the nonlocal parameter a. The value of a is considered as a lattice parameter. The nonlocal continuum elasticity is popular among researchers due to its capability in modeling small-sized structures [6, 7] like nanorods [8, 9, 10], nanotubes [11, 12], nanoplates [13, 14, 15, 16, 17], and nanoshells. Païdoussis [18] discussed elaborately linear and nonlinear dynamics phenomena in pipe-conveying fluid inside [19–23].

DOI: 10.1201/9781003328032-5

Yoon [24] et al. used the Euler-Bernoulli beam model [18] to investigate the effect of fluid inside a carbon nanotube. Yoon [24] et al. investigated the carbon nanotube-conveying fluid using a continuum model [18, 25]. This model consists of deflection $w(x,t)$ as the main function of the carbon nanotube. Recently, the axial-wave propagation of nanorod-conveying fluid has attracted researchers.

Oveissi [21] et al. proposed the carbon nanotube-conveying fluid model in the axial direction as a function of $u(x,t)$. Since carbon nanorod [3] predicts the axial effects alone, it is interesting to investigate the impact of fluid conveying. Oveissi [21] et al. examined the uniform mean flow velocity of nanotube/nanorod is v_u for axial direction effects and defined it as $v_u(x,t) = \dfrac{\partial u(x,t)}{\partial t}$. To distinguish from nanotube notation $v_w(x,t) = \dfrac{\partial w(x,t)}{\partial t}$ for transverse direction effect, the axial direction notation $v_u(x,t)$ is introduced in the chapter. The mean flow velocity of carbon nanotube ranges ranges from 2000 m/s [24, 26] to 50,000 m/s [24, 27]. The value of stiffness of the elastic medium k_u ranges from 1.0 GPa [24] to 12.0 GPa [28]. To the best of the author's knowledge, the axial-wave propagation investigation of carbon nanorods-conveying fluid with group speed study is missing in the literature. For the first time, the chapter addresses the effect of elastic medium support over the carbon nanorods-conveying fluid using wave propagation studies.

5.1.1 MATHEMATICAL MODELLING OF CARBON NANOROD-CONVEYING FLUID

Using Eringen's nonlocal continuum theory [29], the stress-strain relation obtained as [28, 29],

$$\left[1-(e_0a)^2\nabla^2\right]t_{ij} = \lambda e_{rr}\delta_{ij} + 2\mu e_{ij} \tag{5.1}$$

$$u = u(x,t) \tag{5.2}$$

$$e_{xx} = \frac{\partial u}{\partial x} \tag{5.3}$$

$$f = -k_u u \tag{5.4}$$

$$\left[1-(e_0a)^2\frac{\partial^2}{\partial x^2}\right]t_{xx} = Ee_{xx} \tag{5.5}$$

Integrate with $\displaystyle\int_A dA$

$$\left[1-(e_0a)^2\frac{\partial^2}{\partial x^2}\right]\int_A t_{xx}dA = E\int_A e_{xx}dA \tag{5.6}$$

$$\left[1-(e_0a)^2\frac{\partial^2}{\partial x^2}\right]N = EAe_{xx} \tag{5.7}$$

Differentiate w.r.t 'x' and define $N = \int_A t_{xx} dA$.

$$\left[1 - (e_0 a)^2 \frac{\partial^2}{\partial x^2}\right] \frac{\partial N}{\partial x} = EA \frac{\partial^2 u}{\partial x^2} \tag{5.8}$$

The equation of motion for fluid-filled carbon nanorod in the axial direction [21] along with the elastic support [28] k_u takes form of

$$\frac{\partial N}{\partial x} = m_c \frac{\partial^2 u}{\partial t^2} + k_u u + m_f \left(2v_u \frac{\partial^2 u}{\partial x \partial t} + v_u^2 \frac{\partial^2 u}{\partial x^2} + \frac{\partial^2 u}{\partial t^2}\right) \tag{5.9}$$

The mean flow velocity is defined in axial direction [21] as:

$$v_u(x,t) = \frac{\partial u(x,t)}{\partial t} \tag{5.10}$$

Using Eqns. (5.8) and (5.9), the governing equation appears as,

$$EA \frac{\partial^2 u}{\partial x^2} - (m_c + m_f) \frac{\partial^2 u}{\partial t^2} - k_u u - 2m_f v_u \frac{\partial^2 u}{\partial x \partial t} + (e_0 a)^2 2m_f v_u \frac{\partial^4 u}{\partial x^3 \partial t} - m_f v_u^2 \frac{\partial^2 u}{\partial x^2}$$
$$+ (e_0 a)^2 m_f v_u^2 \frac{\partial^4 u}{\partial x^4} + (e_0 a)^2 (m_c + m_f) \frac{\partial^4 u}{\partial x^2 \partial t^2} + (e_0 a)^2 k_u \frac{\partial^2 u}{\partial x^2} = 0 \tag{5.11}$$

The wave function [30] assumes the form of:

$$u(x,t) = \hat{u}(x,\omega) e^{i(\omega t - kx)} \tag{5.12}$$

Substitute Eqn. (5.12) in Eqn. (5.11) and it becomes,

$$(m_c + m_f)\omega^2 [1 + (e_0)^2 a^2] 1 + (e_0 a)^2 k^2 - 2m_f v_u \omega \left(k + (e_0 a)^2 k^3\right)$$
$$+ m_f v_u^2 k^2 [1 + (e_0 a)^2 k^2] - EA k^2 - k_u [1 + (e_0 a)^2 k^2] = 0 \tag{5.13}$$

$$P\omega^2 - Q\omega - R = 0 \tag{5.14}$$

$$P = (m_c + m_f)\left[1 + (e_0 a)^2 k^2\right] \tag{5.15}$$

$$Q = 2km_f v_u \left(1 + (e_0 a)^2 k^2\right) \tag{5.16}$$

$$R = \left(EA k^2 + k_u [1 + (e_0 a)^2 k^2] - m_f v_u^2 k^2 [1 + (e_0 a)^2 k^2]\right) \tag{5.17}$$

The frequency and wavenumber relation is,

$$\omega = \left(\frac{Q \pm \sqrt{Q^2 + 4PR}}{2P} \right) \tag{5.18}$$

The phase speed and group speed are defined as,

$$v^P = \frac{\omega}{k} \tag{5.19}$$

$$v^G = \frac{d\omega}{dk} \tag{5.20}$$

5.2 WAVE PROPAGATION STUDIES OF CARBON NANOROD-CONVEYING FLUID

Table 5.1 refers to the material properties of carbon nanorod conveying fluid inside.

Figure 5.1 shows the impact in the frequency of carbon nanorod is conveying fluid with mean flow velocity $v_u = 2000$ m/s. The dispersion curve shows that more significant influence of flow velocity inside carbon nanorod at higher wavenumbers in the frequency curve with the increase of nonlocal parameters. Though it impacts smaller wavenumbers, it is predominant in higher wavenumbers. It means that an increase in nonlocal parameters decreases the frequency values. The present dispersion curve study considers three different values of v_u as 2000 m/s, 6000 m/s, and 12000 m/s. Figures 5.1–5.3 show the effect of nonlocal parameters over frequencies. For an increase in v_u from Figures 5.1 to 5.3, the frequency increases with the increase in nonlocal parameters. So Figures 5.1–5.3 imply that both mean flow velocity and nonlocal parameters predominantly affect the frequencies at higher wavenumbers. In the absence of fluid flow, the present carbon nanorod model predicts Eringen's results with the value of the nonlocal parameter as $e_0 = 0.39$ [28, 29]. Further, Eringen's model matches lattice dynamics results without fluid flow reasonably well. Also, if the nonlocal parameter attains the value as zero, this model yields classical nanorod/continuum results.

TABLE 5.1
Material Properties of Carbon Nanorod

$E = 1.0\ TPa$ [20, 21]	$h = 1.0\ nm$ [21]	$R_{out} = 3.4\ nm$ [20, 21]
$\rho_c = 2.3 \times 10^3\ kg/m^3$ [20, 21]	$m_c = 8.3529 \times 10^{-14}\ kg/m$ [21]	$\rho_f = 1.0 \times 10^3\ kg/m^3$ [20, 21]
$m_f = 1.8095 \times 10^{-14}\ kg/m$ [21]		

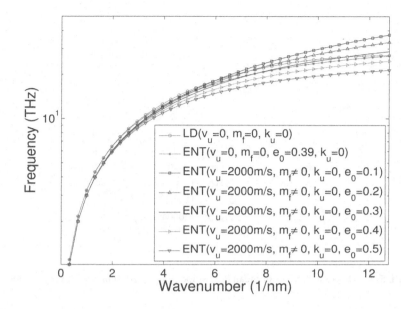

FIGURE 5.1 Dispersion curve of mean flow velocity $v_u = 2000$ m/s with varying nonlocal parameter.

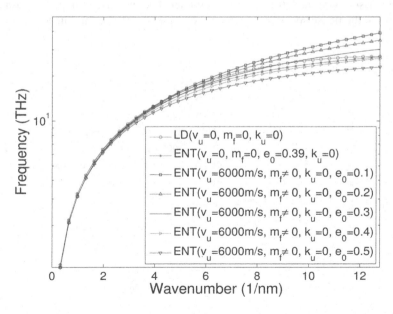

FIGURE 5.2 Dispersion curve of mean flow velocity $v_u = 6000$ m/s with varying nonlocal parameter.

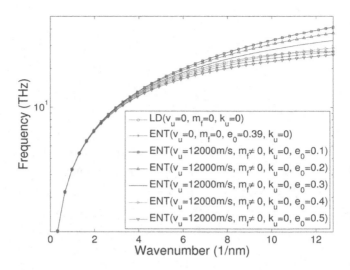

FIGURE 5.3 Dispersion curve of mean flow velocity $v_u = 12000$ m/s with varying nonlocal parameter.

Figures 5.4–5.6 reveal the phase speed study for the fluid-conveying nanorod without elastic support, and the mean velocity v_u ranges between 2000 m/s, 6000 m/s, and 12,000 m/s, respectively. Eringen's nonlocal continuum model captures well the phase speed with the nonlocal parameter value $e_0 = 0.39$ in the absence of fluid flow in comparison with the lattice dynamics model. The increase in nonlocal parameter e_0 impacts the phase speed at a higher wavenumber with the fluid inside. The phase

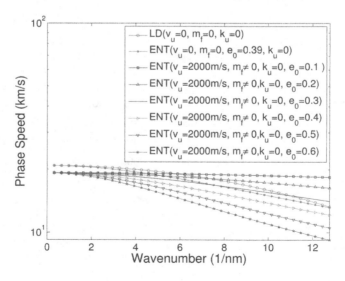

FIGURE 5.4 Phase speed of mean flow velocity $v_u = 2000$ m/s with varying nonlocal parameter.

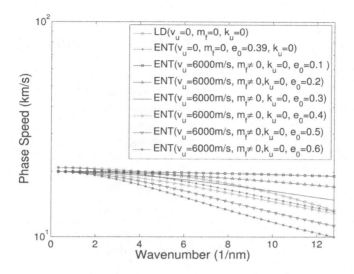

FIGURE 5.5 Phase speed of mean flow velocity $v_u = 6000$ m/s with varying nonlocal parameter.

speed decreases with an increase in nonlocal effects for the v_u values of 2000 m/s. This phenomenon is visible in higher wavenumber in Figure 5.4. But the increase in the mean flow velocity increases the phase speed of the carbon nanorod with the increment in nonlocal parameter. This occurrence is shown in Figures 5.4–5.6 for higher wavenumbers.

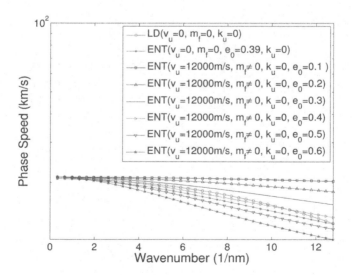

FIGURE 5.6 Phase speed of mean flow velocity $v_u = 12000$ m/s with varying nonlocal parameter.

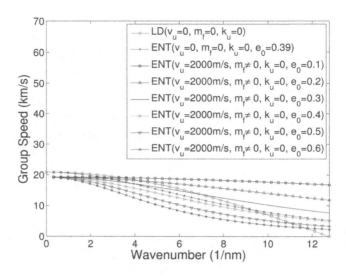

FIGURE 5.7   Group speed of mean flow velocity $v_u = 2000$ *m/s* with varying nonlocal parameter.

The group velocity studies from Figures 5.7 to 5.9 show that the mean flow velocity influences its greater extent at the higher wavenumbers. The increase in nonlocal parameters decreases the group speed at $v_u = 2000$ *m/s*. Further, the increase in mean flow velocity in two specific cases as 6000 m/s and 12,000 m/s increases the group speed. The study consists of a dispersion curve and phase speed, revealing that the mean flow velocity and nonlocal parameters significantly impact higher

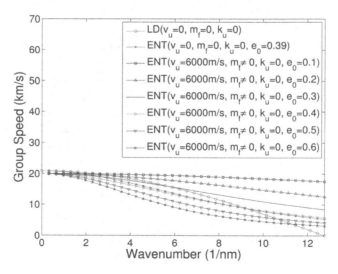

FIGURE 5.8   Group speed of mean flow velocity $v_u = 6000$ *m/s* with varying nonlocal parameter.

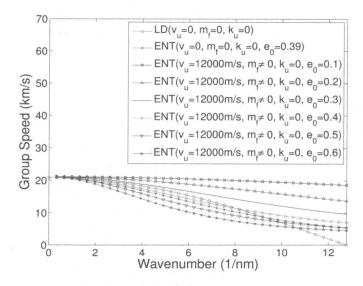

FIGURE 5.9   Group speed of mean flow velocity $v_u = 12000$ m/s with varying nonlocal parameter.

wavenumbers. However, these effects were seen in the small wavenumbers too. The studies deal without the elastic support effect.

5.3   WAVE PROPAGATION STUDIES OF CARBON NANOROD-CONVEYING FLUID WITH ELASTIC SUPPORT

The elastic support in the axial direction of carbon nanorod is influenced by the flexible elastic medium k_u. Aydogdu [29] proposed k_u values range from 1×10^9 N/m^2 to 12×10^9 N/m^2. The present investigation considers the values of k_u as 10 GPa. The dispersion curve in Figure 5.10 with elastic support shows an increase in frequency in comparison with Figure 5.1 without elastic support when the mean flow velocity is at $v_u = 2000$ m/s. A similar phenomenon occurs if the mean velocity increases in the range of $v_u = 6000$ m/s and $v_u = 12000$ m/s for the dispersion with elastic support. Also, the nonlocal effects influence the frequency of fluid-filled carbon nanorods in all these cases, shown in Figures 5.10–5.12.

The phase speed in Figure 5.13 captures the behavior of fluid-filled carbon nanorod with elastic support at mean flow velocity $v_u = 2000$ m/s. The phase speed changes significantly with small wavenumbers and higher wavenumbers. The increase in flow velocity in the case of $v_u = 6000$ m/s and $v_u = 12000$ m/s increases the phase speed. Also, the increase in the nonlocal parameter decreases the phase speed with elastic support, evident in Figures 5.13–5.15.

Figures 5.16–5.18 show the group speed of fluid-filled carbon nanorod with elastic stiffness $k_u = 10$ GPa. The mean flow velocity $v_u = 2000$ m/s affects the group speed for an increase in wavenumber. The group speed shows a significant change with

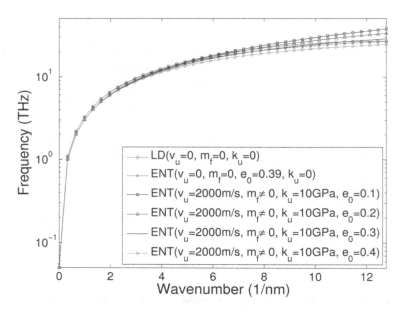

FIGURE 5.10   Dispersion curve of mean flow velocity $v_u = 2000\ m/s$ with varying nonlocal parameter at $k_u = 10\ GPa$.

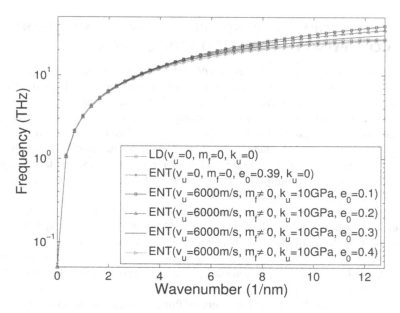

FIGURE 5.11   Dispersion curve of mean flow velocity $v_u = 6000\ m/s$ with varying nonlocal parameter at $k_u = 10\ GPa$.

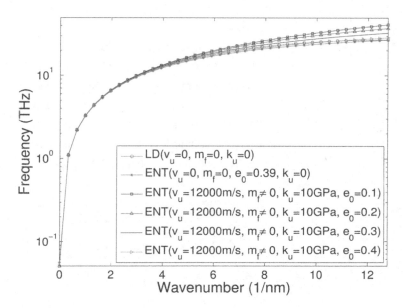

FIGURE 5.12 Dispersion curve of mean flow velocity $v_u = 12000$ m/s with varying nonlocal parameter at $k_u = 10$ GPa.

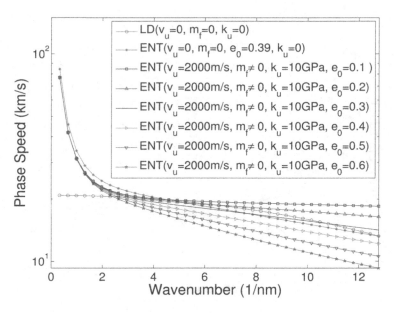

FIGURE 5.13 Phase speed of mean flow velocity $v_u = 2000$ m/s with varying nonlocal parameter at $k_u = 10$ GPa.

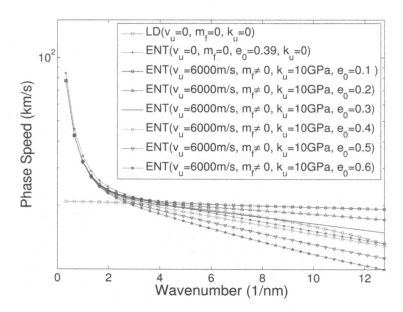

FIGURE 5.14 Phase speed of mean flow velocity $v_u = 6000$ *m/s* with varying nonlocal parameter at $k_u = 10$ *GPa*.

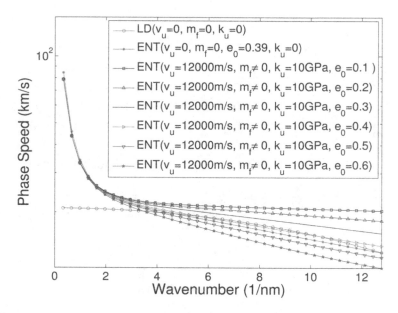

FIGURE 5.15 Phase speed of mean flow velocity $v_u = 12000$ *m/s* with varying nonlocal parameter at $k_u = 10$ *GPa*.

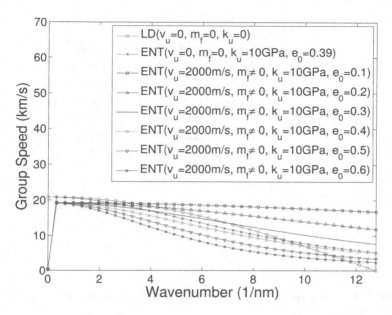

FIGURE 5.16 Group speed of mean flow velocity $v_u = 2000$ m/s with varying nonlocal parameter at $k_u = 10$ GPa.

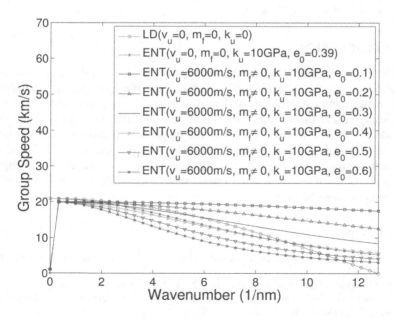

FIGURE 5.17 Group speed of mean flow velocity $v_u = 6000$ m/s with varying nonlocal parameter at $k_u = 10$ GPa.

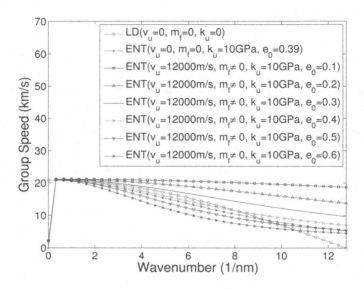

FIGURE 5.18   Group speed of mean flow velocity $v_u = 12000$ m/s with varying nonlocal parameter at $k_u = 10$ GPa.

elastic support for the higher wavenumber. Also, the increase in the flow velocity in the value of $v_u = 6000$ m/s and $v_u = 12000$ m/s shows the increase in the value of group speed. Further, the nonlocal effects decrease the group speed with an elastic stiffness value of $k_u = 10$ GPa.

5.4   CONCLUSION

This chapter investigated the axial-wave propagation of fluid-filled carbon nanorods with elastic support. Eringen's nonlocal continuum theory captured the nanoscale effects using nonlocal parameters. The dispersion curve, phase speed, and group speed study reveal the influence of elastic stiffness and nonlocal parameters. To conclude, the elastic stiffness increases the dispersion, phase, and group speed in the axial propagation study. Further, the nonlocal effects decrease the same. Since the elastic stiffness is in the range of GPa and the wave propagation investigation is influenced, it will be interesting to study the elastic stiffness in the value of TPa, which is not established until now.

REFERENCES

[1] L. Brillouin. 2013. *Wave propagation and group velocity*, volume 8. Academic Press, New York.

[2] S. Gopalakrishnan and S. Narendar. 2013. *Wave propagation in nanostructures: nonlocal continuum mechanics formulations*. Springer Science & Business Media, New York; Dordrecht; London.

[3] D. Karlicic, T. Murmu, S. Adhikari, and M. McCarthy. 2015. *Non-local structural mechanics*. John Wiley & Sons, Hoboken, NJ.

[4] V. Senthilkumar. 2010. Buckling analysis of a single-walled carbon nanotube with nonlocal continuum elasticity by using differential transform method. *Advanced Science Letters*, 3(3): 337–340.

[5] V. Senthilkumar, S.C. Pradhan, and G. Prathap. 2010. Buckling analysis of carbon nanotube based on nonlocal Timoshenko beam theory using differential transform method. *Advanced Science Letters*, 3(4): 415–421.

[6] A. Farajpour, M.H. Ghayesh, and H. Farokhi. 2018. A review on the mechanics of nanostructures. *International Journal of Engineering Science*, 133: 231–263.

[7] J. Peddieson, G.R. Buchanan, and R.P. McNitt. 2003. Application of nonlocal continuum models to nanotechnology. *International Journal of Engineering Science*, 41(3–5): 305–312.

[8] M. Arda. 2021. Axial dynamics of functionally graded Rayleigh-bishop nanorods. *Microsystem Technologies*, 27(1): 269–282.

[9] M. Aydogdu. 2009. Axial vibration of the nanorods with the nonlocal continuum rod model. *Physica E: Low-dimensional Systems and Nanostructures*, 41(5): 861–864.

[10] S. Narendar and S. Gopalakrishnan. 2010. Ultrasonic wave characteristics of nanorods via nonlocal strain gradient models. *Journal of Applied Physics*, 107(8): 084312.

[11] S.K. Jena, S. Chakraverty, and M. Malikan. 2021. Implementation of Haar wavelet, higher order Haar wavelet, and differential quadrature methods on buckling response of strain gradient nonlocal beam embedded in an elastic medium. *Engineering with Computers*, 37(2): 1251–1264.

[12] J.N. Reddy. 2007. Nonlocal theories for bending, buckling and vibration of beams. *International Journal of Engineering Science*, 45(2–8): 288–307.

[13] B. Arash and Q. Wang. 2012. A review on the application of nonlocal elastic models in modeling of carbon nanotubes and graphenes. *Computational Materials Science*, 51(1): 303–313.

[14] S. Chakraverty and L. Behera. 2016. *Static and dynamic problems of nanobeams and nanoplates*. World Scientific, Singapore.

[15] P. Lu, P.Q. Zhang, H.P. Lee, C.M. Wang, and J.N. Reddy. 2007. Non-local elastic plate theories. *Proceedings of the Royal Society A: Mathematical, Physical and Engineering Sciences*, 463(2088): 3225–3240.

[16] H. Thai, T.P. Vo, T. Nguyen, and S. Kim. 2017. A review of continuum mechanics models for size-dependent analysis of beams and plates. *Composite Structures*, 177: 196–219.

[17] C. Wu and H. Hu. 2021. A review of dynamic analyses of single-and multi-layered graphene sheets/nanoplates using various nonlocal continuum mechanics-based plate theories. *Acta Mechanica*, 232(11): 4497–4531.

[18] M.P. Païdoussis. 1998. *Fluid-structure interactions: slender structures and axial flow*, volume 1. Academic Press, California.

[19] S. Oveissi, S.A. Eftekhari, and D. Toghraie. 2016. Longitudinal vibration and instabilities of carbon nanotubes conveying fluid considering size effects of nanoflow and nanostructure. *Physica E: Low-dimensional Systems and Nanostructures*, 83: 164–173.

[20] S. Oveissi and A. Ghassemi. 2018. Longitudinal and transverse wave propagation analysis of stationary and axially moving carbon nanotubes conveying nano-fluid. *Applied Mathematical Modelling*, 60: 460–477.

[21] S. Oveissi, D. Toghraie, and S.A. Eftekhari. 2016. Longitudinal vibration and stability analysis of carbon nanotubes conveying viscous fluid. *Physica E: Low-dimensional Systems and Nanostructures*, 83: 275–283.

[22] S. Oveissi, D.S. Toghraie, and S.A. Eftekhari. 2017. Analysis of transverse vibrational response and instabilities of axially moving cnt conveying fluid. *International Journal of Fluid Mechanics Research*, 44(2): 115–129.

[23] S. Oveissi, D.S. Toghraie, and S.A. Eftekhari. 2018. Investigation on the effect of axially moving carbon nanotube, nanoflow, and Knudsen number on the vibrational behavior of the system. *International Journal of Fluid Mechanics Research*, 45(2): 171–186.

[24] J. Yoon, C.Q. Ru, and A. Mioduchowski. 2005. Vibration and instability of carbon nanotubes conveying fluid. *Composites Science and Technology*, 65(9): 1326–1336.

[25] M.P. Païdoussis and G.X. Li. 1993. Pipes conveying fluid: A model dynamical problem. *Journal of Fluids and Structures*, 7(2): 137–204.

[26] A.I. Skoulidas, D.M. Ackerman, J.K. Johnson, and D.S. Sholl. 2002. Rapid transport of gases in carbon nanotubes. *Physical Review Letters*, 89(18): 185901.

[27] Z. Mao and S.B. Sinnott. 2000. A computational study of molecular diffusion and dynamic flow through carbon nanotubes. *The Journal of Physical Chemistry B*, 104(19): 4618–4624.

[28] M. Aydogdu. 2012. Axial vibration analysis of nanorods (carbon nanotubes) embedded in an elastic medium using nonlocal elasticity. *Mechanics Research Communications*, 43: 34–40.

[29] A.C. Eringen. 1983. On differential equations of nonlocal elasticity and solutions of screw dislocation and surface waves. *Journal of Applied Physics*, 54(9): 4703–4710.

[30] H. Liu, Z. Lv, and Q. Li. 2017. Flexural wave propagation in fluid-conveying carbon nanotubes with system uncertainties. *Microfluidics and Nanofluidics*, 21(8): 1–13.

6 Differential Transformation and Adomian Decomposition Methods for the Radiation Effect on Marangoni Boundary Layer Flow of Carbon Nanotubes

P.K. Ratha, R.S. Tripathy, and S.R. Mishra

CONTENTS

6.1　INTRODUCTION

To improve heat transfer rate, large numbers of strategies are used; low thermal conductivity of traditional fluid inhibits high compression and effectiveness of heat transfer. Enhancing the thermal characteristics of energy transmission fluids might be a way to improve heat transfer. Suspending micro-solid particles in fluids is one of the approaches for the improvement in the thermal conductivities of fluids. Slurries can be made from a variety of powders, including metallic as well as non-metallic along with polymeric particles. The fluids involve solid particles are predicted to have greater thermal conductivities than typical fluids. In recent studies on the flow

DOI: 10.1201/9781003328032-6

phenomenon, it reveals that the heat transfer using nanofluids attains its largest interest because of its enhanced conductivity properties. The term 'nanofluids' was first coined by Choi [1] by suspending nanoparticles in the traditional liquids. Studies show that the thermal conductivity of the base liquids is enhanced by the inclusion of higher conductivity of the nanoparticles like either metals or metal oxides along with carbon nanotubes. Therefore, the performance of heat transfer properties is improved effectively. In several industrial as well as engineering applications like powder energy, vehicle engineering, chemical engineering, aerospace, and biomedical engineering, the use of nanofluid is huge. Xuan [2] discovered that utilizing a suspension of Nano-phase particles and a base liquid, the nanofluid displays a high rate of heat transmission. The particle suspensions have a high significant effect on the nanofluid conductivity, and they also give a comprehensive theory for predicting the conductivity of nanofluids. Sidik et al. [3] consider three ways for preparing the nanofluid without changing its thermophysical properties: sonication, PH-control, and PH-value. Lomascolo et al. [4] reviewed experimental results on the heat transmission capabilities of nanofluids and discussed the effects of various factors. Guo et al. [5] considered buoyancy forces relating to both densities and surface tension showing Marangoni effects depending on temperature on a two-phase model. The proposed model deals with Benard–Marangoni convection mechanism for the two-phase field model, and the modified midpoint method is useful for the solution of the designed model. Lin et al. [6] considered the impact of radiative heat in a steady 2-D Marangoni convection in critical micelle concentration (CMC) nanofluids with base liquid water containing various nanoparticles, including metal like copper (Cu), or oxide like aluminium oxide (Al2O3). However, the surface tension is supposed to change in a linear fashion as the temperature rises. Lin et al. [7] examined the use of nanofluids to provide a constant electrically conducting MHD Marangoni convection bounding surface heat transport through a permeable medium. Jiao et al. [8] investigated the correctness and efficacy of analytical conclusions are validated with earlier solutions, which explored Marangoni convection generated by surface tension, relating to the temperature. On the flow phenomena, the results of the power-law index proposed with temperature and velocity along with permeability, Marangoni number, and heat production parameter are graphically studied.

Nomenclatures

U, V Velocity component along the co-ordinate axes
T_{const} Reference temperature
T Temperature
T_w Wall temperature
T_∞ Ambient temperature
Re   Reynolds number
Ma Marangoni number
Pr   Prandtl number
Nr Thermal radiation
Nu_x Heat transfer rate
q_r Rosseland approximation
C_p Specific heat at constant pressure

Greek Symbols

μ_{nf} Viscosity of carbon nanotubes
ρ_{nf} Nanofluid density
α_{nf} Nanofluid thermal diffusivity
k_{nf} nanofluid thermal conductivity
γ_{nf} Kinematic viscosity
k_f Thermal conductivity of water
ρ_f Density of the water
k_s Thermal conductivity of carbon nanotubes
σ^* Stefan-Boltzmann constant
k^* Mean absorption
σ Surface tension
σ_0 Positive constant

Subscripts

nf Nanoliquid
f Base liquid
s Nanoparticles

Al-Sharafi et al. [9] obtained the droplets created using the carrier fluid, which is a combination of *water* and *carbon nanotubes* (CNT) are taken into account and simulations are run to study fluid flow numerically inside the droplets, as well as the impacts of both convections on the flow field. Hayat et al. [10] demonstrated the Marangoni convection behavior in a CNT–water nanofluid flow and the thermal radiation is also taken into account. The homotopy analysis approach is used to obtain convergent series solutions for the governing equation and n numerous relevant factors and their effects on the Nusselt number are explored. Tiwari et al. [11] investigated radiation along the two types of carbon nanotubes; the Marangoni convection of CNTs within a permeable medium was examined. Using similarity transformations, the resorted governing set of equations are handled numerically for the various parameters were examined using the graphical presentation. Mahanthesh et al. [12] developed a model on the impact of magnetic force and addition of the Marangoni transport of dissipating single and multi-walled carbon nanotubes nanofluids was constructed. To provide a numerical solution, a conventional Runge-Kutta-based shooting procedure is used. Through graphical outputs, the impact graphs of relevant physical factors are detected on rate coefficients and other profiles. Gul et al. [13] consider the CNT blood-based nanofluid in detail using the results of the Marangoni convection against a thin liquid film of Casson fluid for the inclusion of magnetic field. Through graphical depiction, the impact of various model restrictions on flow field, temperature, and other engineering coefficients is examined. Rehman et al. [14] consider the liquid film spray of CNT nanofluid across an unstable expanding surface of a cylinder, the Marangoni convection is considered. Using the optimum homotopy analysis approach, the altered equations of the flow issue were solved (OHAM). The numerical and visual validation of the acquired results for both types of nanofluids was done using the sum of total residual errors. Palanisamy [15] worked on the energy transport and pressure drop of a helically coiled tube heat transfer utilizing *multi-walled carbon nanotubes* (MWCNT) nanofluids, and this experimental study focuses on the thermal and flow behaviour of MWCNT nanofluids in a helically cone coiled tube heat exchanger. Ahmed [16] studied a viscid nanofluid with changing viscosity and a magnetic field through a contracting permeable surface. To represent the flow issue, the boundary-layer approximation is used employing *Navier–Stokes equations* and energy equation; these are solved by the Keller box approach that is used to identify the governing equations' features. The velocity, velocity gradient, and skin friction coefficient graphs have been presented. Anuar et al. [17] investigate the carbon nanotubes and the effects of buoyancy on heat transfer across a permeable vertical plate where the role of suction is vital. There are two types of base liquids used, *single-walled carbon nanotubes* (SWCNT) and *multi-walled carbon nanotubes* (MWCNT). The impact of the characterizing parameters along with particle concentration on the various profiles is graphically analysed.

Following the aforesaid literatures, it aims to investigate the influence of radiative heat transfer on the flow of nanofluid for the inclusion of Marangoni boundary conditions. Both the SWCNT and MWCNT nanoparticles are undertaken in the base liquid water for the preparation of nanofluids. The novelty arises due to the conjunction of Hamilton–Crosser conductivity model for the enrichment in the heat transfer criterion. The approximate analytical techniques such as *Differential Transform Method* (DTM) and *Adomian Decomposition Method* (ADM) are useful to solve the transformed governing equations and the validation with the numerical solution is employed.

6.2   MATHEMATICAL MODEL

A two-dimensional tie-independent carbon nanotube-water nanofluid with Marangoni convection over a flat surface for the inclusion of surface tension gradient in the form of exponential temperature is analysed in the current investigation. Here considering the Cartesian co-ordinate system (X, Y), where X is placed along the plate and Y is placed normal to it, respectively. The governing equations for the flow phenomena are

$$\frac{\partial U}{\partial X} + \frac{\partial V}{\partial Y} = 0 \tag{6.1}$$

$$U\frac{\partial U}{\partial X} + V\frac{\partial U}{\partial Y} = \frac{\mu_{nf}}{\rho_{nf}}\frac{\partial^2 U}{\partial Y^2} \tag{6.2}$$

$$U\frac{\partial T}{\partial X} + V\frac{\partial T}{\partial Y} = \alpha_{nf}\frac{\partial^2 T}{\partial Y^2} - \frac{1}{\left(\rho C_p\right)_{nf}}\frac{\partial\left(q_r\right)}{\partial Y} \tag{6.3}$$

The boundary conditions are

$$\mu_{nf}\left(\frac{\partial U}{\partial Y}\right)_{Y=0} = \left(\frac{\partial \sigma}{\partial X}\right)_{Y=0} = 0, (V)_{Y=0} = 0, (T)_{Y=0} = T_w = T_\infty + T_{const} e^{\frac{-X}{L_0}} \quad as \ Y = 0$$

$$(U)_{Y\to\infty} = 0, \ (T)_{Y\to\infty} = T_\infty \qquad\qquad\qquad as \ Y \to \infty$$

$$(6.4)$$

where (U,V) deployed for the velocity components of the coordinate axes (X,Y) and μ_{nf}, the velocity, ρ_{nf}, the density, and T, the temperature of the nanofluid.

Where ϕ, the nanoparticle concentration, and following [18–22] the physical parameters are:

$$\mu_{nf} = \left(\frac{1}{1-2.5(\phi)}\right)\mu_f = \left(1+2.5\phi+6.25\phi^2 +O\left(\phi^3\right)\right)\mu_f \qquad (6.5)$$

$$\rho_{nf}/\rho_f = 1-\phi+\phi\rho_s/\rho_f \left(\rho C_p\right)_{nf} /\left(\rho C_p\right)_f = 1-\phi+\phi\left(\rho C_p\right)_s /\left(\rho C_p\right)_f \qquad (6.6)$$

$$\alpha_{nf} = \frac{k_{nf}}{\left(\rho C_p\right)_{nf}}, \ \upsilon_{nf} = \frac{\mu_{nf}}{\rho_{nf}}, \qquad (6.7)$$

where ρ_f, density, $\left(\rho C_p\right)_f$, heat capacity of water, as well as ρ_s, density, $\left(\rho C_p\right)_s$, the heat capacity of carbon nanotubes and the nanoparticle shape of the thermal conductivity of carbon nanotubes are assumed by considering the *Hamilton–Crosser* conductivity model [23–31]

$$k_{nf} = \frac{\left(k_s +(m-1)k_f\right)-(m-1)\phi\left(k_f - k_s\right)}{\left(k_s +(m-1)k_f\right)+\phi\left(k_f - k_s\right)} k_f \qquad (6.8)$$

Here, k_f, k_s are the thermal conductivity (water and carbon nanotubes). The radiative heat flux employing the Rosseland approximation is

$$q_r = -\frac{4\sigma^*}{3k^*}\frac{\partial T^4}{\partial Y} \qquad (6.9)$$

where σ^* and k^* denote the *Stefan–Boltzmann constant* and mean absorption, respectively. Considering only up to the first order after expanding T^4 using Taylor series about T_∞, it becomes,

$$T^4 \approx 4T_\infty^3 T - 3T_\infty^4 \qquad (6.10)$$

In Eqn. (6.4), σ is expressed as the surface tension and, therefore, the temperature gradient near the interface is

$$\frac{\partial \sigma}{\partial X} = \frac{\partial \sigma}{\partial T} \cdot \frac{\partial T}{\partial X} \tag{6.11}$$

and it is treated as a linear with respect to temperature, i.e.,

$$\sigma = \sigma_0 - \gamma_T (T - T_\infty), \gamma_T = -\frac{\partial \sigma}{\partial T}, \tag{6.12}$$

where σ_0 is the positive constant when $T = T_\infty$ and γ_T coefficient of the surface tension.

Considering U_0 the unit velocity and L_0 the unit length, the variables in dimensionless form are as follows:

$$u = \frac{U}{U_0}, \quad v = \frac{V}{U_0} \left(\frac{U_0 L_0}{\gamma_f} \right)^{\frac{1}{2}} = \frac{V}{U_0} (\mathrm{Re})^{\frac{1}{2}}, \tag{6.13}$$

$$x = \frac{X}{L_0}, \quad y = \frac{Y}{L_0} \left(\frac{U_0 L_0}{\gamma_f} \right)^{\frac{1}{2}} = \frac{Y}{L_0} (\mathrm{Re})^{\frac{1}{2}}, \tag{6.14}$$

$$t = \frac{T}{T_\infty}, \tag{6.15}$$

$$a = (1 - 2.5\phi) \left[(1 - \phi) + \phi \frac{\rho_s}{\rho_f} \right], \tag{6.16}$$

$$b = \left((1 - \phi) + \phi \frac{(\rho C_p)_s}{(\rho C_p)_f} \right) \left[\frac{(k_s + (m-1)k_f) + \phi(k_f - k_s)}{(k_s + (m-1)k_f) - (m-1)\phi(k_f - k_s)} \right], \tag{6.17}$$

$$c = (1 - 2.5\phi), \tag{6.18}$$

$$\mathrm{Re} = \frac{U_0 L_0}{\gamma_f}, \mathrm{Pr} = \frac{\gamma_f}{\alpha_f}, \gamma_f = \frac{\mu_f}{\rho_f}, Ma = \frac{\gamma_T T_{const} L_0}{\mu_f \alpha_f}, Nr = \frac{16\sigma^* T_\infty^3}{3k^* k_{nf}} \tag{6.19}$$

Here, Re is the Reynolds number, Pr the Prandtl number, Nr the Radiation, and Ma the Marangoni parameter. Equations (6.2)–(6.4) can be transformed into

$$\frac{\partial u}{\partial x} + \frac{\partial v}{\partial y} = 0 \tag{6.20}$$

$$u\frac{\partial u}{\partial x}+v\frac{\partial u}{\partial y}=\frac{1}{a}\left(\frac{\partial^2 u}{\partial y^2}\right) \tag{6.21}$$

$$u\frac{\partial t}{\partial x}+v\frac{\partial t}{\partial y}=\left(\frac{1+Nr}{b\,Pr}\right)\frac{\partial^2 t}{\partial y^2} \tag{6.22}$$

The boundary condition can be expressed as:

$$\frac{T_{const}}{T_\infty}\left(\frac{\partial u}{\partial y}\right)_{y=0}=-\frac{Ma}{Pr}(Re)^{-\frac{3}{2}}c\left(\frac{\partial t}{\partial x}\right)_{y=0}, \quad v=0,\,(t)_{y=0}=1+\frac{T_{const}}{T_\infty}e^{-x} \quad as\ y=0$$

$$(u)_{y\to\infty}=0,\,(t)_{y\to\infty}=1 \qquad as\ y\to\infty$$

$$\tag{6.23}$$

Adopting the standard definition $u=\dfrac{\partial \psi}{\partial y}$ and $v=-\dfrac{\partial \psi}{\partial x}$, the following transformation variables are written in the form of:

$$\psi(x,y)=Fe^{\frac{-x}{3}}f(\eta),\quad \eta=Fe^{\frac{-x}{3}}y,\quad F=\left(\frac{Ma}{Pr}\right)^{\frac{1}{3}}(Re)^{-\frac{1}{2}},\quad t(x,y)=1+\frac{T_{const}}{T_\infty}e^{-x}\theta(\eta)$$

$$\tag{6.24}$$

Therefore, Eqns. (6.21) and (6.22) with Eqn. (6.23) are described as:

$$f'''(\eta)=a\left[\left(\frac{1}{3}\right)f(\eta)f''(\eta)-\left(\frac{2}{3}\right)f'(\eta)^2\right], \tag{6.25}$$

$$\theta''(\eta)=\frac{b\,Pr}{1+Nr}\left\{\left(\frac{1}{3}\right)f(\eta)\theta'(\eta)-f'(\eta)\theta(\eta)\right\}, \tag{6.26}$$

$$\left.\begin{array}{l}f(0)=0,\ f''(0)=c,\ f'(\infty)=0\\[4pt]\theta(0)=1,\theta(\infty)=0\end{array}\right\} \tag{6.27}$$

The X and Y components are

$$U=U_0\left(\frac{Ma}{Pr}\right)^{\frac{2}{3}}\frac{1}{Re}e^{\frac{-2X}{3L_0}}f'(\eta)=\left\{\frac{\gamma_T T_{const}\gamma_f}{\rho_f L_0^2}\right\}^{\frac{2}{3}}\frac{Re}{U_0}e^{\frac{-2X}{3L_0}}f'(\eta) \tag{6.28}$$

$$V=\frac{1}{3}\left\{\frac{\gamma_T T_{const}\gamma_f}{\rho_f L_0^2}\right\}^{\frac{1}{3}}e^{-\frac{X}{3L_0}}(f(\eta)+\eta f'(\eta)) \tag{6.29}$$

$$T = T_\infty + T_{const} e^{-\frac{X}{L_0}} \theta(\eta) \tag{6.30}$$

A quantity of interest is the local Nusselt number Nu_x and is defined as:

$$Nu_x = \frac{X q_w(X)}{k(T)\left[T(X,0) - T(X,\infty)\right]}, \tag{6.31}$$

where $q_w(X)$ is the heat flux as $q_w(X) = -k\{T\}\left(\dfrac{\partial T}{\partial Y}\right)_{Y=0}$. Using Eqns. (6.28)–(6.30), Eqn. (6.31) becomes

$$Nu_x = -\frac{X}{L_0} e^{-\frac{X}{3L_0}} \left\{ \frac{\gamma_T^2 T_{const}^2}{\rho_f^2 U_0^3 L_0 \gamma_f} \right\}^{\frac{1}{6}} \theta'(0) \tag{6.32}$$

So, $-\theta'(0)$ presents the local Nusselt number Nu_x.

6.3 METHODOLOGY

6.3.1 DTM Approximations

Zhou [34], in 1986, introduced the concept of differential transform for the solution of boundary value problems by using a semi-analytical/an approximate analytical technique. The proposed method, i.e., the DTM is based on Taylor series expansion in which the differential equation is converted into a recurrence relation to perform a series in polynomials. First of all, one-dimensional transform method is used to solve IVPs as well as BVPs for ordinary differential equations.

For the analytical function $f(x)$, the differential transform is

$$F(k) = \frac{1}{k!}\left[\frac{d^k f(x)}{dx^k}\right]_{x=0} \tag{6.33}$$

Here, $f(x)$ is the considered function and the transformed function is $F(k)$.

However, the inverse transform of $F(k)$ is

$$f(x) = \sum_{k=0}^{\infty} F(k)(x - x_0)^k \tag{6.34}$$

By substituting Eqn. (6.33) in Eqn. (6.34),

$$f(x) = \sum_{k=0}^{\infty} \frac{(x - x_0)^k}{k!}\left(\frac{d^k}{dx^k} f(x)\right)_{x=x_0} \tag{6.35}$$

By taking the differential transforms of Eqns. (6.25)–(6.27), we get

$$
(k+3)(k+2)(k+1)F[k+3] = \frac{a}{3}\sum_{r=0}^{k}(k-r+1)(k-r+2)F[r]F[k-r+2]
$$

$$
\left.\begin{array}{r} -\frac{2a}{3}\sum_{r=0}^{k}(r+1)(k-r+1)F[r+1]F[k-r+1] = 0, \end{array}\right\}
$$

(6.36)

$$
(k+2)(k+1)\Theta[k+2] = \frac{b\,\text{Pr}}{1+Nr}\left(\left(\begin{array}{c} \dfrac{1}{3}\sum_{r=0}^{k}(k-r+1)F[r]\Theta[k-r+1] \\[2mm] -\sum_{r=0}^{k}(k-r+1)\Theta[r]F[k-r+1] \end{array}\right)\right).
$$

(6.37)

where $f(t)$ and $\theta(t)$ represent the inverse transforms of $F(k)$ and $\Theta(k)$, respectively. Also, the boundary conditions are transformed into

$$
F[0] = 0, F[1] = A, F[2] = c, \Theta[0] = 1, \Theta[1] = B. \tag{6.38}
$$

The recursive values are obtained using $k = 0,1,2,--- = 0$.

The numerical treatment is useful for the evaluation of the unknowns A and B.

6.3.2   ADOMIAN DECOMPOSITION METHOD (ADM)

The coupled nonlinear ODEs (6.25) and (6.26) with (6.27) are also solved using Adomian Decomposition Method (ADM). The procedure for the said problem is described in details as follows:

To re-write Eqns. (6.25) and (6.26), let us introduce the operators $\Re_1 = \dfrac{d^3}{d\eta^3}(\bullet)$

and $\Re_2 = \dfrac{d^2}{d\eta^2}(\bullet)$ with inverse operators $\Re_1^{-1}(\bullet) = \displaystyle\int_{-1}^{\eta}\int_{-1}^{\eta}\int_{-1}^{\eta}(\bullet)d\eta\,d\eta\,d\eta$ and

$\Re_2^{-1}(\bullet) = \displaystyle\int_{-1}^{\eta}\int_{-1}^{\eta}(\bullet)d\eta\,d\eta$.

Therefore, Eqns. (6.25) and (6.26) can be expressed as:

$$
\Re_1(f) = a\left(\frac{1}{3}ff_{\eta\eta} - \frac{2}{3}f_{\eta}^{2}\right) \tag{6.39}
$$

$$
\Re_2(\theta) = \frac{b\,\text{Pr}}{1+Nr}\left(\frac{1}{3}f\theta_{\eta} - f_{\eta}\theta\right) \tag{6.40}
$$

Hence, taking the inverse of above equations, we get

$$f(\eta) = \Re_1^{-1}\left(a\left(\frac{1}{3}ff_{\eta\eta} - \frac{2}{3}f_\eta^2\right)\right) \tag{6.41}$$

$$\theta(\eta) = \Re_2^{-1}\left(\frac{b\,\mathrm{Pr}}{1+Nr}\left(\frac{1}{3}f\theta_\eta - f_\eta\theta\right)\right) \tag{6.42}$$

The linear as well as nonlinear terms of RHS of these equations are assumed in the form of a series each such as,

$$\sum_{m=0}^{\infty}A_m = f,\ \sum_{m=0}^{\infty}B_m = f_\eta,\ \sum_{m=0}^{\infty}C_m = f_{\eta\eta},\ \sum_{m=0}^{\infty}D_m = f\theta_\eta,\ \sum_{m=0}^{\infty}E_m = f_\eta\theta$$

Introducing the initial conditions

$$f(0) = 0, f_\eta(0) = p, f_{\eta\eta}(0) = c, \theta(0) = 1, \theta_\eta(0) = q$$

The initial guess solutions as well as recursive expression are as follows:

$$f_0(\eta) = p\eta + \frac{c}{2}\eta^2 \tag{6.43}$$

$$\theta_0(\eta) = 1 + q\eta \tag{6.44}$$

$$f_{k+1}(\eta) = \Re_1^{-1}\left(a\left(\frac{1}{3}A_k - \frac{2}{3}B_k\right)\right) \tag{6.45}$$

$$\theta_{k+1}(\eta) = \Re_2^{-1}\left(\frac{b\,\mathrm{Pr}}{1+Nr}\left(\frac{1}{3}C_k - D_k\right)\right) \tag{6.46}$$

For $k = 0, 1, 2, \ldots$ and evaluating the unknowns numerically, we get the approximate analytical solutions and the solutions for velocity and temperature are obtained.

6.4   RESULTS AND DISCUSSION

Two-dimensional flow of nanofluid with Marangoni boundary condition over a flat heat surface is investigated in the current literature. The impact of thermal radiation enriches the flow and heat transport properties. The enrichment in the flow characteristics is due to the interaction of Hamilton–Crosser conductivity model. The novelty arises due to the involvement of carbon nanotube for both SWCNT and MWCNT with water as the base liquid (Thermophysical properties are presented in Table 6.1). The analysis of the contributing parameters is obtained by solving the

TABLE 6.1

Physical Properties of the Regular Fluid, Nanoparticle, SWCNTs, and MWCNTs

Physical Properties	SWCNTs	MWCNTs	Water
$C_p(J)(Kg\ K)^{-1}$	425	796	4179
$\rho(Kg)(m^3)^{-1}$	2600	1600	997.1
$\kappa(W/mK)$	6600	3000	0.613
$\sigma(S/m)$	10^6 to 10^7	10^6 to 10^7	5.5×10^{-6}

Source: From Upreti et al. [32, 33].

modelled equations using approximate analytical technique for the specified values of the parameters. The fundamentals of DTM are listed in Table 6.2. These are the volume concentration, surface condition parameter, thermal radiation, and Prandtl number. Figure 6.1 illustrates the behaviour of the particle concentration on the velocity distribution. For $\phi = 0$, it gives rise to the case of pure fluid, whereas the non-zero values indicate the role of SWCNT and WCNT-H2O nanofluids. The volume concentration is the amount of particles submerged into the base liquid. The heavier density of the particles lowers down the velocity profiles and this is due to the clogging of the particles near the sheet region. The observation reveals that for the case of pure fluid, the profile augments with higher magnitude. Figure 6.2 is deployed for the variation of the surface condition parameter on the fluid velocity. The result indicates that the attenuation in the velocity is due to the continuous enhancement in the parametric growth. The role of the nanoparticles on the fluid temperature distribution

TABLE 6.2

Transformation Rules

Standard Function	Standard Transformed
$f(x) = u(x) \pm v(x)$	$F(t) = U(t) \pm V(t)$
$f(x) = \lambda u(x)$	$F(t) = \lambda U(t)$
$f(x) = x^m$	$F(t) = \sigma(t - m)$
$f(x) = \dfrac{d^n u(x)}{dx^n}$	where if $t = m$, $\sigma(t - m) = 1$ and 0 otherwise $F(t) = (t + 1)(t + 1).....(t + n)F(t + n)$
$f(x) = u(x)v(x)$	$F(t) = \displaystyle\sum_{n=0}^{t} U(n)V(t - n)$
$f(x) = u'(x)v'(x)$	$F(t) = \displaystyle\sum_{n=0}^{t} (n + 1)(t - n + 1)U(n + 1)V(t - n + 1)$
$f(x) = u(x)v'(x)$	$F(t) = \displaystyle\sum_{n=0}^{t} (t - n + 1)U(n)V(t - n + 1)$

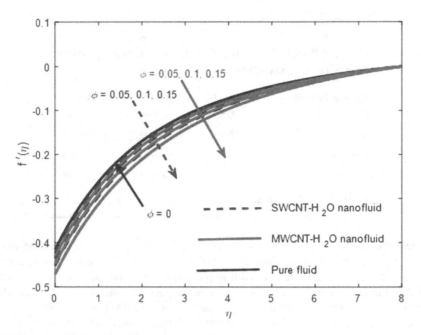

FIGURE 6.1   Variation of particle concentration on fluid velocity.

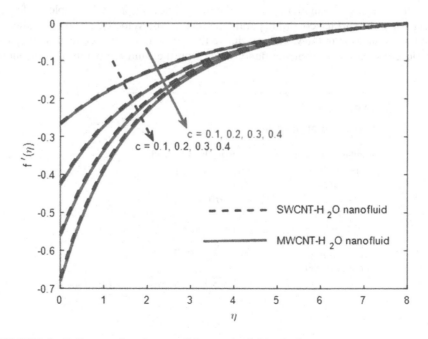

FIGURE 6.2   Influence of surface condition on the fluid velocity.

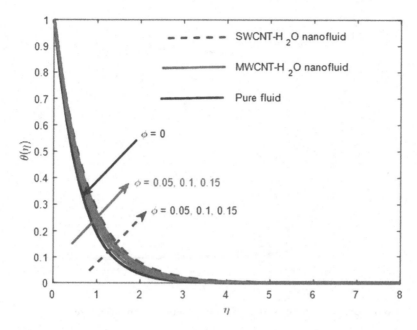

FIGURE 6.3   Variation of particle concentration on the fluid temperature.

is important since the thermophysical property depends upon the conductivity of the particles and that is exhibited in Figure 6.3. It reveals that the augmentation in the particle concentration enriches the fluid temperature for both the case of SWCNT and MWCNT-water nanofluids in comparison to the base fluid. The amount of particles deposited in the sheet region encourages the increase in the fluid temperature. In a comparative study, it was concluded that the case of SWCNT-water nanofluid exhibits higher temperature throughout the domain. Figure 6.4 exhibits the impact of surface condition parameter on the fluid temperature and result shows a smooth retardation in the entire domain. Figure 6.5 portrays the behaviour of thermal radiation for the variation in fluid temperature. Thermal radiation measures the amount of emission of the electromagnetic radiation from the fluid element. This transforms the heat radiation to the electromagnetic radiation. Furthermore, the augmented radiation encourages the fluid temperature to boost up and, therefore, the temperature profile enhances. The nature of the profile is asymptotic to meet the requisite boundary condition. Figure 6.6 renders the significant contribution of the Prandtl number on the fluid temperature. Prandtl number is expressed as the ratio of the kinematic viscosity with the thermal diffusivity. An increase in Pr suggests a strong deceleration in the diffusion and it favours to control the fluid temperature significantly. Figure 6.7 elaborates the behaviour of shape factor on the fluid temperature. Here, various shapes like spherical, hexahedron, tetrahedron, column, and lamina are presented in this figure. The observation shows that the shape of the particles is in the order of increasing as spherical < hexahedron < tetrahedron < column < lamina (as mentioned in the figure). With this it is seen that as shape increases the fluid temperature

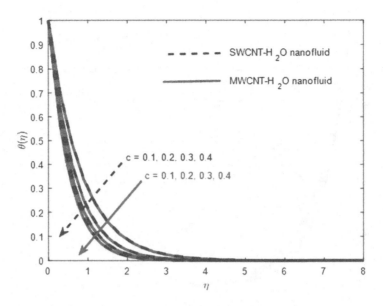

FIGURE 6.4   Behaviour of the surface condition on the fluid temperature.

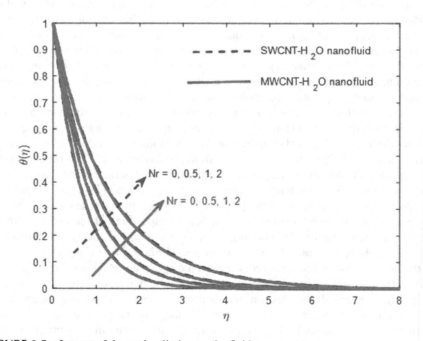

FIGURE 6.5   Impact of thermal radiation on the fluid temperature.

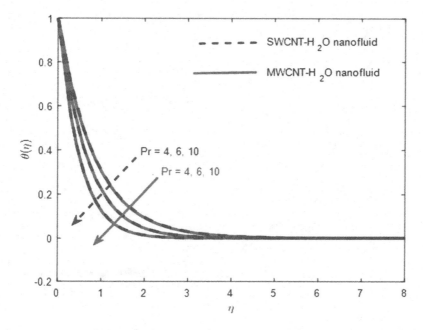

FIGURE 6.6   Impact of Prandtl number on the fluid temperature.

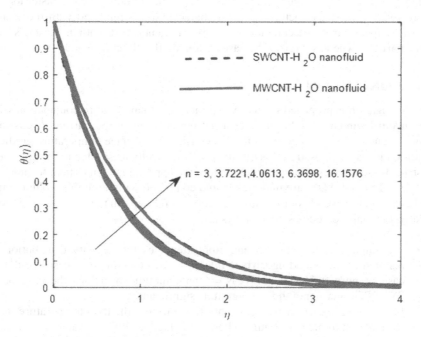

FIGURE 6.7   Impact of shape factor on the fluid temperature.

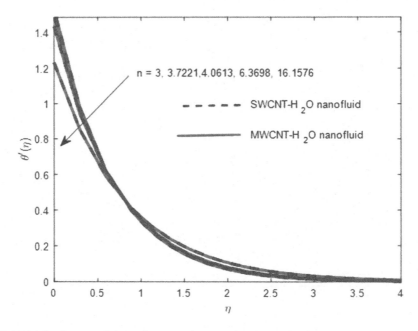

FIGURE 6.8   Impact of shape factor on the rate of heat transfer.

enhances subsequently. Figure 6.8 portrays the profiles of rate of heat transfer for the various values of shape factor. The dual characteristic is rendered with increasing particle shape. Near the sheet region, the rate coefficient retards with increasing shape of the particles whereas with an increase in domain the effect reversed.

6.5   CONCLUSION

An investigation is proposed for the two-dimensional nanofluid in conjunction with Marangoni boundary conditions over a flat hot surface. In the current discussion, SWCNT and MWCNT-water nanofluid is considered. Surface temperature gradient condition is also proposed in this discussion. The novelty arises due to the consideration of the shape fact of nanoparticles. Both the approximate analytical techniques such as DTM and ADM are undertaken and compared with numerical solution and the behaviour of several components are deliberated briefly in the discussion section. Finally, the major outcomes are displays as:

- The enhanced particle concentration attenuates the velocity distribution resulted in the bounding surface thickness increases in comparison to the case of pure fluid. The MWCNT-water nanofluid overrides the case of SWCNT-water nanofluid in magnitude significantly.
- Inclusion of higher-particle concentration enriches the fluid temperature in comparison to the base liquid, whereas the fact of SWCNT-water nanofluid enhances more in magnitude.

- The Marangoni boundary condition ceases to decelerate the bounding surface thickness of the temperature profiles.
- The radiative heat energy and increase in particle shape correspondingly for spherical, hexahedron, tetrahedron, column, and lamina augments the nanofluid temperature irrespective of the type of particles.
- The rate of heat transfer attenuates smoothly for the increasing in shape near the surface region, and it behaves in reverse direction as the boundary domain increases.

REFERENCES

[1] S.U.S. Choi. Enhancing thermal conductivity of fluids with nanoparticles. *ASME Publ. Fed.*, 231 (1995): 99–106.

[2] Y.M. Xuan and Q. Li. Heat transfer enhancement of nanofluids. *Int. J. Heat Fluid Flow*, 21 (2000): 58–64.

[3] N.A.C. Sidik, H.A. Mohammed, O.A. Alawi, and S. Samion. A review on preparation methods and challenges of nanofluids. *Int. Commun. Heat Mass Transfer*, 54 (2014): 115–125.

[4] M. Lomascolo, G. Colangelo, M. Milanese, and A.D. Risi. Review of heat transfer in nanofluids: Conductive, convective and radiative experimental results. *Renew Sust. Energ. Rev.*, 43 (2015): 1182–1198.

[5] Z.L. Guo, P. Lin, and Y.F. Wang. Continuous finite element schemes for a phase model in two-layer fluid Bénard-Marangoni convection computations. *Comput. Phys. Commun.*, 185 (2014): 63–78.

[6] Y.H. Lin, L.C. Zheng, and X.X. Zhang. Radiation effects on Marangoni convection flow and Heat transfer in pseudo-plastic non-Newtonian nanofluids with variable thermal conductivity. *Int. J. Heat Mass Transf.*, 77 (2014): 708–716.

[7] Y.H. Lin, L.C. Zheng, and X.X. Zhang. MHD Marangoni boundary layer flow and heat transfer of pseudo-plastic nanofluids over a porous medium with a modified model. *Mech. Time-Depend. Mater.*, 19 (2015): 519–536.

[8] C.R. Jiao, L.C. Zheng, Y.H. Lin, L.X. Ma, and G. Chen. Marangoni abnormal convection heat transfer of power-law fluid driven by temperature gradient in porous medium with heat generation. *Int. J. Heat Mass Transf.*, 92 (2016): 700–707.

[9] A. Al-Sharafi, A.Z. Sahin, B.S. Yilbas, and S.Z. Shuja. Marangoni convection flow and heat transfer characteristics of water–CNT nanofluid droplets. *Numer. Heat Transfer, Part A*, 69 (7) (2016): 763–780.

[10] T. Hayat et al. Impact of Marangoni convection in the flow of carbon–water nanofluid with thermal radiation. *Int. J. Heat Mass Transfer*, 106 (2017): 810–815.

[11] A.K. Tiwari, F. Raza, A.K. Jahan. Mathematical model for marangoni convection MHD flow of carbon nanotubes through a porous medium, *IAETSD J. Adv. Res. Appl. Sci.*, 4 (7) (2017): 216–222.

[12] B. Mahanthesh, B.J. Gireesha, N.S. Shashikumar, and S.A. Shehzad. Marangoni convective MHD flow of SWCNT and MWCNT nano liquids due to a disk with solar radiation and irregular heat source. *Physica E.*, 94 (2017): 25–30.

[13] T. Gul, R. Akbar, Z. Zaheer, and I.S. Amiri. The impact of the Marangoni convection and magnetic field versus blood-based carbon nanotube nanofluids. *Proc. Inst. Mech.*, 234 (1–2) (2020): 37–46.

[14] A. Rehman, T. Gul, Z. Salleh, S. Mukhtar, F. Hussain, K.S. Nisar, and P. Kumam. Effect of the Marangoni convection in the unsteady thin film spray of CNT nanofluids. *Processes*, 7 (2019): 392.

[15] K. Palanisamy and P.C. Mukesh Kumar. Experimental investigation on convective heat transfer and pressure drop of cone helically coiled tube heat exchanger using carbon nanotubes/water nanofluids. *Heliyon*, 5 (5) (2019): e01705.

[16] Z. Ahmed, S. Nadeem, S. Saleem, and R. Ellahi. Numerical study of unsteady flow and heat transfer CNT-based MHD nanofluid with variable viscosity over a permeable shrinking surface. *Int. J. Numer. Methods Heat Fluid Flow*, 29 (12) (2019): 4607–4623.

[17] N.S. Anuar, N. Bachok, N. MdArifin, and H. Rosali. Role of multiple solutions in flow of nanofluids with carbon nanotubes over a vertical permeable moving plate. *Alexandria Eng. J.*, 59 (2) (2020): 763–773.

[18] A. Malvandi, F. Hedayati, and D.D. Ganji. Slip effects on unsteady stagnation point flow of a nanofluid over a stretching sheet. *Powder Technol.*, 253 (2014): 377–384.

[19] M. Sheikholeslami, D.D. Ganji, and H.R. Ashorynejad. Investigation of squeezing unsteady nanofluid flow using ADM. *Powder Technol.*, 239 (2013): 259–265.

[20] M. Sheikholeslami, M. Gorji-Bandpy, D.D. Ganji, and S. Soleimani. Effect of a magnetic field on natural convection in an inclined half-annulus enclosure filled with Cuwater nanofluid using CVFEM. *Adv. Powder Technol.*, 24 (2013): 980–991.

[21] E. Sourtiji, M. Gorji-Bandpy, D.D. Ganji, and S.F. Hosseinizadeh. Numerical analysis of mixed convection heat transfer of Al2O3-water nanofluid in a ventilated cavity considering different positions of the outlet port. *Powder Technol.*, 262 (2014): 71–81.

[22] E. Sourtiji, D.D. Ganji, and S.M. Seyyedi. Free convection heat transfer and fluid flow of Cu-water nanofluids inside a triangular-cylindrical annulus. *Powder Technol.*, 277 (2015): 1–10.

[23] H.U. Kang, S.H. Kim, and J.M. Oh. Estimation of thermal conductivity of nanofluid using experimental effective particle volume. *Exp. Heat Transfer*, 19 (2006): 181–191.

[24] Q. Li, Y.M. Xuan, and J. Wang. Experimental investigations on transport properties of magnetic fluids. *Exp. Thermal Fluid Sci.*, 30 (2005): 109–116.

[25] Q. Li and Y.M. Xuan. Enhanced heat transfer behaviors of new heat carrier for spacecraft thermal management. *J. Spacecr. Rocket*, 43 (3) (2006): 687–689.

[26] F.M. Abbasi, T. Hayat, F. Alsaadi, A.M. Dobai, and H.J. Gao. MHD peristaltic transport of spherical and cylindrical magneto-nanoparticles suspended in water. *AIP Adv.*, 5 (077104) (2015): 1–12.

[27] N.S. Akbar and M.T. Mustafa. Ferromagnetic effects for nanofluid venture through composite permeable stenosed arteries with different nanosize particles. *AIP Adv.*, 5 (077102) (2015): 1–9.

[28] N.S. Akbar. A new thermal conductivity model with shaped factor ferromagnetism nanoparticles study for the blood flow in non-tapered stenosed arteries. *IEEE Trans. Nanobiosci.*, 14 (7) (2015): 780–789.

[29] N.S. Akbar. Biofluidics study in digestive system with thermal conductivity of shape nanosize H2O + Cu nanoparticles. *J. Bionic Eng.*, 12 (2015): 656–663.

[30] N.S. Akbar. Ferromagnetic CNT suspended H2O + Cu nanofluid analysis through composite stenosed arteries with permeable wall. *Phys. E.*, 72 (2015): 70–76.

[31] N.S. Akbar and A.W. Butt. Ferromagnetic effects for peristaltic flow of Cu-water nanofluid for different shapes of nanosize particles. *Appl. Nanosci.*, 6 (2016): 379–385.

[32] H. Upreti, K.S. Rawat, and M. Kumar. Radiation and non-uniform heat sink/source effects on 2D MHD flow of CNTs-H_2O nanofluid over a flat porous plate. *Multidisp. Model. Mater. Struct.*, 16 (4) (2019): 791–809.

[33] H. Upreti, A.K. Pandey, M. Kumar, and O.D. Makinde. Ohmic heating and non-uniform heat source/sink roles on 3D Darcy-Forchheimer flow of CNTs nanofluids over a stretching surface. *Arab. J. Sci. Eng.*, 45 (2020): 7705–7717.

[34] J.K. Zhou. *Differential Transformation and Its Applications for Electrical Circuits*. Wuhan: Huazhong University Press (1986).

7 Min-Max Game Theory for Coupled Partial Differential Equation Systems in Fluid Structure

S. M. Chithra

CONTENTS

DOI: 10.1201/9781003328032-7

7.1   INTRODUCTION

Game theory is widely regarded as a rapidly developing branch in applied mathematics and is applied to a broad variety of fields of study, which include economy, public administration, national guard scientific knowledge, neuroscience, and ecosystems. In this chapter, we take a look at the min-max game issue that occurs in fluid dynamic interactions and is devised for couples PDE systems. Also, we take into account two vastly distinct designs for the framework: adjustable strap and visco-elastic. Although the first elastic prototype is prominent in engineering in which the solid is engrossed in a fluid, the initial visco-elastic model is commonly used during biomedical activities where the framework displays some visco-elastic reaction. This section's primary objective is to perceive game theory findings in the context of a highly nonlinear fluid dynamic model theory with direct authority and disruption trying to act just at the interaction. It turns out that systemic object's stretchy or visco-elastic reaction has huge ramifications for the asserted final match of mathematical formalism outcomes.

7.2   GAME THEORY STRUCTURE

Let $p = \{p_1, \ldots p_n\}$ falsely claim we're in a game. We'll $w = w_1 \times w_2$ presume it's a distinct strategic $p_i \in p$ planning scope. That is, we could indeed assign a distinct range of strategies w_i to every actor.

7.2.1   Definition: Normal Form

Let p be a team of players, $w = w_i \times \ldots \times w_n$ a strategic planning space and $y : w \to \Re^n$ a strategic planning kickback feature. The quadruple $G = \{p, w, y\}$ is a basic game.

Remark: If $G = \{p, w, y\}$ and the match is of the normalized form $y_i : w \to \Re$, the role is the risk-reward role for the golfer p_i and i^{th} gets back the portion of the function of y.

7.2.2   Definition: Constant/General Sum Game

Let $G = \{p, w, y\}$ plays a regular game. A continual zero sums are $C \in \Re$, where there would be a fixed value $(\sigma_1, \ldots \sigma_n) \in w$ such that with all item sets we have $\sum_{i=1}^{n} y_i(\sigma_1, \ldots \sigma_n) = C$. Public sum refers to any game G which is not a continual sum.

7.2.2.1   Strategic Form Games

The $m \times n$ data structure is a data sequence of numbers with values $\Re$. The matrix $A \in \Re^{m \times n}$ is made up of multiple rows. The component in the ith rows or j^{th} columns of A is indicated by A_{ij} and the line can be composed of A_j as, in which the • is construed as a valuation numbered i from 1 to m. Therefore, the ith row A could be authored as A_i. When $m = n$, the framework A is made up of a rectangular shape.

7.2.3 DEFINITION: STRATEGIC FORM – 2 PLAYER GAMES

Let $G = \{p, w, y\}$ performs a simple form of game with $p = \{p_1, p_2\}$ and $w = w_1 \times w_2$. If $w_i (i = 1, 2)$ the schemes are arranged in such a way that $w_i = \{\sigma_1^i, ... \sigma_{n_i}^i\} (i = 1, 2)$. There is a $A_i \in \Re^{m_1 \times n_2}$ framework for every participant so that element (r, c) of A_i is given by $w_i(\sigma_r^1, \sigma_c^2)$ then tuple $G = \{p, w, A_1, A_2\}$ is a two-player game strategy.

7.2.4 DEFINITION: UNIT AND ZERO VECTORS

The tensor $e \in \Re^n$ is a single template $e = (1, 1, ... 1)$. Likewise, the zero-game theory $o = (0, 0.... 0) \in \Re^n$. We predict that the distance of e and o would be ascertained.

Theorem 7.1: Let $G = \{p, w, A, B\}$ be just two different complicated games with $w_1 = \{\sigma_1^1, ... \sigma_m^1\}$ and $w_2 = \{\sigma_1^2, ... \sigma_n^2\}$. If the game p_1 started tactic σ_r^1 as well as the player p_2 selects strategic plan σ_c^2 then $w_1(\sigma_r^1, \sigma_c^2) = e_r^T A e_c$ and $w_2(\sigma_r^1, \sigma_c^2) = e_r^T B e_c$.

7.2.5 DEFINITION: SYMMETRIC AND EQUILIBRIUM GAME

  i. **Symmetric Game:**
     If $G = \{p, w, A, B\}$, then G is made reference to as a synchronous game $A = B^T$.
 ii. **Equilibrium:**
     Let $G = \{p, w, A, B\}$ perform two key tournaments with $w = w_1 \times w_2$. If and only if the steady state strategic approach, $(\sigma_i^1, \sigma_j^2) \in w_1 \times w_2$ the phrasings $e_i^T A e_j \geq e_k^T A e_j \forall k \neq i$ and $e_i^T B e_j \geq e_i^T B e_l \forall l \neq j$ hold.

Theorem 7.2: Let $G = \{p, w, A\}$ be two different games with a negligible eventual result. A strategic planning duo (e_i, e_j) is an equitable tactic if and only if $e_i^T A e_j = \max_{k \in \{1, 2, ...m\}} \min_{l \in \{1, 2, ...n\}} A_{kl} = \min_{l \in \{1, 2, ...n\}} \max_{k \in \{1, 2, ...m\}} A_{kl}$ then the value of the game is $V_G = e_i^T A e_j$.

7.3 THE FLUID-STRUCTURE INTERACTION MODEL

The aforementioned fluid-structure connectivity model includes both stretchy and fluid aspects into consideration.

7.3.1 CONTROL/DISTURBANCE PDE MODEL

The schematic upon which fluid-particle communication adapts will be the unifica-tion $\Omega_f \cup \Omega_s$, where the "fluid site" Ω_f is a constrained subset of $\Re^n$, $n \geq 2$ and indeed the geometry's "sturdy field" is a delimited small segment of $\Re^n$ that is also engrossed in Ω_f. Let Γ_f be this same site's southern boundary and Γ_s be the boundary of Ω_s, both the regions that also verge the outer area Ω_f and are where the systems interact.

As a consequence, the features of both components Γ_s are formed. Let $\upsilon(x)$ be the property's unit outer regular vertex Ω_f and, therefore, inward with respect to Ω_s and u becomes an m-dimensional vector function of Ω_f, going to represent the fluid's kinetic energy, and the particle feature chosen to $\wp$ represent the stress. Let v and u be the dimensional space scattering and speed characteristics of the sturdy Ω_s. The leaping control $p \in L_2 \left\{ 0,T; \left(L_2 \left(\Gamma_s \right) \right)^n \right\}$ has been depicted and is active today on boundary Γ_s. We start by presuming that the predetermined disruption $w = \{w_1, w_2\}$ acts on one element w_1 in Ω_f and one element w_2 in Ω_s. We support our work on the assumption of comparatively tiny but quick fluctuations of the rigid body, such that the functionality Γ_s can be presumed to be stationary for further modeling work.

7.3.2 MATHEMATICAL MODEL

Let the boundary control $p \in L_2 \{0,T; A\}$, where $A = \left(L_2 \left(\Gamma_s \right) \right)^n$ and the inner deterministic trouble $w = \{w_1, w_2, w_3\} \in L_2 \{0,T; U \times V \times A\}$, where $U = \left(L_2 \left(\Omega_f \right) \right)^n$, $V = \left(L_2 \left(\Omega_s \right) \right)^n$ and $A = \left(L_2 \left(\Gamma_s \right) \right)^n$. The unknowns' parabolic-hyperbolic coupled PDE system $\{u, v, v_t, p\}$ is

$$u_t - \Delta u + L_f u + \nabla p = w_1 \text{ where } Q_f \equiv \Omega_f \times (0,T] \tag{7.1}$$

$$divu = 0 \text{ where } Q_f \equiv \Omega_f \times (0,T] \tag{7.2}$$

$$v_{tt} - div\sigma(v) - \rho div\sigma(v_t) = w_2 \text{ where } Q_s \equiv \Omega_s \times (0,T] \tag{7.3}$$

$$v_t = u + p_0 \text{ where } \Pi_s \equiv \Gamma_s \times (0,T] \tag{7.4}$$

$$u = 0 \text{ if } \Pi_f \equiv \Gamma_f \times (0,T] \tag{7.5}$$

$$\sigma(v + \rho v_t) \bullet v = \in(u) \bullet \upsilon - pv - p_1 + w_3 \text{ if } \Pi_s \equiv \Gamma_s \times (0,T] \tag{7.6}$$

$$u(0,\bullet) = u_0 \text{ where } \Omega_f \tag{7.7}$$

$$v(0,\bullet) = v_0, v_t(0,\bullet) = v_t \text{ where } \Omega_s \tag{7.8}$$

The elastic strain tensor

$$\in_{ij}(u) = \in_{ji}(u) = \frac{1}{2} \left(\frac{\partial u_i}{\partial x_j} + \frac{\partial u_j}{\partial x_i} \right) \tag{7.9}$$

and the stress tensor

$$\sigma_{ij}(u) = \sigma_{ji}(u) = \lambda \left(\sum_{k=1}^{n} \in_{kk}(u) \right) \delta_{ij} + 2\mu \in_{ij}(u) \tag{7.10}$$

where $\lambda > 0$ and $\mu > 0$ are the lame constants. Observably, we have:

$$|\in(u)| \leq |\nabla u| \text{ and}$$

$$|\sigma(u)| \leq 2 \max\{\lambda, 2\mu\}|\in(u)| \tag{7.11}$$

$$|\sigma(u)| \leq 2 \max\{\lambda, 2\mu\}|\nabla u|$$

Here, Lu is defined as $Lu = (u \bullet \nabla)y_e + (y_e \bullet \nabla)u$
If $div\ y_e = 0$,

$$y_e / \Gamma_f = 0 \tag{7.12}$$

7.4 MIN-MAX GAME THEORY STRUCTURE

7.4.1 ABSTRACT OF DYNAMICS

Allow Hilbert Spaces to be delegated to U = command, Y = jurisdiction, Z = monitoring, and V = perturbation. The dynamics-governed governmental formula is

$$y_t = My + Np + Zw \text{ on } \left[D(M^*) \right]' y_0 \in Y \tag{7.13}$$

which is subject to certain conditions.

7.4.1.1 Hypothesis

H_1: $M : Y \supset D(M) \to Y$ is the minuscule originator of a semi-conductor semi-group e^{Mt} on $Y, t \geq 0$.

H_2: N is a rectilinear operative on $U = D(N) \to \left[D(M^*) \right]'$, gratifying the state-run $R(\lambda, M)N \in \ell(U, Y)$, for certain $\lambda \in \rho(M)$, anywhere $R(\lambda, M)$ and $\rho(M)$ is the resolution of M.

H_3: Let Z be a lined operative in $V = D(Z) \to \left[D(M^*) \right]'$ satisfying the nationwide $R(\lambda, M)Z \in \ell(V, Y)$, for certain $\lambda \in \rho(M)$.

H_4: If $R : D(R) \to \Psi$ is a lined worker, then

$$D(R) \supset \{ e^{Mt}NU, 0 < t \leq T \} \cup \{ e^{Mt}ZV, 0 < t \leq T \} \cup Y. \tag{7.14}$$

H_5: If the tripartite $\{M, N, R\}$ fulfills the subsequently output notable approximation disorder, then $0 < \alpha < 1$ is continuous in $C_T > 0$ if

$$\left\| Re^{Mt} N \right\|_{\ell(U:\Psi)} = \left\| N^* e^{M^*t} R^* \right\|_{\ell(\Psi:U)} \leq \frac{C_T}{t^\alpha}, 0 < t \leq T$$

$$Re^{Mt}Np \in {}_\alpha C([0,T]; \Psi), \forall p \in U, \tag{7.15}$$

where

$$(Np, y)_Y = (p, N^* y)_U, \forall p \in V \tag{7.16}$$

$y \in D(N^*) \supset D(M^*)$. The intention liberty is distinct as $_\alpha C([0,T]; \bullet)$ $_r C([s,T]; X)$ is defined as:

$$\left\{ f \in C([s,T]; X) : \|f\|_{r C([s,T];X)} \right\}$$
$$= \underset{s < t \le T}{Sup} (t-s)^r \|f(t)\|_X \tag{7.17}$$

H_6: With R as in H_4, the triple $\{A, G, R\}$ satisfies the following output Singular Estimate Condition:

$$\Rightarrow \left\| \text{Re}^{Mt} Z \right\|_{\ell(V:\Psi)} = \left\| Z^* e^{M^* t} R^* \right\|_{\ell(\Psi:V)} \le \frac{C_T}{t^\alpha}, 0 < t \le T \tag{7.18}$$

$$\text{Re}^{Mt} Zw \in {}_\alpha C([0,T]; \Psi), \forall w \in V \tag{7.19}$$

$$\&(Zw, y)_\gamma = (w, Z^* y)_v, w \in v$$

H_7: R in the hypotheses of $H_4 \Rightarrow R \in \ell(Y, \Psi)$, so that Re^{Mt} is continuous in

$$Y \to C([0,T]; \Psi) \tag{7.20}$$

7.4.2 Finite Time Interval of Game Structure

If $0 < T < \infty$ and $\gamma > 0$, we show a relationship with $y_t = My + Np + Zw$ on $\left[D(M^*) \right]'$. If $y_0 \in Y$, the cost functional

$$\Phi(p, w; y_0) = \Phi(p, w; y(p, w); y_0)$$
$$= \int_0^T \left[\|R_y(t)\|_\Psi^2 + \|g(t)\|_U^2 - \gamma^2 \|w(t)\|_V^2 \right] dt$$

where $y(t) = y(t; y_0)$ is the solution of (7.1) due to $p(t)$ and $w(t)$. Thus, the following game theory problem is $\underset{w}{Sup} \underset{u}{Inf} \Phi(p, w; y_0)$ where the Infimum is taken over all $p \in L_2(0,T;U)$ for $w \in L_2(0,T;V)$ fixed, and the Supremum is then taken over all $w \in L_2(0,T;V)$.

7.4.3 In the Semi-Group

If $y_t = My + Np + Zw$ on $\left[D(M^*) \right]'$ to $y_0 \in Y$ in $\{y(t), p(t), w(t)\}$ where the states, controls, and disturbances have been depicted. Here, M is the power source of a resolutely constant fuzzy set on the Hilbert state space. It is a linear, extremely unbounded operator larger than the power transfer, so that $N : U \rightarrow \left[D(M^*) \right]$ or $M^{-1}N \in \ell(U, Y)$ and correspondingly Z over the perturbation space V. In addition, at the early stages, together fulfil the basic solitary guesstimate.

$$\left\| e^{Mt} Np \right\|_{Y} \leq \frac{C_T}{t^{\alpha}} \|u\|_{U},$$

$$\left\| e^{Mt} Zw \right\|_{Y} \leq \frac{C_T}{t^{\alpha}} \|w\|_{V}, \text{ where } 0 < \alpha < 1, 0 < t \leq T \tag{7.21}$$

thereby, the remedy to the min-max game theory predicament for trajectories is (7.19). $\left\{ y*(t, y_0), p*(t, y_0), w*(t, y_0) \right\}$ on $(0, T)$, such that

$$\underset{w \in V}{Sup} \underset{\ell \in p}{Inf} \int_{0}^{T} \left[\left\| R_y(t) \right\|_{Z}^{2} + \left\| g(t) \right\|_{U}^{2} - \gamma^{2} \left\| w(t) \right\|_{V}^{2} \right] dt \tag{7.22}$$

where R is a favorable continual Z and γ is a piecewise delimited monitoring provider Y from one Hilbert space to the other.

7.5 MIN-MAX GAME THEORY IN FLEXIBLE AND VISCO-ELASTIC

7.5.1 ABSTRACT MODEL

The abstract spaces of the coupled PDE problem of (7.1) are $H^* \equiv \Pi \times \left(H^l(\Omega_s)^n \times (L_2(\Omega_s))^n \right)$ for $\{u, v, v_t\}$ where,

$$\Pi \equiv \left\{ u \in \left(L_2(\Omega_f) \right)^n : div\ u = 0, u \bullet v|_{\Gamma_f} = 0 \right\} \tag{7.23}$$

and the interplanetary

$$E \equiv \left\{ u \in \left(H^l(\Omega_f) \right)^n : div\ u = 0, u|_{\Gamma_f} = 0 \right\} \tag{7.24}$$

The area Ω_s and the consistent border Γ_s, the L_2-inner crops are symbolized by

$$(u_1, u_2) = \int_{\Omega_s} u_1 \bullet u_2 d\Omega_s,$$

$$(u_1, u_2) = \int_{\Gamma_s} u_1 \bullet u_2 d\Gamma_s \tag{7.25}$$

On the domain Ω_f the space E is topologized with respect to the inner product given by

$$(u_1 \bullet u_2)_E \equiv \int_{\Omega_f} \in (u_1) \bullet \in (u_2) d\Omega_f \tag{7.26}$$

$\therefore H^l\left(\Omega_f\right) \equiv \|\bullet\|_{1,\Omega_f}$ from Koran's and Poincare's inequity.

$$\|u\|_{1,\Omega_f} = \left(\int_{\Omega_f} |\in(u)|^2 \, d\Omega_f\right)^{1/2} \tag{7.27}$$

The storage $H^l\left(\Omega_s\right)$ of configurations in regard to the internal product is defined as:

$$(v,\Psi)_{1,s} \equiv \int_{\Omega_s} v \bullet \Psi d\Omega_s + \int_{\Omega_s} \sigma(v) \bullet \in(\Psi)d\Omega_s \tag{7.28}$$

$$\therefore \|\bullet\|_{1,\Omega_s} \equiv H^l\left(\Omega_s\right) \Rightarrow \|v\|_{1,\Omega_f}^2 = \int_{\Omega_s} \sigma(v) \bullet \in(v)d\Omega_s + |v|_{0,\Omega_s}^2 \equiv H^l\left(\Omega_s\right) \tag{7.29}$$

with $T > 0$, the irrational modes $y_t = My + Np + Zw$ on $\left[D\left(M^*\right)\right]'$ originating $y_0 \in Y$ from the free flowing model theory (7.1) with regulation p and the predetermined disruption w we link up for a fixed $\gamma > 0$ this same ratio used to measure integral is

$$\Phi(p,w,y_0) = \int_0^T \left[\|R_y\|_\Psi^2 + \|p\|_U^2 - \gamma^2 \|w\|_V^2\right]dt \tag{7.30}$$

where y is the remedy of (7.29) due to (p,w,y_0) and R fulfills the presumption.

$$\left\|\operatorname{Re}^{Mt} Np\right\|_\Psi \le \frac{C_T}{t^\alpha}\|p\|_U, 0 < t \le T, 0 < \alpha < 1$$

Therefore the min-max game problem is then

$$\underset{w \in L_2(0,T;\upsilon)}{Inf} \underset{p \in L_2(0,T;u)}{Sup} \Phi(p,w,y_0), y_0 \in H^* \tag{7.31}$$

and now it is

$$\Phi(u,v,p,w) = \int_0^T \left(|p(t)|_{L_2(\Gamma_s)}^2 + |u(t)|_{L_2(\Omega_f)}^2 + |v(t)|_{H^l(\Omega_s)}^2 + |v_t(t)|_{L_2(\Omega_s)}^2\right)dt - \gamma^2 \tag{7.32}$$

$$\int_0^T \left(|w_1(t)|_{L_2(\Omega_f)}^2 + |w_2(t)|_{L_2(\Omega_s)}^2\right)dt$$

The min-max game theory corresponds to Eqn. (7.32), i.e.

$$\underset{w \in L_2(\Omega_f) \times L_2(\Omega_s)}{Sup} \underset{p \in L_2(\Gamma_s)}{Inf} \Phi(u,v,p,w).$$

7.5.2 ELASTIC MODEL

If $\rho = 0$ and the min-max game problem $\underset{w \in L_2(\Omega_f) \times L_2(\Omega_s)}{Sup} \underset{p \in L_2(\Gamma_s)}{Inf} \Phi(u, v, p, w)$, there exists a critical $\gamma_c > 0$, for all first disorder in H^*, that is, $y_0 = (v_0, v_1, u_0) \in H'(\Omega_s) \times L_2(\Omega_s)$ the subsequent circumstances are content.

  i. If $0 < \gamma < \gamma_c$, $\Phi(u, v, p, w)$ will grow to infinity as we take supermoms over $w = (w_1, w_2)$. Thus, there is no finite solution for game problem for any initial condition $y_0 \in H^*$.
  ii. If $\gamma > \gamma_c$, then for each initial condition $y_0 \in H^*$ there exist a unique control $\left(w_0^*, w_1^*\right) \in C\left([0, T], L_2(\Omega_f) \times L_2(\Omega_s)\right)$ and corresponding optimal state $y^*(t) = \left(v^*(t), v_t^*(t), u^*(t)\right) \in C\left([0, T], H'(\Omega_s) \times L_2(\Omega_s) \times H\right)$ such that $\Phi(u*, v*, p*, w*) = \underset{w \in L_2(\Omega_f) \times L_2(\Omega_s)}{Sup} \underset{p \in L_2(\Gamma_s)}{Inf} \Phi(u, v, p, w)$

7.5.3 VISCO-ELASTIC MODEL

If $\rho > 0$, then the inclusion of visco-elastic effects has two advantages.

  i. It follows the applicability of a much richer class of controls.
  ii. It does not have the need of an incremental deal successfully with hypothesis requirement on the remark of R.

Once $\sigma = 0$, then Eqn. (7.1) can take an account for a Dirichlet type of control added to the interface, the resulting model is the following PDE:

$$u_t - \Delta u + L_f u + \nabla p = w_1 \text{ where } Q_f \equiv \Omega_f \times (0, T] \qquad (7.33)$$

$$div\, u = 0 \text{ then } Q_f \equiv \Omega_f \times (0, T] \qquad (7.34)$$

$$v_{tt} - div\sigma(v) - \rho div\sigma(v_t) = w_2 \text{ then } Q_s \equiv \Omega_s \times (0, T] \qquad (7.35)$$

$$v_t = u + p_0 \text{ if } \Pi_s \equiv \Gamma_s \times (0, T] \qquad (7.36)$$

$$u = 0 \text{ then } \Pi_f \equiv \Gamma_f \times (0, T] \qquad (7.37)$$

$$\sigma(v + \rho v_t) \bullet v = \in (u) \bullet v = p\upsilon - p_1 \text{ if } \Sigma_s \equiv \Gamma_s \times (0, T] \qquad (7.38)$$

$$u(0, \bullet) = u_0 \text{ in } \Omega_f \qquad (7.39)$$

$$v(0, \bullet) = v_0, v_t(0, \bullet) = v_1 \text{ in } \Omega_s \qquad (7.40)$$

The significant difference between Eqn. (7.1) and model (7.33) is that in (7.36), and one more control g_0 is added and $\rho > 0$ must be positive. This difference changes the entire setup. The overall fluid-structure dynamics will satisfy the singular estimate.

7.6   CONCLUSION

In this chapter, we discussed the min-max game theory dilemma of a sequential fluid-structure model theory, by means of regulation and disruption stand-in on the crossing point of two models, stretchy and visco-elastic, again for system comprising of coupled PDE framework.

REFERENCES

1. J. Von Neumann. Zur Theorie det Gesell's chats spieled. *Math Ann*, 100(1), 1928, 295–320.
2. R. Myerson. *Game Theory: Analysis of Conflict*. Managerial and Decision Economics, Vol 13, No.4 (July- Aug. 1992) page no. 369.
3. I. Lasiecka. Roberto trigging and Jing Zhang. *Open Appl Math J*, 2, 2013, 7–17.
4. J. Zhang. Virginia State University, Min-max Game Theory for coupled PDE Systems–Abstract Theory with Applications to Fluid Structure Interactions, January 2013.
5. R. Triggiani. Min-max game theory and non-standard differential Riccati equations for abstract hyperbolic-like equations. *J Nonlinear Anal*. Series A, Special Issue, 75, 2012, 1572–1591.
6. P. Bernhard. Linear-quadratic, two-person, zero-sum differential games: necessary and sufficient conditions. *J Opto Theory Appl*, 27, 1979, 51–69.
7. V. Barbu, Z. Grujic, I. Lasiecka and A. Tuffaha. Smoothness of weak solutions to a nonlinear fluid-structure interaction model. *Indian J Math*, 57(3), 2008, 1173–1207.
8. G. Avalosabd and T. Triggiani. Uniform stabilization of a coupled PDE system arising in fluid-structure interaction with boundary dissipation at the interface. *Discr Contin Dyn Syst*, 22(4), 2008, 817–833.
9. F. Bucci and I. Lasiecka. Optimal boundary control with critical penalization for a PDE model of fluid-solid interactions. *Calc Var Partial Differ Equations*, 37(1–2), 2010, 217–235.
10. P. Consstantin and C. Foias. *Navier Stokes Equations*. Chicago Lectures in Mathematics, 1988.

8 Numerical Simulation for Time Fractional Integro Partial Differential Equations Arising in Viscoelastic Dynamical System

Jugal Mohapatra and Sudarshan Santra

CONTENTS

8.1 INTRODUCTION

The study on fractional calculus gains more attention of many researchers in recent times, due to its immense applicability to define various models, such as viscoelastic damped structure [1], the model due to radiative transfer [2], the theory of linear transport [3], and the mathematical structure due to kinetic energy of gases [4]. A detailed investigation about the application of fractional differential as well as fractional integro-differential equation is available in [5–7]. The general form of a fractional derivative viscoelastic models can be written as:

$$X(t) + \sum_{m=1}^{M} a_m \mathcal{D}_t^{\alpha_m} X(t) = E_0 Y(t) + \sum_{n=1}^{N} E_n \mathcal{D}_t^{\beta_n} Y(t), \qquad (8.1)$$

where the time-dependent stress field $X(t)$ is related to time-dependent strain field $Y(t)$ through a series of derivatives of fractional order. E_i's denote the Young's modulus

DOI: 10.1201/9781003328032-8

and a_i's are some parameters. α_m, β_n for $m = 1, 2, \cdots, M; n = 1, 2, \cdots, N$ lie between zero and one. Many viscoelastic materials can be modeled by retaining only the first fractional derivative term in each series in Eqn. (8.1) and the resulting model becomes

$$X(t) + a\mathcal{D}_t^{\alpha} X(t) = E_0 Y(t) + E_1 \mathcal{D}_t^{\beta} Y(t). \tag{8.2}$$

Here, $0 < \alpha < \beta < 1$. Model (Eqn. [8.2]) is equivalent to the following integro-differential equation including a Volterra integral operator arising due to viscoelasticity.

$$D_t^{\delta} Y(t) + \frac{E_0}{E_1 \Gamma(\alpha)} \int_{s=0}^{t} (t-s)^{\alpha-1} Y(s) ds = F(t),$$

where $\delta = \beta - \alpha \in (0, 1)$ and $F(t) = \dfrac{a}{E_1} X(t) + \dfrac{1}{E_1 \Gamma(\alpha)} \displaystyle\int_{s=0}^{t} (t-s)^{\alpha-1} X(s) ds$. Further, $X(t)$ is a known function with homogeneous initial condition and $Y(t)$ is considered to be unknown. The basic difference between the fractional-ordered derivative and the classical ordinary derivative is that the ordinary derivative is defined on a very small neighborhood of a given point, whereas the fractional derivative of a function is defined by an integral over the entire region. This chapter devotes to the numerical solution for time fractional parabolic integro partial differential equations (PDEs) which play an important role in physical phenomena.

Let $\Omega := [0,1] \times [0,1]$. The model which involves in heat flow in materials with memory and linear viscoelastic mechanics in dynamical system can be written mathematically in a general form as:

$$\left\{ \begin{aligned} &L\mathcal{W}(x,t) := \mathcal{D}_t^{\alpha} \mathcal{W}(x,t) - \mathcal{L}\mathcal{W}(x,t) + \int_0^t \mathcal{K}(x,t-s)\mathcal{W}(x,s)ds = f(x,t), \\ &(x,t) \in \Omega, \text{ satifying the following conditions:} \\ &\mathcal{W}(0,t) = \psi_1(t) \text{ and } \mathcal{W}(1,t) = \psi_2(t) \; \forall t \in (0,1], \\ &\mathcal{W}(x,0) = \varphi(x) \; \forall x \in [0,1]. \end{aligned} \right. \tag{8.3}$$

Here, $\mathcal{L}\mathcal{W}(x,t) := p(x)\dfrac{\partial^2 \mathcal{W}}{\partial x^2}$ with $p(x)$ is continuous on $[0,1]$ satisfying $p(x) \geq p_0 > 0 \; \forall x \in [0,1]$. $\mathcal{D}_t^{\alpha}$ denotes the fractional Caputo derivative of order α, $\alpha \in (0,1)$. Further, φ, ψ_1, ψ_2 are smooth functions on $[0,1]$. The function f and the kernel $\mathcal{K}$ are continuous on $\bar{\Omega}$.

Many numerical methods are available in the literature for solving an integro-differential equation (IDE). For instance: A sharp error bound was shown in [8] for the numerical solution of a linear Fredholm–Volterra integro-differential equation (VIDE). The readers might be interested in various numerical techniques for solving singularly perturbed VIDEs. Tari and Shahmorad in [9] examined a two-dimensional linear VIDE, where they used the matrix Tau method to solve the model numerically. People are working not only on IDEs but also there is a great deal of interest for developing many numerical techniques to work on partial integro-differential equations (PIDEs). To mention a few, the Laplace transform method was introduced in [10] to determine an analytical approximate solution of a linear PIDE with a convolution kernel. From numerical point of view, a second-order finite difference scheme was constructed in [11] for the numerical solution of a PIDE having weakly singular kernel. Li and Da in [12] studied on the numerical solution of parabolic PIDEs where they used the combined finite difference and finite element approximation. For further studies, one may refer to [13–15]. Finding analytical solution of an IDE itself is a challenging task and it is more critical when the model problem involves a fractional-order differential operator. The readers can find the existence and uniqueness properties of a fractional integro-differential equation (FIDE) in [16, 17]. Several numerical techniques, such as Adomian decomposition method (ADM), Laplace decomposition method (LDM), Homotopy perturbation method (HPM), finite difference method, and finite element method, are developed to solve FIDEs. To mention a few, Hamoud and Ghadle in [18] considered a nonlinear FIDE involving both Volterra and Fredholm operator where they used the ADM and modified LDM to solve the model. An efficient numerical scheme was constructed in [19] by using ADM for solving fourth-order FIDE. A novel finite difference scheme was constructed in [20, 21] for the numerical solution of FIDEs. For further studies, one may refer to [22–25] and references therein. Very few articles are available in the literature for studying integro PDEs involving fractional derivatives.

One can find the existence and uniqueness asymptotic stability properties in [26], to solve time fractional integro PDEs involving a finite delay. Our main intention in this chapter is to develop an efficient finite difference scheme to study the behavior of the numerical solution of time fractional integro PDEs arising in viscoelastic dynamical system. The solution of such problem has a weak singularity at $t = 0$. To deal with such singular behavior, the L1 discretization is used to approximate the temporal derivative. A second-order central difference scheme is applied to approximate the spatial derivative and a repeated quadrature rule is employed to dealt with the integral part. The convergence of the given numerical scheme is being analyzed and a sharp error bound for the given numerical solution is estimated. A detailed analysis proves that the proposed scheme gives $O(\tau^\alpha)$ rate of convergence over the entire region but on any subdomain away from the origin, it shows $O(\tau)$ rate of convergence. τ denotes the mesh parameter towards time direction. Finally, the proposed method is applied on a more general integro PDE involving multi-term time fractional differential operators. It is also shown to be effective with the desired accuracy for such multi-term time fractional integro PDEs. Some examples are given to prove the theoretical claim.

8.2 SOME PRELIMINARIES

In this segment, we discuss some well-known definitions of various fractional derivatives and integrals and some of their properties. One may refer to [27, 28].

Definition 8.1: *For a continuous function u(x,t), The Riemann-Liouville (RL) fractional integral of order $\delta \in \mathbb{R}^+$ is defined as:*

$$\mathcal{J}^\delta u(x,t) = \frac{1}{\Gamma(\delta)} \int\limits_{\zeta=0}^{t} (t-\zeta)^{\delta-1} u(x,\zeta)d\zeta, \ (x,t) \in \bar{\Omega}.$$

$\mathbb{R}^+$ *denotes the set of all positive real numbers.*

Definition 8.2: *The fractional Caputo derivative of u(x,t) can be defined in terms of RL fractional integral as:*

$$\mathcal{D}_t^\delta u(x,t) = \left[\mathcal{J}^{m-\delta}\left(\frac{\partial^n u}{\partial t^n} \right) \right](x,t) \ \text{for}\,(x,t) \in \Omega.$$

Here, $m = \delta$ is the smallest integer which is greater or equals to δ. Further, we have:

1. $\mathcal{D}^\delta u = 0$ for any constant function u.
2. For $m \in \mathbb{N}, \gamma \in \mathbb{R}$, one has

$$\mathcal{D}^\delta t^\gamma = \begin{cases} \dfrac{\Gamma(\gamma+1)}{\Gamma(\gamma-\delta+1)} t^{\gamma-\delta} & \text{if } m-1 < \delta < m, \gamma > m-1, \\ 0 & \text{if } m-1 < \delta < m, \gamma \leq m-1. \end{cases}$$

3. Linearity property: $\mathcal{J}^\delta\{c_1 u \pm c_2 v\} = c_1\mathcal{J}^\delta u \pm c_2\mathcal{J}^\delta v$ and $\mathcal{D}_t^\delta\{c_1 u \pm c_2 v\} = c_1\mathcal{D}_t^\delta u \pm c_2\mathcal{D}_t^\delta v$, c_1, c_2 are some constants.
4. $\mathcal{D}_t^\delta \mathcal{J}^\delta u(x,t) = u(x,t)$ but $\mathcal{J}^\delta\mathcal{D}_t^\delta u(x,t) = u(x,t) - u(x,0^+)$, $\delta \in (0,1)$.

Definition 8.3: *Consider the mesh function $\mathcal{V}_m^n$, $m = 0,1, \cdots, M$; $n = 0,1, \cdots, N$ corresponding to a continuous function $\mathcal{V} : \Omega \subset \mathbb{R}^2 \to \mathbb{R}$, then one can define*

$$\|\mathcal{V}\| := \max_{(x,t)\in\Omega} |\mathcal{V}(x,t)| \ and \ \|\mathcal{V}^n\| := \max_{0\leq m\leq M} |\mathcal{V}_m^n|.$$

Definition 8.4: *A linear operator $\mathcal{T} : \mathcal{X} \to \mathcal{Y}$ is said to be bounded if $\exists \tilde{c} > 0$, such that*

$$\|\mathcal{T}x\|_y \leq \tilde{c}\|x\|_x \ \forall x \in \mathcal{X}.$$

$\mathcal{X}, \mathcal{Y}$ *are some normed vector spaces over $\mathbb{R}$ or $\mathbb{C}$. $\mathcal{T}$ is bounded if $\mathcal{T}$ is continuous everywhere in $\mathcal{X}$.*

Definition 8.5: *A mapping* $F : D(\subseteq \mathbb{R}^2) \to D$ *be such that*

$$\|F(x) - F(y)\| \le \hat{c}\|x - y\| \quad \forall x, y \in D,$$

then F *is called contraction with* $0 \le \hat{c} < 1$. *If* D *be complete then* F *has a fixed point* x^* *in* D *which is unique.*

Further, we assume that the derivatives of the solution for Eqn. (8.3) satisfies the following bounds:

$$\begin{cases} \left| \dfrac{\partial^{k_1} \mathcal{W}}{\partial x^{k_1}} \right| \le C \quad \text{for } k_1 = 0,1,2,3,4, \\[4mm] \left| \dfrac{\partial^{k_2} \mathcal{W}}{\partial t^{k_2}} \right| \le C(1 + t^{\alpha - k_2}) \quad \text{for } k_2 = 0,1,2 \end{cases} \tag{8.4}$$

for $(x,t) \in \Omega$. Notice that the kernel $\mathcal{K}$, defined on $\bar{\Omega}$ is continuous. So, there exists $\Re > 0$ such that $\|\mathcal{K}\| \le \Re$.

Lemma 8.1: *Assume that the partial derivatives of the solution of Eqn. (8.3) with respect to space are continuous. The sufficient condition for existing unique* $\mathcal{W}(x,t) \in \bar{\Omega}$ *is*

$$\frac{(\alpha + 1)C\|p\| + \Re}{\Gamma(\alpha + 2)} < 1,$$

where the generic constant $C(> 0)$ *can be different at different places.*

Proof. The idea behind the proof of this lemma is available in [16]. Apply $\mathcal{J}^\alpha$ in Eqn. (8.3) to get $\mathcal{W}(x,t) = \mathcal{H}\mathcal{W}(x,t)$, $(x,t) \in \bar{\Omega}$. The operator $\mathcal{H}\mathcal{W}$ is defined as:

$$\mathcal{H}\mathcal{W}(x,t) = \varphi(x) + \mathcal{J}^\alpha[f(x,t)] + p(x)\mathcal{J}^\alpha\left[\frac{\partial^2 \mathcal{W}}{\partial x^2}\right] - \mathcal{J}^\alpha\left[\int_0^t \mathcal{K}(x,t-s)\mathcal{W}(x,s)ds\right].$$

For $\mathcal{W}_1, \mathcal{W}_2 \in C(\bar{\Omega})$, we have

$$\left\| \mathcal{H}\mathcal{W}_1 - \mathcal{H}\mathcal{W}_2 \right\| \le \left\| \frac{p(x)}{\Gamma(\alpha)} \int_0^t (t-s)^{\alpha-1} \frac{\partial^2}{\partial x^2} (\mathcal{W}_1 - \mathcal{W}_2)(x,s)ds \right\|$$

$$+ \left\| \frac{1}{\Gamma(\alpha)} \int_0^t (t-\rho)^{\alpha-1} \int_0^\rho \mathcal{K}(x,\rho-s)(\mathcal{W}_1 - \mathcal{W}_2)(x,s)ds\, d\rho \right\|$$

$$\le \frac{\|p\|}{\Gamma(\alpha)} \int_0^t (t-s)^{\alpha-1} \left\| \frac{\partial^2}{\partial x^2} (\mathcal{W}_1 - \mathcal{W}_2)(x,s) \right\| ds$$

$$+ \frac{1}{\Gamma(\alpha)} \int_0^t (t-\rho)^{\alpha-1} \int_0^\rho \|\mathcal{K}(x,\rho-s)\| \|(\mathcal{W}_1 - \mathcal{W}_2)(x,s)\| ds\, d\rho.$$

Now, by definition 8.4 and using $\|\mathcal{K}\| \le \Re$, we obtain

$$\|\mathcal{HW_1} - \mathcal{HW_2}\| \le \frac{C\|p\|}{\Gamma(\alpha+1)} \|\mathcal{W_1} - \mathcal{W_2}\|$$

$$+ \frac{\Re}{\Gamma(\alpha)} \int\limits_0^t (t-\rho)^{\alpha-1} \int\limits_0^\rho \|(\mathcal{W_1} - \mathcal{W_2})(x,s)\| ds\, d\rho$$

$$\le \left[\frac{(\alpha+1)C\|p\| + \Re}{\Gamma(\alpha+2)}\right] \|\mathcal{W_1} - \mathcal{W_2}\| < \|\mathcal{W_1} - \mathcal{W_2}\|.$$

This proves that $\mathcal{H}$ is a contraction mapping. Since $(C(\bar{\Omega}), \|.\|)$ is a Banach space, then the proof follows from definition 8.5.

8.3   NUMERICAL DISCRETIZATION

Let $\mathcal{M}, \mathcal{N} \in \mathbb{N}$ be fixed. The domain $\bar{\Omega} = [0,1] \times [0,1]$ is discretized uniformly as: $\{(x_m, t_n): m = 0,1,\cdots,\mathcal{M}, n = 0,1,\cdots,\mathcal{N}\}$, where $x_m = mh$ for $m = 0,1,\cdots,\mathcal{M}$ and $t_n = n\tau$ for $n = 0,1,\cdots,\mathcal{N}$. $h = 1/\mathcal{M}, \tau = 1/\mathcal{N}$. Let $\{\mathcal{W}(x_m,t_n)\}_{m=0,n=0}^{\mathcal{M},\mathcal{N}}$ be the exact solution and denote $\{\mathcal{W}_m^n\}_{m=0,n=0}^{\mathcal{M},\mathcal{N}}$ as the approximate solution at each (x_m,t_n) of Eqn. (8.3). The Caputo derivative of $\mathcal{W}$ at each (x_m,t_n) is defined as:

$$\mathcal{D}_t^\alpha \mathcal{W}(x_m,t_n) = \frac{1}{\Gamma(1-\alpha)} \sum_{j=0}^{n-1} \int\limits_{s=t_j}^{t_{j+1}} (t_n-s)^{-\alpha} \frac{\partial \mathcal{W}}{\partial s}(x_m,s)\, ds.$$

The L1 approximation is given by:

$$\mathcal{D}_t^\alpha \mathcal{W}(x_m,t_n) \approx \mathcal{D}_N^\alpha \mathcal{W}_m^n := \frac{1}{\Gamma(1-\alpha)} \sum_{j=0}^{n-1} \frac{\mathcal{W}_m^{j+1} - \mathcal{W}_m^j}{\tau} \int\limits_{s=t_j}^{t_{j+1}} (t_n-s)^{-\alpha}\, ds$$

$$= \frac{1}{\tau^\alpha \Gamma(2-\alpha)} \sum_{j=0}^{n-1} (\mathcal{W}_m^{j+1} - \mathcal{W}_m^j) d_{n-j}, \tag{8.5}$$

where, $d_j = j^{1-\alpha} - (j-1)^{1-\alpha}$, $j \ge 1$. The readers can refer to [20, 21, 29–31] for detailed study on L1 discretization. The operator $\dfrac{\partial^2 \mathcal{W}}{\partial x^2}$ is discretized as:

$$\frac{\partial^2 \mathcal{W}}{\partial x^2}(x_m,t_n) \approx \delta_x^2 \mathcal{W}_m^n := \frac{\mathcal{W}_{m+1}^n - 2\mathcal{W}_m^n + \mathcal{W}_{m-1}^n}{h^2}. \tag{8.6}$$

Then, Eqn. (8.3) reduces to

$$
\left\{
\begin{aligned}
& L_{\mathcal{M},\mathcal{N}}\mathcal{W}(x_m,t_n) := \mathcal{D}_N^{\alpha}\mathcal{W}(x_m,t_n) - p(x_m)\delta_x^2\mathcal{W}(x_m,t_n) \\
& \qquad + \int_0^{t_n}\mathcal{K}(x_m,t_n-s)\mathcal{W}(x_m,s)ds = f(x_m,t_n) +^{(1)}\mathcal{R}_m^n +^{(2)}\mathcal{R}_m^n \\
& \text{for } 1 \le m \le \mathcal{M}-1, 1 \le n \le \mathcal{N}, \text{ with} \\
& \mathcal{W}(x_0,t_n) = \psi_1(t_n) \text{ and } \mathcal{W}(x_{\mathcal{M}},t_n) = \psi_2(t_n) \text{ for } 0 < n \le \mathcal{N}, \\
& \mathcal{W}(x_m,t_0) = \varphi(x_m) \text{ for } 0 \le m \le \mathcal{M},
\end{aligned}
\right.
\tag{8.7}
$$

where, $^{(1)}\mathcal{R}_m^n = (\mathcal{D}_N^{\alpha} - \mathcal{D}_t^{\alpha})\mathcal{W}(x_m,t_n)$ and $^{(2)}\mathcal{R}_m^n = p(x_m)\left(\dfrac{\partial^2}{\partial x^2} - \delta_x^2\right)\mathcal{W}(x_m,t_n)$. The integral part is replaced by the composite trapezoidal rule as:

$$
\int_0^{t_n}\mathcal{K}(x_m,t_n-s)\mathcal{W}(x_m,s)\,ds = \sum_{j=0}^{n-1}\int_{t_j}^{t_{j+1}}\mathcal{K}(x_m,t_n-s)\mathcal{W}(x_m,s)\,ds
$$

$$
= \frac{\tau}{2}\sum_{j=0}^{n-1}[\mathcal{K}(x_m,t_n-t_{j+1})\mathcal{W}(x_m,t_{j+1}) + \mathcal{K}(x_m,t_n-t_j)\mathcal{W}(x_m,t_j)] + {}^{(3)}\mathcal{R}_m^n,
\tag{8.8}
$$

where, $\ {}^{(3)}\mathcal{R}_m^n = \sum\limits_{j=0}^{n-1}\int_{t_j}^{t_{j+1}}(t_{j+1/2}-s)\dfrac{d}{ds}[\mathcal{K}(x_m,t_n-s)\mathcal{W}(x_m,s)]ds$. Then from Eqn. (8.7), we get

$$
\left\{
\begin{aligned}
& L_{\mathcal{M},\mathcal{N}}\mathcal{W}(x_m,t_n) := \mathcal{D}_N^{\alpha}\mathcal{W}(x_m,t_n) - p(x_m)\delta_x^2\mathcal{W}(x_m,t_n) \\
& \ + \frac{\tau}{2}\sum_{j=0}^{n-1}[\mathcal{K}(x_m,t_n-t_{j+1})\mathcal{W}(x_m,t_{j+1}) + \mathcal{K}(x_m,t_n-t_j)\mathcal{W}(x_m,t_j)] = f(x_m,t_n) + \mathcal{R}_m^n \\
& \text{for } 1 \le m \le \mathcal{M}-1, 1 \le n \le \mathcal{N}, \text{ with} \\
& \mathcal{W}(x_0,t_n) = \psi_1(t_n) \text{ and } \mathcal{W}(x_{\mathcal{M}},t_n) = \psi_2(t_n) \text{ for } 0 < n \le \mathcal{N}, \\
& \mathcal{W}(x_m,t_0) = \varphi(x_m) \text{ for } 0 \le m \le \mathcal{M},
\end{aligned}
\right.
\tag{8.9}
$$

where,

$$
\mathcal{R}_m^n = {}^{(1)}\mathcal{R}_m^n + {}^{(2)}\mathcal{R}_m^n + {}^{(3)}\mathcal{R}_m^n = \left(\mathcal{D}_N^{\alpha} - \mathcal{D}_t^{\alpha}\right)\mathcal{W}(x_m,t_n) + p(x_m)\left(\frac{\partial^2}{\partial x^2} - \delta_x^2\right)\mathcal{W}(x_m,t_n)
$$

$$
+ \sum_{j=0}^{n-1}\int_{t_j}^{t_{j+1}}(t_{j+1/2}-s)\frac{d}{ds}[\mathcal{K}(x_m,t_n-s)\mathcal{W}(x_m,s)]ds.
\tag{8.10}
$$

Now neglecting $\mathcal{R}_m^n$, we have the following discrete problem:

$$
\left\{
\begin{aligned}
& L_{\mathcal{M},\mathcal{N}} \mathcal{W}_m^n := \mathcal{D}_N^\alpha \mathcal{W}_m^n - p(x_m)\delta_x^2 \mathcal{W}_m^n \\
& \quad + \frac{\tau}{2} \sum_{j=0}^{n-1} [\mathcal{K}(x_m, t_n - t_{j+1})\mathcal{W}_m^{j+1} + \mathcal{K}(x_m, t_n - t_j)\mathcal{W}_m^j] = f(x_m, t_n) \\
& \text{for } 1 \leq m \leq \mathcal{M} - 1, 1 \leq n \leq \mathcal{N}, \text{ with} \\
& \mathcal{W}_0^n = \psi_1(t_n) \text{ and } \mathcal{W}_{\mathcal{M}}^n = \psi_2(t_n) \text{ for } 0 < n \leq \mathcal{N}, \\
& \mathcal{W}_m^0 = \varphi(x_m) \text{ for } 0 \leq m \leq \mathcal{M}.
\end{aligned}
\right.
\tag{8.11}
$$

By using Eqns. (8.5) and (8.6) into Eqn. (8.11), we have:

$$
\left\{
\begin{aligned}
& A_m \mathcal{W}_{m-1}^n + B_m \mathcal{W}_m^n + C_m \mathcal{W}_{m+1}^n = D(m; 0, 1, \cdots, n-1) \\
& \text{for } 1 \leq m \leq \mathcal{M} - 1, 1 \leq n \leq \mathcal{N}, \text{ with} \\
& \mathcal{W}_0^n = \psi_1(t_n) \text{ and } \mathcal{W}_{\mathcal{M}}^n = \psi_2(t_n) \text{ for } 0 < n \leq \mathcal{N}, \\
& \mathcal{W}_m^0 = \varphi(x_m) \text{ for } 0 \leq m \leq \mathcal{M},
\end{aligned}
\right.
$$

where,

$$
\left\{
\begin{aligned}
& A_m = C_m = -\frac{p(x_m)}{h^2}, \; B_m = \frac{d_1}{\tau^\alpha \Gamma(2-\alpha)} + \frac{2p(x_m)}{h^2} + \frac{\tau}{2}\mathcal{K}(x_m, t_0), \\
& D(m; 0, 1, \cdots, n-1) = \frac{d_1}{\tau^\alpha \Gamma(2-\alpha)}\mathcal{W}_m^{n-1} - \frac{1}{\tau^\alpha \Gamma(2-\alpha)} \sum_{j=0}^{n-2} \left(\mathcal{W}_m^{j+1} - \mathcal{W}_m^j\right)d_{n-j} \\
& \quad - \frac{\tau}{2}\mathcal{K}(x_m, t_n - t_{n-1})\mathcal{W}_m^{n-1} - \frac{\tau}{2}\sum_{j=0}^{n-2}\left[\mathcal{K}(x_m, t_n - t_{j+1})\mathcal{W}_m^{j+1} + \mathcal{K}(x_m, t_n - t_j)\mathcal{W}_m^j\right] \\
& \quad + f(x_m, t_n), \hspace{3cm} m = 1, 2, \cdots, \mathcal{M} - 1; n = 1, 2, \cdots, \mathcal{N}.
\end{aligned}
\right.
$$

At any time label, we have the following tridiagonal system in the unknowns $\mathcal{W}_1^n, \mathcal{W}_2^n, \cdots, \mathcal{W}_{\mathcal{M}-1}^n$ as:

$$
\begin{bmatrix}
B_1 & C_1 & & & & \\
A_2 & B_2 & C_2 & & & \\
& \ddots & \ddots & \ddots & & \\
& & & A_{\mathcal{M}-2} & B_{\mathcal{M}-2} & C_{\mathcal{M}-2} \\
& & & & A_{\mathcal{M}-1} & B_{\mathcal{M}-1}
\end{bmatrix}
\begin{bmatrix}
\mathcal{W}_1^n \\
\mathcal{W}_2^n \\
\vdots \\
\mathcal{W}_{\mathcal{M}-2}^n \\
\mathcal{W}_{\mathcal{M}-1}^n
\end{bmatrix}
=
\begin{bmatrix}
D(1; 0, 1, \cdots, n-1) - A_1 \mathcal{W}_0^n \\
D(2; 0, 1, \cdots, n-1) \\
\vdots \\
D(\mathcal{M}-2; 0, 1, \cdots, n-1) \\
D(\mathcal{M}-1; 0, 1, \cdots, n-1) - C_{\mathcal{M}-1}\mathcal{W}_{\mathcal{M}}^n
\end{bmatrix}.
$$

The restriction on $p(x) \geq p_0 > 0 \, \forall x \in [0,1]$ ensures the stability of the coefficient matrix corresponding to the operator $L_{\mathcal{M},\mathcal{N}}$, the entries of which are in correct sign pattern.

8.4 CONVERGENCE ANALYSIS

In this section, we analyze the convergence of the proposed scheme. The following three lemmas reveal the truncation error bounds for $\mathcal{D}_N^\alpha$, δ_x^2 and trapezoidal approximation followed by the main convergence result.

Lemma 8.2: *Suppose the solution of Eqn. (8.3) satisfies Eqn. (8.4), Then for each* $(x_m, t_n) \in \Omega$, *one has*

$$\left\| {}^{(1)}\mathcal{R}_m^n \right\| \leq Cn^{-\min\{2-\alpha, \alpha+1\}}.$$

Proof. Similar result is available in [29].

Lemma 8.3: *For each* $(x_m, t_n) \in \Omega$, *the operator* δ_x^2 *satisfies the following bounds.*

$$\left\| {}^{(2)}\mathcal{R}_m^n \right\| \leq Ch^2.$$

Proof. The Taylor series expansion gives $\left\| \left(\dfrac{\partial^2}{\partial x^2} - \delta_x^2 \right) \mathcal{W}(x_m, t_n) \right\| \leq Ch^2$. Notice that $p(x)$ is a continuous function on a closed interval $[0,1]$, hence bounded. So

$$\left\| {}^{(2)}\mathcal{R}_m^n \right\| = \left\| p(x_m) \left(\frac{\partial^2}{\partial x^2} - \delta_x^2 \right) \mathcal{W}(x_m, t_n) \right\| \leq \| p(x_m) \| \left\| \left(\frac{\partial^2}{\partial x^2} - \delta_x^2 \right) \mathcal{W}(x_m, t_n) \right\| \leq Ch^2.$$

Lemma 8.4: *For* ${}^{(3)}\mathcal{R}_m^n$, *one can prove*

$$\left\| {}^{(3)}\mathcal{R}_m^n \right\| \leq C\mathcal{N}^{-1}, \quad 0 \leq m \leq \mathcal{M}, 0 \leq n \leq \mathcal{N}.$$

Proof.

$$\left\| {}^{(3)}\mathcal{R}_m^n \right\| = \left\| \sum_{j=0}^{n-1} \int_{t_j}^{t_{j+1}} (t_{j+1/2} - s) \frac{d}{ds} [\mathcal{K}(x_m, t_n - s) \mathcal{W}(x_m, s)] ds \right\|$$

$$\leq \sum_{j=0}^{n-1} \int_{t_j}^{t_{j+1}} (t_{j+1/2} - s) \left\| \frac{d}{ds} [\mathcal{K}(x_m, t_n - s) \mathcal{W}(x_m, s)] \right\| ds$$

$$\leq \sum_{j=0}^{n-1} \int_{t_j}^{t_{j+1}} (t_{j+1/2} - s) \left\| \frac{\partial}{\partial s} [\mathcal{K}(x_m, t_n - s) \mathcal{W}(x_m, s)] \right.$$

$$\left. + \frac{\partial}{\partial \mathcal{W}} [\mathcal{K}(x_m, t_n - s) \mathcal{W}(x_m, s)] \frac{\partial \mathcal{W}}{\partial s}(x_m, s) \right\| ds$$

$$\leq C\tau \int_0^{t_n} \left(1 + \frac{\partial \mathcal{W}}{\partial s}(x_m, s) \right) ds \leq C\tau \leq C\mathcal{N}^{-1}.$$

For each $(x_m, t_n) \in \Omega$, denote the maximum error as $\left\| \mathfrak{e}_m^n \right\| = \left\| \mathcal{W}(x_m, t_n) - \mathcal{W}_m^n \right\|$. The error equation can be obtained by subtracting Eqn. (8.11) from Eqn. (8.9) as:

$$
\left\{
\begin{aligned}
&L_{\mathcal{M},\mathcal{N}} \mathfrak{e}_m^n := \mathcal{D}_N^\alpha \mathfrak{e}_m^n - p(x_m) \delta_x^2 \mathfrak{e}_m^n \\
&\qquad + \frac{\tau}{2} \sum_{j=0}^{n-1} \left[\mathcal{K}(x_m, t_n - t_{j+1}) \mathfrak{e}_m^{j+1} + \mathcal{K}(x_m, t_n - t_j) \mathfrak{e}_m^j \right] = \mathcal{R}_m^n \\
&\text{for } 1 \le m \le \mathcal{M} - 1, 1 \le n \le \mathcal{N}, \text{ with} \\
&\quad \mathfrak{e}_0^n = \mathfrak{e}_{\mathcal{M}}^n = 0 \qquad \text{for } 0 < n \le \mathcal{N}, \\
&\qquad \mathfrak{e}_m^0 = 0 \qquad \text{for } 0 \le m \le \mathcal{M}.
\end{aligned}
\right.
$$

Lemma 8.5: *The discrete problem in Eqn. (8.11) satisfies*

$$
\left\| \mathcal{W}(x_m, t_n) - \mathcal{W}_m^n \right\| \le \tau^\alpha \Gamma(2 - \alpha) \sum_{j=1}^n \theta_{n-j} \left\| \mathcal{R}_m^j \right\|, \quad (x_m, t_n) \in \Omega,
$$

where, $\theta_0 := 0, \ \theta_i := \sum_{k=1}^i (d_k - d_{k+1}) \theta_{i-k}$ *for* $i = 1, 2, \cdots$, *and* $\mathcal{R}_m^j$ *stands for*

$$
\mathcal{R}_m^j = {}^{(1)}\mathcal{R}_m^j + {}^{(2)}\mathcal{R}_m^j + {}^{(3)}\mathcal{R}_m^j = (\mathcal{D}_N^\alpha - \mathcal{D}_t^\alpha) \mathcal{W}(x_m, t_j) + p(x_m) \left(\frac{\partial^2}{\partial x^2} - \delta_x^2 \right) \mathcal{W}(x_m, t_j)
$$

$$
+ \sum_{k=0}^{j-1} \int_{t_k}^{t_{k+1}} (t_{k+1/2} - s) \frac{d}{ds} [\mathcal{K}(x_m, t_j - s) \mathcal{W}(x_m, s)] ds.
$$

Proof. A detailed proof of this lemma is available in [32].

Theorem 8.1: *If* $\left\{ \mathcal{W}(x_m, t_n) \right\}_{m,n=0}^{\mathcal{M},\mathcal{N}}$ *and* $\left\{ \mathcal{W}_m^n \right\}_{m,n=0}^{\mathcal{M},\mathcal{N}}$ *be the continuous and the corresponding discrete solutions of Eqn. (8.3), then one has*

$$
\left\| \mathfrak{e}_m^n \right\| \le C \left[\tau t_n^{\alpha-1} + \tau + h^2 \right], \quad (x_m, t_n) \in \Omega.
$$

Proof. Lemma 8.5 yields

$$
\left\| \mathfrak{e}_m^n \right\| = \left\| \mathcal{W}(x_m, t_n) - \mathcal{W}_m^n \right\| \le \tau^\alpha \Gamma(2 - \alpha) \sum_{j=1}^n \theta_{n-j} \left\| \mathcal{R}_m^j \right\|
$$

$$
\le C\tau^\alpha \sum_{j=1}^n \theta_{n-j} \left\| {}^{(1)}\mathcal{R}_m^j \right\| + C\tau^\alpha \sum_{j=1}^n \theta_{n-j} \left\| {}^{(2)}\mathcal{R}_m^j \right\| + C\tau^\alpha \sum_{j=1}^n \theta_{n-j} \left\| {}^{(3)}\mathcal{R}_m^j \right\|.
$$

Now, applying Lemma 8.2, Lemma 8.3, and Lemma 8.4, the above inequality reduces to

$$\left\|e_m^n\right\| \leq C\tau^\alpha \sum_{j=1}^n \theta_{n-j} j^{-\min\{2-\alpha,\alpha+1\}} + C\tau^\alpha \sum_{j=1}^n \theta_{n-j} h^2 + C\tau^\alpha \sum_{j=1}^n \theta_{n-j} \mathcal{N}^{-1}. \qquad (8.12)$$

The inequality $\tau^\alpha \sum_{k=1}^n k^{-\varsigma}\theta_{n-k} \leq C\mathcal{N}^{-1}t_n^{\alpha-1}$ holds if $\varsigma > 1$ but $\varsigma \leq \alpha$ implies $\tau^\alpha \sum_{k=1}^n k^{-\varsigma}\theta_{n-k} \leq \dfrac{\mathcal{N}^{-\varsigma}}{1-\alpha}$ for each $n = 1, 2, \cdots, \mathcal{N}$. For more details, one can refer to [29, 32]. Then Eqn. (8.12) yields

$$\left\|e_m^n\right\| \leq C\left(\mathcal{N}^{-1}t_n^{\alpha-1} + \frac{h^2}{1-\alpha} + \frac{\mathcal{N}^{-1}}{1-\alpha}\right) \leq C\left[\tau t_n^{\alpha-1} + \tau + h^2\right],$$

which is the desired result.

Remark 8.1: *Suppose $t_n \geq K$, $n = 1, 2, \cdots, \mathcal{N}$ for some $K > 0$. Then we have:*

$$\left\|e_m^n\right\|_{t_n \geq K > 0} \leq C\left[\tau + h^2\right] \text{ for } m = 0, 1, \cdots, \mathcal{M}.$$

8.5 MULTI-TERM TIME FRACTIONAL INTEGRO PDES

In this section, we consider a generalized version of the time fractional integro PDEs in which more than one fractional differential operators are involved. It is proved that the proposed scheme is also effective to solve such more general models. The model can be described as:

$$\begin{cases} \left[\mathcal{D}_t^{\alpha_1} + \sum_{k=2}^r \eta_k \mathcal{D}_t^{\alpha_k}\right]\mathcal{W}(x,t) - p(x)\dfrac{\partial^2 \mathcal{W}}{\partial x^2} + \displaystyle\int_0^t \mathcal{K}(x,t-s)\mathcal{W}(x,s)ds = f(x,t) \\ \text{for } (x,t) \in \Omega, \text{with} \\ \mathcal{W}(0,t) = \psi_1(t) \text{ and } \mathcal{W}(1,t) = \psi_2(t) \, \forall t \in (0,1], \\ \mathcal{W}(x,0) = \varphi(x) \, \forall x \in [0,1]. \end{cases} \qquad (8.13)$$

where, $0 < \alpha_r < \alpha_{r-1} < \cdots < \alpha_2 < \alpha_1 < 1$ and η_k's are given positive constants. Using Eqns. (8.5), (8.6), and (8.8), we have the following discrete version of Eqn. (8.13).

$$
\begin{cases}
\sum_{k=1}^{r} \eta_k \mathcal{D}_N^{\alpha_k} \mathcal{W}_m^n - p(x_m)\delta_x^2 \mathcal{W}_m^n + \dfrac{\tau}{2} \sum_{l=0}^{n-1} \Big[\mathcal{K}(x_m, t_n - t_{l+1}) \mathcal{W}_m^{l+1} \\
\qquad\qquad\qquad\qquad\qquad + \mathcal{K}(x_m, t_n - t_l) \mathcal{W}_m^l \Big] = f(x_m, t_n) \\
\text{for } 1 \le m \le \mathcal{M} - 1, 1 \le n \le \mathcal{N}, \text{with} \\
\mathcal{W}_0^n = \phi_1(t_n) \text{ and } \mathcal{W}_\mathcal{M}^n = \phi_2(t_n) \text{ for } 0 < n \le \mathcal{N}, \\
\mathcal{W}_m^0 = \psi(x_m) \text{ for } 0 \le m \le \mathcal{M}.
\end{cases}
\tag{8.14}
$$

Remark 8.2: *If $\{\mathcal{W}(x_m, t_n)\}_{m,n=0}^{\mathcal{M},\mathcal{N}}$ be the exact solution of Eqn. (8.13) and $\{\mathcal{W}_m^n\}_{m,n=0}^{\mathcal{M},\mathcal{N}}$ be the numerical solution of Eqn. (8.13) by using the scheme in Eqn. (8.14), then by using similar argument of Theorem 8.1, one has:*

$$
\left\| \mathfrak{e}_m^n \right\| \le C \left[\tau t_n^{\alpha_1 - 1} + \tau + h^2 \right], \ (x_m, t_n) \in \Omega.
$$

Further, for any $\tilde{K} > 0$ satisfying $t_n \ge \tilde{K}$, we have

$$
\left\| \mathfrak{e}_m^n \right\|_{t_n \ge \tilde{K} > 0} \le C \left[\tau + h^2 \right] \quad \text{for } n = 1, 2, \cdots, \mathcal{N}; \ m = 0, 1, \cdots, \mathcal{M}.
$$

8.6   NUMERICAL EXPERIMENTS

Several tests are performed in this section to show the effectiveness of the proposed method.

Example 8.1:

Consider the following time fractional integro PDE of the form:

$$
\begin{cases}
\mathcal{D}_t^\alpha \mathcal{W}(x,t) - \dfrac{\partial^2 \mathcal{W}}{\partial x^2} + \int_0^t \mathcal{K}(x, t-s)\mathcal{W}(x,s)ds = f(x,t), \ (x,t) \in [0,1] \times (0,1], \\
\mathcal{W}(0,t) = t + t^\alpha, \ \mathcal{W}(1,t) = 0 \ \forall t \in (0,1], \\
\mathcal{W}(x,0) = 0 \ \forall x \in [0,1],
\end{cases}
$$

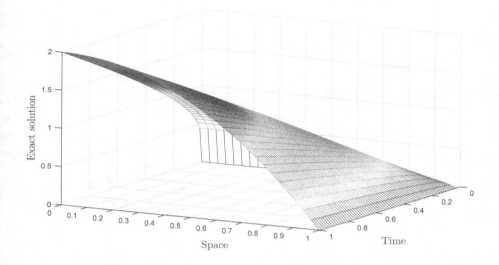

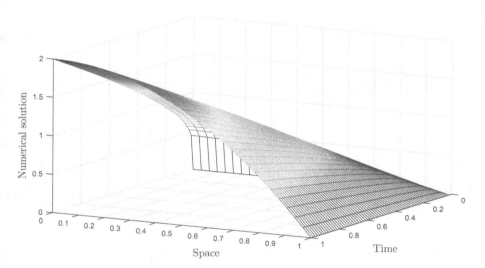

FIGURE 8.1   (a) Exact solution (b) Numerical solution for Example 8.1 with $\alpha = 0.2$.

where $\alpha \in (0,1)$, $\mathcal{K}(x, t - s) = x(t - s)$. Choose $f(x,t)$ such that the exact solution of Example 8.1 is $\mathcal{W}(x,t) = (1 - x^2)(t + t^{\alpha})$. The maximum error $E_{\mathcal{M},\mathcal{N}}$ and the rate of convergence $P_{\mathcal{M},\mathcal{N}}$ are defined by:

$$E_{\mathcal{M},\mathcal{N}} = \max_{(x_m,t_n)\in\bar{\Omega}}\left|\mathcal{W}(x_m,t_n) - \mathcal{W}_m^n\right|, \quad P_{\mathcal{M},\mathcal{N}} = \log_2\left(\frac{E_{\mathcal{M},\mathcal{N}}}{E_{2\mathcal{M},2\mathcal{N}}}\right).$$

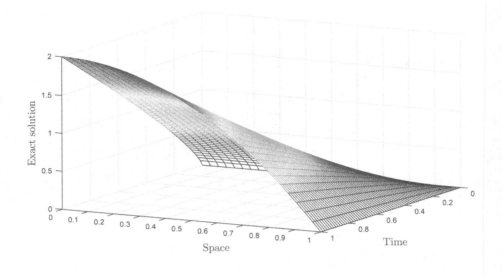

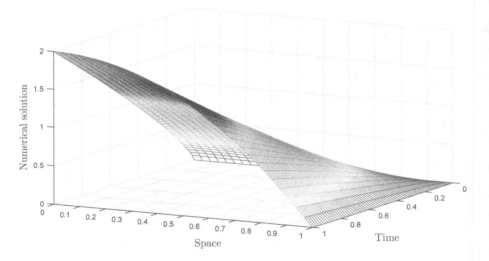

FIGURE 8.2   (a) Exact solution (b) Numerical solution for Example 8.1 with $\alpha = 0.7$.

The exact solution and the corresponding discrete solution with $\alpha = 0.2, \mathcal{M} = 32$, $\mathcal{N} = 64$ are shown in Figure 8.1 for Example 8.1. Similarly, Figure 8.2 displays the same for $\alpha = 0.7$. The graphical representation of the log-log plot is shown in Figure 8.3 for different values of α. $E_{\mathcal{M},\mathcal{N}}$ and $P_{\mathcal{M},\mathcal{N}}$ are shown in Table 8.1 over $\bar{\Omega}$. Tables 8.2 and 8.3 represent $E_{\mathcal{M},\mathcal{N}}, P_{\mathcal{M},\mathcal{N}}$ at $t = 0.1$ and $t = 1$, respectively. One can see that the scheme is of order $O(\tau^{\alpha})$ over $\bar{\Omega}$ but it gives $O(\tau)$ rate of convergence away from $t = 0$.

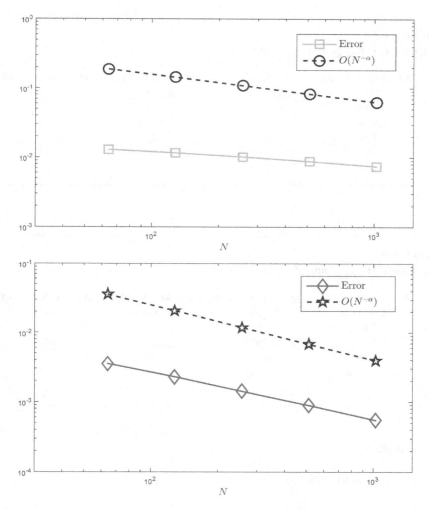

FIGURE 8.3 Log-log plots with (a) $\alpha = 0.4$ (b) $\alpha = 0.8$ for Example 8.1.

TABLE 8.1

Example 8.1: $E_{\mathcal{M},\mathcal{N}}$ and $P_{\mathcal{M},\mathcal{N}}$ on $\overline{\Omega}$

α	$\mathcal{M} = 128, \mathcal{N} = 128$	$\mathcal{M} = 256, \mathcal{N} = 256$	$\mathcal{M} = 512, \mathcal{N} = 512$	$\mathcal{M} = 1024, \mathcal{N} = 1024$
0.3	1.2327e-2	1.1474e-2	1.0575e-2	9.6459e-3
	0.103	0.118	0.133	
0.5	9.5084e-3	7.7324e-3	6.1274e-3	4.7477e-3
	0.298	0.336	0.368	
0.7	4.2892e-3	2.9268e-3	1.9406e-3	1.2602e-3
	0.551	0.593	0.623	

TABLE 8.2

Example 8.1: $E_{\mathcal{M},\mathcal{N}}$ and $P_{\mathcal{M},\mathcal{N}}$ at $t = 0.1$

α	$\mathcal{M} = 128, \mathcal{N} = 128$	$\mathcal{M} = 256, \mathcal{N} = 256$	$\mathcal{M} = 512, \mathcal{N} = 512$	$\mathcal{M} = 1024, \mathcal{N} = 1024$
0.4	5.1782e-4	2.1614e-4	9.3358e-5	4.1479e-5
	1.260	1.211	1.170	
0.6	7.5129e-4	3.3047e-4	1.5006e-4	6.9758e-5
	1.185	1.139	1.105	
0.8	9.0364e-4	4.4693e-4	2.2228e-4	1.1103e-4
	1.016	1.008	1.001	

TABLE 8.3

Example 8.1: $E_{\mathcal{M},\mathcal{N}}$ and $P_{\mathcal{M},\mathcal{N}}$ at $t = 1$

α	$\mathcal{M} = 128, \mathcal{N} = 128$	$\mathcal{M} = 256, \mathcal{N} = 256$	$\mathcal{M} = 512, \mathcal{N} = 512$	$\mathcal{M} = 1024, \mathcal{N} = 1024$
0.3	1.3067e-5	5.9026e-6	2.6970e-6	1.2448e-6
	1.147	1.130	1.115	
0.5	2.0851e-6	1.2357e-6	7.0308e-7	3.8627e-7
	0.755	0.814	0.864	
0.7	9.5306e-6	3.7294e-6	1.4316e-6	5.3962e-7
	1.354	1.381	1.408	

Example 8.2:

Consider the following problem:

$$
\begin{cases}
\mathcal{D}_t^\alpha \mathcal{W}(x,t) - \dfrac{\partial^2 \mathcal{W}}{\partial x^2} + \displaystyle\int_0^t e^{xt}\mathcal{W}(x,s)ds = f(x,t), \ (x,t) \in [0,1] \times (0,1], \\
\mathcal{W}(0,t) = 0, \ \mathcal{W}(1,t) = 0 \ \forall t \in (0,1], \\
\mathcal{W}(x,0) = 0 \ \forall x \in [0,1],
\end{cases}
$$

where $f(x,t) = \Gamma(1+\alpha)\sin(\pi x) + \pi^2 t^\alpha \sin(\pi x) + \dfrac{t^{1+\alpha}}{1+\alpha} e^{xt} \sin(\pi x)$. The exact solution
is given by $\mathcal{W}(x,t) = t^\alpha \sin(\pi x)$. Taking $\mathcal{M} = 32, \mathcal{N} = 64$, the exact solution of Example 8.2
is displayed in Figure 8.4a for $\alpha = 0.3$. The numerical solution is given in Figure 8.4b.
Similarly, for $\alpha = 0.8$, the exact and the numerical solution is shown in Figure 8.5.
The log-log plot for Example 8.2 is given in Figure 8.6. Table 8.4 shows $E_{\mathcal{M},\mathcal{N}}$ and
$P_{\mathcal{M},\mathcal{N}}$ on $\overline{\Omega}$. The results are in agreement with the theoretical analysis.

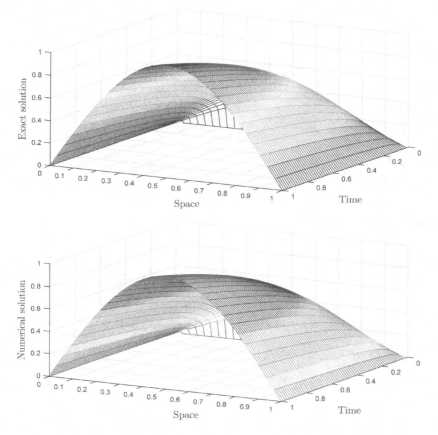

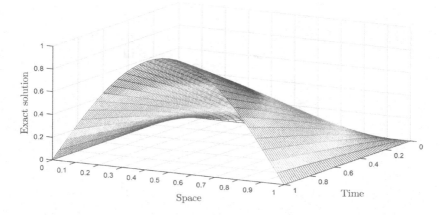

FIGURE 8.4   (a) Exact solution (b) Numerical solution for Example 8.2 with $\alpha = 0.3$.

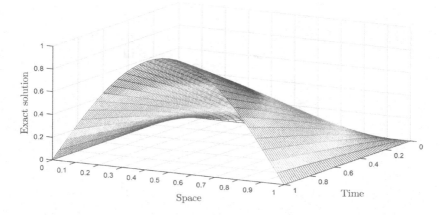

FIGURE 8.5   (a) Exact solution (b) Numerical solution for Example 8.2 with $\alpha = 0.8$.

(Continued)

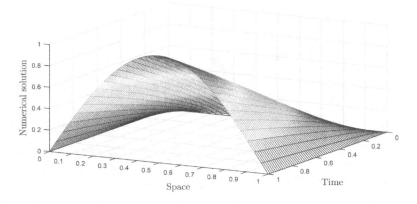

FIGURE 8.5   *(Continued)*

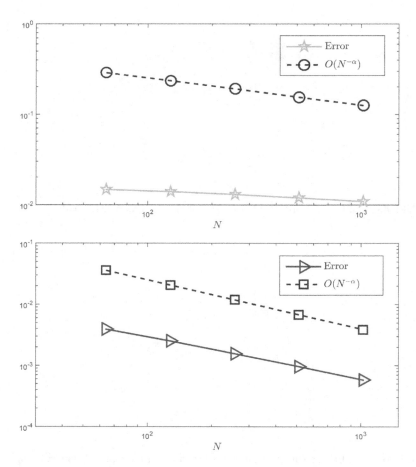

FIGURE 8.6   Log-log plots with (a) $\alpha = 0.3$ (b) $\alpha = 0.8$ for Example 8.2.

TABLE 8.4

Example 8.2: $E_{M,N}$ and $P_{M,N}$ on $\bar{\Omega}$

α	$M = 64, N = 128$	$M = 128, N = 256$	$M = 256, N = 512$	$M = 512, N = 1024$
0.5	1.0684e-2	8.6689e-3	6.8394e-3	5.2665e-3
	0.301	0.342	0.377	
0.7	4.7639e-3	3.2166e-3	2.1052e-3	1.3483e-3
	0.567	0.612	0.643	
0.9	1.0580e-3	6.4045e-4	3.7850e-4	2.1863e-4
	0.724	0.759	0.792	

Example 8.3:

Consider another text problem for $\alpha \in (0,1)$:

$$\begin{cases} \mathcal{D}_t^\alpha \mathcal{W}(x,t) - \frac{\partial^2 \mathcal{W}}{\partial x^2} + \int_0^t xt\mathcal{W}(x,s)ds = f(x,t), \ (x,t) \in [0,1] \times (0,1], \\ \mathcal{W}(0,t) = 0, \ \mathcal{W}(1,t) = t^\alpha \cos(1) \ \forall t \in (0,1], \\ \mathcal{W}(x,0) = 0 \ \forall x \in [0,1]. \end{cases}$$

Here

$$f(x,t) = \Gamma(1+\alpha)\sin\left(x+\frac{\pi x}{2}\right) + \left(1+\frac{\pi}{2}\right)^2 t^\alpha \sin\left(x+\frac{\pi x}{2}\right)$$
$$+ \left(\frac{1}{1+\alpha}\right) xt^{2+\alpha} \sin\left(x+\frac{\pi x}{2}\right).$$

The exact solution is $\mathcal{W}(x,t) = t^\alpha \sin\left(x+\frac{\pi x}{2}\right)$. The graphical representation of the exact and the numerical solution is shown in Figure 8.7 with $\alpha = 0.4$. Figure 8.8a and b displays the maximum errors for different values of N. It is clear that the error plot decreases as N increases. This proves the convergence of the scheme. Table 8.5 represents $E_{M,N}$ and $P_{M,N}$ on $\bar{\Omega}$ and Table 8.6 shows the same at $t = 0.5$. It is proved that the rate of convergence is very high away from the origin compared to the entire region which satisfies the theoretical analysis.

The following example is the generalized form of an integro PDE where the model contains more than one fractional Caputo derivative.

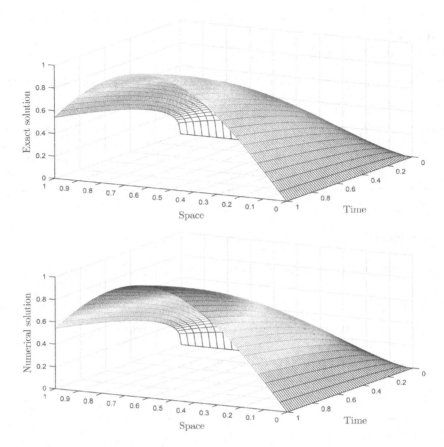

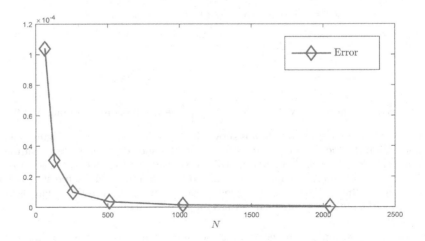

FIGURE 8.7   (a) Exact solution (b) Numerical solution for Example 8.3 with $\alpha = 0.4$.

FIGURE 8.8   Error plots with (a) $\alpha = 0.4$ (b) $\alpha = 0.6$ for Example 8.3.   *(Continued)*

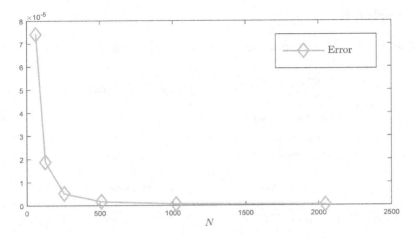

FIGURE 8.8   *(Continued)*

TABLE 8.5

Example 8.3: $E_{\mathcal{M},\mathcal{N}}$ and $P_{\mathcal{M},\mathcal{N}}$ on $\overline{\Omega}$

α	$\mathcal{M} = 256, \mathcal{N} = 256$	$\mathcal{M} = 512, \mathcal{N} = 512$	$\mathcal{M} = 1024, \mathcal{N} = 1024$	$\mathcal{M} = 2048, \mathcal{N} = 2048$
0.3	1.3359e-2	1.2302e-2	1.1209e-2	1.0103e-2
	0.119	0.134	0.150	
0.5	8.9231e-3	7.0309e-3	5.4065e-3	4.0726e-3
	0.344	0.379	0.409	
0.7	3.2956e-3	2.1496e-3	1.3710e-3	8.6157e-4
	0.614	0.649	0.670	

TABLE 8.6

Example 8.3: $E_{\mathcal{M},\mathcal{N}}$ and $P_{\mathcal{M},\mathcal{N}}$ at $t = 0.5$

α	$\mathcal{M} = 200, \mathcal{N} = 200$	$\mathcal{M} = 400, \mathcal{N} = 400$	$\mathcal{M} = 800, \mathcal{N} = 800$	$\mathcal{M} = 1600, \mathcal{N} = 1600$
0.3	1.0324e-5	5.7479e-6	2.8799e-6	1.3852e-6
	0.845	0.997	1.056	
0.5	1.8216e-5	9.0335e-6	4.3414e-6	2.0713e-6
	1.012	1.057	1.068	
0.7	3.2538e-5	1.5566e-5	7.3358e-6	3.4471e-6
	1.064	1.085	1.090	

Example 8.4:

Examine the following multi-term time fractional integro PDE:

$$\begin{cases} \left[\mathcal{D}_t^{\alpha_1} + \sum_{k=2}^{3} \mathcal{D}_t^{\alpha_k} \right] \mathcal{W}(x,t) - \dfrac{\partial^2 \mathcal{W}}{\partial x^2} + \int_0^t \mathcal{K}(x,t-s)\mathcal{W}(x,s)ds = f(x,t), \\[2mm] (x,t) \in [0,1] \times (0,1], \, with \\[1mm] \mathcal{W}(0,t) = 0, \; \mathcal{W}(1,t) = t^{\alpha_1} \; \forall t \in (0,1], \\[1mm] \mathcal{W}(x,0) = 0 \; \forall x \in [0,1], \end{cases}$$

where $0 < \alpha_3 < \alpha_2 < \alpha_1 < 1$. $\mathcal{K}(x,t-s) = e^{xt}$ and $f(x,t)$ is given by:

$$f(x,t) = x \left[\Gamma(\alpha_1+1) + \frac{\Gamma(\alpha_1+1)}{\Gamma(\alpha_1-\alpha_2+1)} t^{\alpha_1-\alpha_2} + \frac{\Gamma(\alpha_1+1)}{\Gamma(\alpha_1-\alpha_3+1)} t^{\alpha_1-\alpha_3} \right] + \frac{xe^{xt}}{\alpha_1+1} t^{\alpha_1+1}.$$

$\mathcal{W}(x,t) = xt^{\alpha_1}$ is the exact solution for Example 8.4. The surface represented in Figure 8.9a shows the exact solution with $\alpha_1 = 0.3, \alpha_2 = 0.2, \alpha_3 = 0.1$ for Example 8.4 with $\mathcal{M} = 32, \mathcal{N} = 64$. The numerical surface plot is shown in Figure 8.9b. One

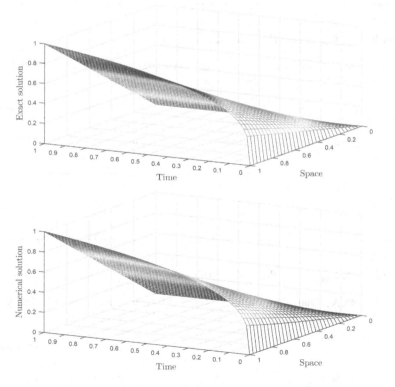

FIGURE 8.9 (a) Exact solution (b) Numerical solution with $\alpha_1 = 0.3, \alpha_2 = 0.2, \alpha_3 = 0.1$ for Example 8.4.

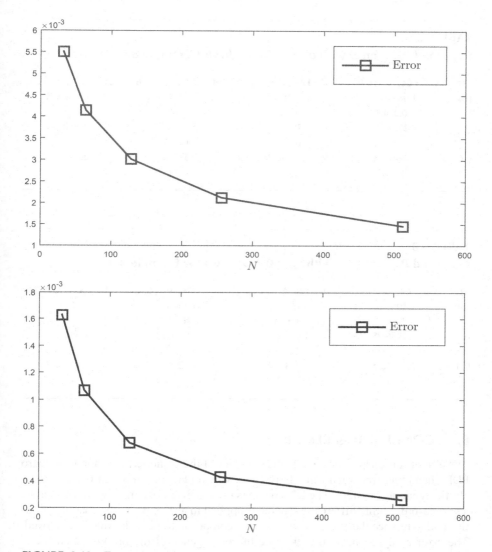

FIGURE 8.10 Error plots with (a) $\alpha_1 = 0.7$ (b) $\alpha_1 = 0.9$ and fixed $\alpha_2 = 0.3, \alpha_3 = 0.1$ for Example 8.4.

can see that the numerical solution converges to the exact solution. The maximum error plot is displayed in Figure 8.10 with fixed $\alpha_2 = 0.3, \alpha_3 = 0.1$ and different values of α_1 which proves the consistency of the proposed scheme. It is clear that the error decreases as the number of mesh points increases. The maximum error and the rate of convergence are shown in tabulated form in Table 8.7 over $\bar{\Omega}$ and in Table 8.8 at $t = 1$ with fixed $\alpha_2 = 0.3, \alpha_3 = 0.1$ and different α_1. It is shown that the rate of convergence is of order $O(\tau^{\alpha_1})$ over entire region $\bar{\Omega}$, but it is of order $O(\tau)$ on any subdomain away from $t = 0$ which satisfies the theoretical arguments (see Remark 8.2).

TABLE 8.7

$E_{\mathcal{M},\mathcal{N}}$ and $P_{\mathcal{M},\mathcal{N}}$ on $\bar{\Omega}$ with $\alpha_2 = 0.3, \alpha_3 = 0.1$ for Example 8.4

α_1	$\mathcal{M} = 64, \mathcal{N} = 64$	$\mathcal{M} = 128, \mathcal{N} = 128$	$\mathcal{M} = 256, \mathcal{N} = 256$	$\mathcal{M} = 512, \mathcal{N} = 512$
0.4	1.0895e-2	9.3910e-3	7.9949e-3	6.7289e-3
	0.214	0.232	0.249	
0.5	8.2706e-3	6.7871e-3	5.4734e-3	4.3440e-3
	0.285	0.310	0.333	
0.6	6.0844e-3	4.7224e-3	3.5779e-3	2.6536e-3
	0.366	0.400	0.431	

TABLE 8.8

$E_{\mathcal{M},\mathcal{N}}$ and $P_{\mathcal{M},\mathcal{N}}$ at $t = 1$ with $\alpha_2 = 0.3, \alpha_3 = 0.1$ for Example 8.4

α_1	$\mathcal{M} = 64, \mathcal{N} = 64$	$\mathcal{M} = 128, \mathcal{N} = 128$	$\mathcal{M} = 256, \mathcal{N} = 256$	$\mathcal{M} = 512, \mathcal{N} = 512$
0.4	7.1060e-5	3.1988e-5	1.4508e-5	6.6393e-6
	1.151	1.141	1.128	
0.5	4.2736e-5	1.9760e-5	9.2183e-6	4.3420e-6
	1.113	1.100	1.086	
0.6	2.4271e-5	1.2182e-5	6.1221e-6	3.0794e-6
	0.994	0.993	0.991	

8.7   CONCLUDING REMARKS

The aim of this chapter is to study the behavior of the numerical solution for integro PDEs including fractional time derivatives, which play an important role in viscoelastic dynamical system. The L1 scheme is used to discretize the Caputo derivative, a second-order finite difference scheme is applied to approximate the spatial derivative and to replace the Volterra operator, the composite trapezoidal rule is employed. The error analysis shows that the rate of convergence of the proposed scheme is higher away from the origin compared to the whole region. Further, the scheme is applied on a more general multi-term time fractional partial integro-differential equation and it is proved that the scheme is also effective to solve such models. In addition, several tests are performed in order to show the desired claim.

REFERENCES

[1] R. L. Bagley and P. J. Torvik. 1985. Fractional calculus in the transient analysis of viscoelastically damped structures. *AIAA J.*, 23(6): 918–925.
[2] J. Caballero, A. B. Mingarelli, and K. Sadarangani. 2006. Existence of solutions of an integral equation of Chandrasekhar type in the theory of radiative transfer. *Electron. J. Differential Equations*, 2006(57): 1–11.

[3] K. M. Case and P. F. Zweifel. 1967. *Linear transport theory.* Reading, MA: Addison-Wesley.

[4] S. Hu, M. Khavani, and W. Zhuang. 1989. Integral equations arising in the kinetic theory of gases. *Appl. Anal.*, 34(3–4): 361–266.

[5] I. Aziz and I. Khan. 2017. Numerical solution of partial integro-differential equations of diffusion type. *Math. Probl. Eng.* 2017: 1–11.

[6] P. Das, S. Rana, and H. Ramos. 2020. A perturbation based approach for solving fractional order Volterra-Fredholm integro differential equations and its convergence analysis. *Int. J. Comput. Math.*, 97(10): 1994–2014.

[7] J. M. Sanz-Serna. 1988. A numerical method for a partial integro-differential equation. *SIAM J. Numer. Anal.*, 25(2): 319–327.

[8] S. Shahmorad. 2005. Numerical solution of the general form linear Fredholm–Volterra integro-differential equations by the Tau method with an error estimation. *Appl. Math. Comput.*, 167(2): 1418–1429.

[9] A. Tari and S. Shahmorad. 2008. A computational method for solving two-dimensional linear Fredholm integral equations of the second kind. *ANZIAM J.*, 49(4): 543–549.

[10] J. Thorwe and S. Bhalekar. 2012. Solving partial integro-differential equations using Laplace transform method. *Am J. Comput. Appl. Math.*, 2(3): 101–104.

[11] M. Dehghan. 2006. Solution of a partial integro-differential equation arising from viscoelasticity. *Int. J. Comput. Math.*, 83(1): 123–129.

[12] W. Li and X. Da. 2010. Finite central difference/finite element approximations for parabolic integro-differential equations. *Computing*, 90: 89–111.

[13] F. Fakhar-Izadi and M. Dehghan. 2014. Space–time spectral method for a weakly singular parabolic partial integro-differential equation on irregular domains. *Comput. Math. Appl.*, 67(10): 1884–1904.

[14] F. Shakeri and M. Dehghan. 2013. A high order finite volume element method for solving elliptic partial integro-differential equations. *Appl. Numer. Math.*, 65: 105–118.

[15] J. Zhou, D. Xu, and X. Dai. 2019. Weak Galerkin finite element method for the parabolic integro-differential equation with weakly singular kernel. *Comput. Appl. Math.*, 38(2): 1–12.

[16] A. A. Hamoud, K. P. Ghadle, M. B. Issa, and Giniswamy. 2018. Existence and uniqueness theorems for fractional Volterra-Fredholm integro-differential equations. *Int. J. Appl. Math.*, 31(3): 333–348.

[17] M. M. Matar. 2009. Existence and uniqueness of solutions to fractional semilinear mixed Volterra-Fredholm integro-differential equations with nonlocal conditions. *Electron. J. Differ. Eq.*, 2009(155): 1–7.

[18] A. A. Hamoud and K. P. Ghadle. 2018. Modified Laplace decomposition method for fractional Volterra-Fredholm integro-differential equations. *J. Math. Model.*, 6(1): 91–104.

[19] S. Momani and M. A. Noor. 2006. Numerical methods for fourth-order fractional integro-differential equations. *Appl. Math. Comput.*, 182(1): 754–760.

[20] S. Santra and J. Mohapatra. 2021. A novel finite difference technique with error estimate for time fractional partial integro-differential equation of Volterra type. *J. Comput. Appl. Math.*, 400: 1–13.

[21] S. Santra and J. Mohapatra. 2021. Numerical analysis of Volterra integro-differential equations with Caputo fractional derivative. *J. Sci. Technol. Trans. A Sci.*, 45(5): 1815–1824.

[22] A. Panda, S. Santra, and J. Mohapatra. 2021. Adomian decomposition and homotopy perturbation method for the solution of time fractional partial integro-differential equations. *J. Appl. Math. Comput.*, 68(3): 2065–2082.

[23] E. A. Rawashdeh. 2006. Numerical solution of fractional integro-differential equations by collocation method. *Appl. Math. Comput.*, 176(1): 1–6.

[24] S. Santra, A. Panda, and J. Mohapatra. 2021. A novel approach for solving multi-term time fractional Volterra-Fredholm partial integro-differential equations. *J. Appl. Math. Comput.*, 68(5): 3545–3563.

[25] S. Shahmorad, M. H. Ostadzad, and D. Baleanu. 2020. A Tau-like numerical method for solving fractional delay integro-differential equations. *Appl. Numer. Math.*, 151: 322–336.

[26] S. Abbas, M. Benchohra, and T. Diagana. 2013. Existence and attractivity results for some fractional order partial integro-differential equations with delay. *Afr. Diaspora J. Math.*, 15(2): 87–100.

[27] K. Diethelm. 2010. *The analysis of fractional differential equations: An application-oriented exposition using differential operators of Caputo type.* Berlin: Springer-Verlag.

[28] I. Podlubny. 1999. *Fractional differential equations: An introduction to fractional derivatives, fractional differential equations, to methods of their solution and some of their applications.* San Diego, CA: Mathematics in Science and Engineering, Academic Press, Inc.

[29] J. L. Gracia, E. O'Riordan, and M. Stynes. 2018. Convergence in positive time for a finite difference method applied to a fractional convection-diffusion problem. *Comput. Methods Appl. Math.*, 18(1): 33–42.

[30] A. Jhinga and V. Daftardar-Gejji. 2019. A new numerical method for solving fractional delay differential equations. *Comput. Appl. Math.*, 38(4): 1–18.

[31] S. Santra and J. Mohapatra. 2020. Analysis of the L1 scheme for a time fractional parabolic-elliptic problem involving weak singularity. *Math. Methods Appl. Sci.*, 44(2): 1529–1541.

[32] M. Stynes, E. O'Riordan, and J. L. Gracia. 2017. Error analysis of a finite difference method on graded meshes for a time-fractional diffusion equation. *SIAM J. Numer. Anal.*, 55(2): 1057–1079.

9 From Continuous Time Random Walk Models to Human Decision-Making Modelling

A Fractional Perspective

Amir Hosein Hadian Rasanan,
Mohammad Mahdi Moayeri,
Jamal Amani Rad, and Kourosh Parand

CONTENTS

9.1   INTRODUCTION

Decision-making is a high-level cognitive function that has some effects on all aspects of our daily life. Choosing food in a restaurant, choosing a vaccine for the

DOI: 10.1201/9781003328032-9

Covid-19 virus, choosing a friend, and many other choices are examples of different types of decisions that we are dealing with in our daily life. Thus, understanding this cognitive process can be very important and also has many applications in different fields of science including computer science (Lehmann et al., 2019), clinical science (Huang-Pollock et al., 2017; Pedersen et al., 2017), economy (Rieskamp, 2008; Rieskamp et al., 2006), and so on. Consequently, different kinds of models have been developed for exploring the underlying mechanisms of decision-making (Brown & Heathcote, 2008; Eric-Jan Wagenmakers et al., 2007; Hawkins & Heathcote, 2021; Krajbich et al., 2012; Ratcliff, 1978; Usher & McClelland, 2001; van Ravenzwaaij et al., 2019).

Consider a scenario in which you are instructed to select the most luminant option among two presenting options. What does happen in the brain to answer this question? The Diffusion Decision Model (DDM) (Ratcliff, 1978; Ratcliff & McKoon, 2008) provides a comprehensive answer to this question. Based on this model, a decision is made when the accumulated information to the benefit of each option exceeds a threshold. So, this model considers an accumulator which is fluctuating between two absorbing boundaries. Therefore, the accumulator can be formulated using the following stochastic differential equations (Smith, 2000):

$$\frac{d}{dt} X(t) = v + \xi(t) \tag{9.1}$$

where $X(t)$ is the accumulator and shows the amount of accumulated information at time t, v is the drift rate that indicates the speed of information processing, and $\xi(t)$ is the normal Gaussian noise that adds some fluctuations to the information accumulation process (Ratcliff & Smith, 2004; Voss et al., 2004, 2019). The process is started from a point between $-\theta$ and θ (i.e., $X(0) = z$) and is stopped whenever $|X(t)| \geq \theta$. Thus, z, v, and θ are the parameters of the DDM and each has a real psychological meaning. In fact, parameter z stands for bias which means the decision-maker is more willing to answer one option. So, if $z > 0$, then the decision-maker is biased to the upper boundary's option and vice versa. Also, parameter v stands for the speed of information processing and defines the difficulty level of the cognitive task (i.e., high value of drift rate parameter implies low level of difficulty and high speed of information accumulation, and low value of drift rate implies high level of difficulty and low speed of information accumulation). Furthermore, parameter θ stands for the amount of information that is needed for making a decision and determines the strategy of making a decision. To put it simply, when the threshold has a high value, the decision-maker needs more information for making a decision. So, the decision process takes more time and is more accurate (i.e., decision strategy is caution). Also, a low-threshold value implies less information is needed for making a decision and the decision process takes less time and is less accurate (i.e., decision strategy is radical). In addition to these parameters, there exists one extra parameter for non-decision components of the response time which is the non-decision time parameter, t_0, and is defined as the summation of the encoding time and motor time (Voss et al., 2004). The whole process of making a decision in the DDM is illustrated in Figure 9.1.

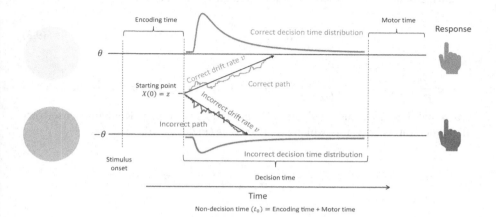

FIGURE 9.1 Schematic view of the diffusion decision model for the luminance discrimination task.

The standard diffusion process can be tracked by the following diffusion equation (Mainardi et al., 2008; Shinn et al., 2020; Voss & Voss, 2008):

$$\frac{\partial}{\partial t} u(x, t) = \frac{\partial^2}{\partial x^2} u(x, t) \tag{9.2}$$

in which $u(x, t)$ is the distribution of the location of the accumulator (i.e., $u(x, t) = P(X(t) = x)$. There are some extensions to this equation that can be interpreted as a psychological process (Ratcliff et al., 2016). For example, the process can be explored in a time-dependent boundary domain which yields a moving boundary problem (Smith, 2000; Smith & Ratcliff, 2021; Voskuilen et al., 2016). The other extension of the diffusion process for decision-making is obtained by considering a Lèvy noise instead of Gaussian noise (Voss et al., 2019) which causes some sudden jumps in the accumulation process. Some research works have been done on the psychological interpretation of jumps of accumulation process and it is clear that the heavy-tailed α-stable noise is a more realistic noise distribution (Hadian Rasanan et al., 2021; Wieschen et al., 2020). Mathematically, evidence accumulation in this kind of decision model can be written in the following discrete form:

$$X(t + \Delta t) = X(t) + v\Delta t + e\Delta t^{\frac{1}{\alpha}}, \ e \sim \text{Stable}\left(\alpha, \beta = 0, \gamma = \frac{1}{\sqrt{2}}, \delta = 0\right)$$

where $\alpha \in [0,2]$ is the stability index (Lèvy index), $\beta \in [-1,1]$ is the skewness parameter, $\gamma > 0$ is the scale parameter, and δ is the shift parameter. By considering this type of noise for the accumulation process, the diffusion equation is converted to the following double-sided fractional form (Hadian Rasanan et al., 2021):

$$\frac{\partial}{\partial t} u(x,\, t) = \frac{\partial^{\alpha}}{\partial |x|^{\alpha}} u(x,\, t) \tag{9.3}$$

where $\dfrac{\partial^{\alpha}}{\partial |x|^{\alpha}} f(x) = \dfrac{1}{2\cos(\alpha\pi/2)} \left({}_{-\infty}^{R}D_{x}^{\alpha} f(x) + {}_{x}^{L}D_{\infty}^{\alpha} f(x) \right)$, $1 < \alpha \leq 2$, is the order of fractional derivative and ${}_{-\infty}^{R}D_{x}^{\alpha}$ and ${}_{x}^{L}D_{\infty}^{\alpha}$ are the right and left Riemann sense fractional derivative operators which are defined as (Padash et al., 2019):

$$_{-\infty}^{R}D_{x}^{\alpha} = \frac{1}{\Gamma(2-\alpha)} \frac{d^2}{dx^2} \int_{-\infty}^{x} \frac{f(\xi)}{(x-\xi)^{\alpha-1}} d\xi \tag{9.4}$$

$$_{x}^{L}D_{\infty}^{\alpha} = \frac{1}{\Gamma(2-\alpha)} \frac{d^2}{dx^2} \int_{x}^{\infty} \frac{f(\xi)}{(\xi-x)^{\alpha-1}} d\xi \tag{9.5}$$

In addition to the Lèvy process, there exist some other processes which are able to capture jumps of the information accumulation process such as the jump-diffusion model. But one of the most general and powerful models for exploration of jumping during the accumulation process is the distributed-order Lèvy walk model which yields a distributed-order fractional diffusion model as follows (Magdziarz & Teuerle, 2017; Mainardi et al., 2008):

$$\int_{0}^{1} \phi(p) \, {}_{0}^{C}D_{t}^{p} u(x,\, t)\, dp = \frac{\partial^2}{\partial x^2} u(x,\, t) \tag{9.6}$$

where $\phi(p)$ is the density function of p over the interval $[0,\, 1]$ and ${}_{0}^{C}D_{t}^{p}$ is the Caputo sense fractional derivative which is defined as follows:

$$_{0}^{C}D_{t}^{p} f(t) = \frac{1}{\Gamma(k-p)} \int_{0}^{t} \frac{f^{(k)}(\tau)}{(t-\tau)^{\alpha+1-k}} d\tau,\ k-1 < p < k \tag{9.7}$$

This equation is applicable in different fields of science (for more applications, interested readers can see Ding et al. [2021]).

Despite the potential benefits of using the Lèvy decision model as a general form of the drift-diffusion model, this model has a fundamental problem, i.e., the lack of its likelihood function which makes it impossible to use this model to fit on behavioural data. Also, this limitation makes it impossible to use it in various applications, including clinical applications. Moreover, it is clear that the likelihood of the Lèvy model can be obtained by solving the fractional differential equation mentioned before. Therefore, the main problem in this chapter is to solve this equation, that is, to find a suitable solution for the joint probability distribution of the Lèvy flight model so that in future research, researchers can use it to fit the model on human behaviour data. However, solving this equation is quite challenging in the field of scientific computing. Besides, in recent years, the use of machine learning grew noticeably around the world, and it is utilized for solving various problems in diverse fields, such as computer science, neuroscience (Carlson et al., 2018), cognitive science (Sadeghi et al., 2017), mechanics (Brunton et al., 2020), astronomy (Kremer et al., 2017), and many others. The high-speed growth attracted scientists' attention to this subject and it resulted in obtaining many new applications for it. One of the fields that captivated researchers and scientists lately is machine learning in dynamic systems. Various machine learning tools have been applied in simulating dynamical systems, for instance, artificial neural network (ANN) (Chakraverty & Mall, 2017), support vector machine (SVM) (Parand et al., 2021), and the Gaussian process (Raissi et al., 2017).

Orthogonal networks or functional link neural networks were first proposed by Chakraverty and his collaborators (Chakraverty & Mall, 2014), and it is used by many other researchers for solving ordinary differential equations (ODEs) and partial differential equations (PDEs). For example, they presented a Chebyshev neural network model to solve Lane–Emden equations (Mall & Chakraverty, 2014) and also used Chebyshev neural networks for nonlinear singular initial value problems of Emden–Fowler type (Mall & Chakraverty, 2015), the elliptic PDEs (Mall & Chakraverty, 2017) and many other titles (Jeswal & Chakraverty, 2018; Mall & Chakraverty, 2016b, 2016a). In particular, they utilized gradient descent as the optimizer and orthogonal polynomials as hidden layer activation functions. Moreover, Hadian and his collaborators presented a fractional orthogonal neural network based on fractional-order Legendre functions for solving different types of Lane–Emden equations (Hadian-Rasanan et al., 2020). In addition, there is another similar and slightly different approach in this branch, and that is the metaheuristic stochastic neural networks. In fact, metaheuristic stochastic neural networks build the ordinary/partial differential equations through the network and by using the metaheuristic optimizers such as genetic algorithm, they can solve various equations. It can be said that the initiators of this approach are Raja et al., who have several fruitful papers on this topic, for example, they solved the MHD Jeffery–Hamel problem using such a neural network (Raja & Samar, 2014), or one-dimensional Bratu equation (Raja & Ahmad, 2014). For more information, interested readers can refer to Ahmad et al., 2017; Masood et al., 2017; Raja et al., 2019. Despite the goodness of these approaches, a more powerful and new research direction in this area that has received much attention recently is the use of deep learning to simulate linear/nonlinear ODE and PDE solutions. In general, it can be said that in the deep learning paradigm, the neural network is considered as a universal function approximator and it has usually more than five hidden layers. In the field of scientific computing, recently Raissi et al. (2019) proposed this approach, which can be helpful for those who are interested to discover more about the application of deep neural networks in differential equations. Another probabilistic method is the Gaussian process approach and it can be used to solve linear ODEs and PDEs. For using this method for a nonlinear problem, first, a linearization procedure should be applied to the problem; for more guidance, Archambeau et al. (2007) and Raissi et al. (2018) can be helpful.

As mentioned, one of the useful tools for solving differential equations are neural networks which have some advantages. The first is its demonstrative ability which means even with a small neural network, the solution of a PDE can be approximated pretty accurately. Another merit is that it can be executed without any great effort and, of course, it should be mentioned since the algorithm is aided by drawing points randomly from a domain, it can be extended to arbitrary domains especially in the higher dimensions. As is obvious, some of the classical methods for solving PDEs have unquestionable restrictions but with neural networks, we have freedom in the approximation spaces. The last good feature would be the sensor data in which the information of the sensors can be combined by adding a term to the loss functions.

Despite the significant benefits of neural networks, some of them have few disadvantages. The first drawback is because of the optimization issues, it cannot indicate that the errors are decreasing and also when the neural network becomes bigger, it is even harder to demonstrate the convergence theoretically. Additionally, it cannot be speculated how well the neural networks can scale to more difficult problems. The last demerit would be that most of the classical methods are completely deterministic but, on the contrary, due to different random initialization, it can obtain different results by different initial value.

In this chapter, we tend to propose a new neural network framework based on a novel family of Lagrange functions that are called *Developed Lagrange Functions* for solving the following distributed-order fractional ordinary and partial differential equations:

1. Distributed-order fractional ordinary differential equations (DFODEs)

$$\int_a^b G_1\big(p,\, D^p u(t)\big)dp + G_2\big(t,\, u(t),\, D^{\alpha_i} u(t)\big) = F(t),\, t \in [0,\, T] \qquad (9.8)$$

with the following initial condition:

$$u^{(h)}(0) = u_h$$

where $\alpha_i \geq 0$ and $h = 0, 1, \ldots, \lfloor \max\{b,\, \alpha_i\}\rfloor$ for $i = 1, 2, \ldots$ Furthermore, G_1 and G_2 are continuous (non)linear functions and $D^p u(t) = {}_0^C D_t^p u(t)$.

2. Distributed-order fractional partial differential equations (DFPDEs)

$$\int_0^1 G\big(p,\, D^p u(x,t)\big)dp = \frac{\partial^2 u(x,t)}{\partial x^2} + F(x,t),\, (x,t) \in [x_0,\, x_1] \times [0,\, T] \qquad (9.9)$$

with the following initial and boundary conditions:

$$u(x,0) = f(x) \qquad (9.10)$$

$$u(x_0,t) = g_0(t),\, u(x_1,t) = g_1(t) \qquad (9.11)$$

where F, g_0, g_1 and f are continuous functions and $D^p u(t,\, x) = {}_0^C D_t^p u(t,\, x)$.

Many researchers worked on the following topic which we have presented in the next section.

9.1.1 A Brief Review of Other Methods Existing in the Literature

In 2019, Sabermahani et al. introduced a general formulation for fractional-order general Lagrange scaling functions that were used for solving a class of fractional differential equations and also a particular class of fractional delay differential equations (Sabermahani et al., 2019; 2020). Sabermahani et al. (2019) provided an algorithm which is based on fractional-order Lagrange polynomials and collocations method for solving delay fractional optimal control problems. Also, Delkhosh and Parand (2019) proposed a new method called the Generalization Pseudospectral Method (GPM) for solving various types of differential equations. On the other hand, for solving a class of FDEs, Sabermahani et al. (2018) presented a new set of fractional functions which is based on Lagrange polynomials. In fact, they first proposed a new representation of Lagrange polynomials and then the fractional-order Lagrange polynomials was introduced. For solving steady-state and time-dependent fractional PDEs, Zayernouri and Karniadakis (2014) proposed an exponentially accurate fractional spectral collocation method, which is called the fractional Lagrange interpolants. Another research work undertaken in this direction is by Esmaeili and Shamsi (2011). In fact, they generalized the pseudospectral method with initial conditions. For achieving this purpose, they presented an appropriate representation of the solution and also derived the pseudospectral differentiation matrix of fractional order. Also, Khosravian-Arab et al. (2017) have tried to provide exponentially accurate Galerkin, Petrov–Galerkin, and pseudospectral methods for FDEs on a semi-infinite interval. It can be said in more detail that they first introduced two new non-classical Lagrange basis functions based on two new families of the Generalized Associated Laguerre Functions: GALFs-1 and GALFs-2, which is obtained recently by Khosravian-Arab et al. (2017) called Non-classical Lagrange Basis Functions, i.e., NLBFs-1 and NLBFs-2 (Khosravian-Arab et al., 2015). Ejlali and Hosseini (2017) used a direct pseudospectral method for solving the fractional optimal control, which is based on the Lagrange interpolating functions with fractional power terms. By considering that the most applied fractional problems have solutions in terms of the fractional power, they understand that by using an appropriate characteristic nodal-based functions with suitable power can obtain more accurate pseudospectral approximations of the solution.

In order to focus on specific fractional models of this chapter, it can be said that Katsikadelis (2014) presented a numerical algorithm based on the finite difference method for approximating both linear and nonlinear DFODEs. Also, a numerical method is represented for the solution of a DFODEs of a general form by Diethelm and Ford (2009). Recently, Mashayekhi and Razzaghi (2016) introduced a method based on hybrid functions approximation and also its properties consisting of block-pulse functions. More recently, Razzaghi and Groza (2019) proposed a technique on the polynomial approximation for solving DFODEs in two independent variables. Also, a numerical algorithm for solving linear and nonlinear DFODEs in which its fractional derivative is described in the Caputo sense is demonstrated in Yuttanan and Razzaghi (2019). On the other hand, Dehestani et al. represented a method for solving various kinds of DFODEs based on Genocchi hybrid functions (Dehestani et al., 2019). With an analytical look, Najafi et al. (2011) analyzed the stability of three classes of DFODE by considering the non-negative density function, and they discovered a robust stability condition for these systems. Also, Atanacković et al. (2007) studied the uniqueness and existence of mild and classical solutions for equations of the form $y^{(2)}(t) + \int_0^2 \phi(\alpha) y^{(\alpha)}(t) d\alpha = f(y(t), t)$. Meerschaert et al. (2011) represented explicit strong solutions and stochastic analogues for distributed-order time-fractional diffusion equations for bounded domains with the help of Dirichlet boundary conditions. Sun et al. (2011) presented a new concept of random-order fractional differential equation model which includes a noise term in the fractional order.

Using artificial intelligence methods for solving diverse types of differential equations became widespread very quickly. The orthogonal neural network, radial basis function networks, and deep neural networks are some examples of artificial neural network frameworks which are used to simulate a dynamical system for solving PDEs and ODE and also FDEs. By using the merger of neural network minimization and finite difference method, Lee and Kang (1990) achieved the ability to solve differential equations. Meade and Fernandez (1994b, 1994a) generated a feed-forward neural network based on spline functions and utilized it for solving linear ODEs and also other related works. Dissanayake and Phan-Thien (1994) proposed a multilayer neural network algorithm to solve PDEs, and then Lagaris et al. (1998) employed the gradient descent approach for solving diverse ODEs and PDEs. Three years later, Mai-Duy and Tran-Cong (2001) developed a radial neural network to solve the mentioned equations. Also, a neural network approach that is utilized to solve fuzzy differential equations is presented by Effati and Pakdaman, (2010). Li Jianyu et al. (2003) introduced a radial basis neural network to solve the elliptic PDEs. It is noteworthy that neural network techniques are applied for solving the integral equations and fuzzy integral equations (Asari et al., 2019; Effati & Buzhabadi, 2012; Golbabai & Seifollahi, 2006; Golbabai et al., 2009). Another significant point is that Mall and Chakraverty (2014) proposed the orthogonal neural networks and utilized them to solve ODEs with the help of Chebyshev basis polynomials. In fact, they were also able to solve Lane–Emden and elliptic differential equations by using artificial neural networks aided by the mentioned Chebyshev functions (Mall & Chakraverty, 2014, 2015, 2016a, 2016b, 2017). Raja et al. (2011) used computational intelligence techniques to solve FDEs. For instance, they proposed a feed-forward neural network in which it applied the genetic algorithm as an optimizer for solving the fractional-order system of the Bagley–Torvik equation. He and his collaborators used artificial neural networks and sequential quadratic programming for optimizing the network weights and proposed a computational intelligence framework that can approximate the solution of nonlinear quadratic Riccati fractional differential equations (Raja et al., 2015). Additionally, they developed an optimizer for solving a system of Bagley–Torvik equations which utilized the interior points (Raja et al., 2017). Pakdaman et al. (2017) utilized the Quasi–Newton method as an optimizer and the sigmoidal neural network for solving the fractional initial value problems.

Also, for solving FDEs of variable order with Mittag–Leffler kernel, Zuniga-Aguilar et al. (2017) introduced a three-layer artificial neural network that can be optimized by the Levenberg–Marquardt algorithm. Recently, the high-order linear fractional differential equations are solved with the help of artificial neural networks (Rostami & Jafarian, 2018). We can also utilize the artificial neural networks for solving some other networks such as fractional delay differential equations (Zúñiga-Aguilar et al., 2018) and fractional optimal control problems (Javad Sabouri et al., 2017).

The rest of this chapter is organized as follows: Section 9.2 presents the required preliminaries. In this section, the definition of Lagrange polynomials is recalled and a new generation of Lagrange functions, namely the Developed Lagrange Functions (DLFs), is introduced. In Section 9.3, our suggested method is illustrated, first by presenting its neural network architecture for solving DFODE and then DFPDE. In Section 9.4, for indicating the efficiency of the mentioned method, seven examples are provided of which four of them are DFODE and the remaining examples are DFPDE and, additionally, these examples are compared with the other related works to demonstrate the accuracy of the method. Further, the convergence of the proposed method is illustrated numerically in this section. Finally, in Section 9.5, a conclusion is assigned for summing up the chapter.

9.2   PRELIMINARIES AND NOTATIONS

In this section, first, by remarking some properties of Lagrange polynomials, we introduce developed Lagrange functions. Then, some basics of fractional calculus and fractional derivative with Caputo sense are recalled.

9.2.1   LAGRANGE POLYNOMIALS

There are many equations in different fields that cannot be solved analytically; therefore, we use numerical and semi-analytical methods for solving them and due to this, we are going to introduce the Lagrange interpolation and polynomials. Suppose we have $n+1$ data points as $\{x_i\}_{i=0}^{n}$, in this case, the Lagrange interpolation of our function would be as follows:

$$f_n(x) = \sum_{j=0}^{n} L_j(x) f(x_j) \tag{9.12}$$

in which $L_j(x)$ is the Lagrange polynomial and it stands for

$$L_j(x) = \prod_{i=0,\, i \neq j}^{n} \frac{x - x_i}{x_j - x_i} \tag{9.13}$$

and also $j \in \{0, 1, \ldots, n\}$. As mentioned before, Sabermahani et al. (2019) proposed a new set of Lagrange functions which is based on the Lagrange polynomials as well as this; it can be used for solving a class of FDEs and it is documented in Sabermahani et al. (2018). Suppose $L_j(x)$, for all $j = 0, 1, \ldots, n$, are Lagrange polynomials on the set of nodes $x_j \in [0,1]$. The Lagrange polynomials with respect to these points are described as follows:

$$L_j(x) = \frac{1}{\prod_{i=0,\, i \neq j}^{n} x_j - x_i} \left(x^n - \sum_{k_1=0,\, i \neq k}^{n} x_{k_1} x^{n-1} + \sum_{k_2=k_1+1}^{n} \sum_{\substack{k_1=0 \\ i \neq k_1}}^{n-1} x_{k_1} x_{k_2} x^{n-2} - \cdots + (-1)^n \right.$$

$$\left. \sum_{\substack{k_n=k_{n-1}+1 \\ i \neq k_n \neq k_{n-1} \neq \cdots \neq k_1}}^{n} \sum_{k_{n-1}=k_{n-2}+1}^{n-1} \cdots \sum_{k_1=0}^{1} \prod_{r=1}^{n} x_{k_r} \right) \tag{9.14}$$

By using Eqn. (9.14), each of the $L_j(x)$ can be rewritten as follows:

$$L_j(x) = \sum_{s=0}^{n} \alpha_{sj} x^{n-s}, \quad j = 0, 1, \ldots, n \tag{9.15}$$

in which

$$\alpha_{j0} = \frac{1}{\prod_{j=0,\, j \neq i}^{n} (x_j - x_i)} \tag{9.16}$$

$$\alpha_{js} = \frac{1}{\prod_{j=0,\, j \neq i}^{n} (x_j - x_i)} \sum_{k_s=k_{s-1}+1}^{n} \cdots \sum_{k_1=0}^{n-s+1} \prod_{r=1}^{s} x_{k_r} \tag{9.17}$$

where $s = 1, 2, \ldots, n$ and also, $j \neq k_1 \neq \cdots \neq k_s$.

9.2.2　Developed Lagrange Functions

In this section, primarily, a new class of Lagrange functions, namely the developed Lagrange functions, is introduced, and some of their properties are proposed. By considering the arbitrary points of $\{x_i\}_{i=0}^{n}$ that are located in the domain Ω, and choosing the functions $\phi_i(x)$ such that they are sufficiently differentiable on Ω and satisfying the following two conditions:

1. $\phi_i(x_j) \neq \phi_j(x_i)$ for all $i \neq j$
2. $\phi_i'(x_i) \neq 0$ for any i

We introduce the Developed Lagrange Functions (DLFs) as a new class of functions for the interpolation methods as follows (Delkhosh et al., 2019):

$$L_j^{\phi_k}(x) = L_j(\phi_0, \phi_1, \ldots, \phi_n, x) = \prod_{\substack{i=0 \\ i \neq j}}^{n} \frac{\phi_i(x) - \phi_i(x_i)}{\phi_i(x_j) - \phi_i(x_i)} \tag{9.18}$$

If we suppose that

$$\omega^{\phi_k}(x) = \prod_{i=0}^{n} (\phi_i(x) - \phi_i(x_i)) \tag{9.19}$$

we can easily obtain

$$\frac{d}{dx}\omega^{\phi_k}(x)\Big|_{x=x_j} = \left(\omega^{\phi_k}(x)\right)'\Big|_{x=x_j} = \phi_j'(x_j) \prod_{\substack{i=0 \\ i \neq j}}^{n} (\phi_i(x) - \phi_i(x_i)) \tag{9.20}$$

Thus, Eqn. (9.18) can be rewritten as:

$$L_j^{\phi_k}(x) = \mu_j \frac{\omega^{\phi_k}(x)}{\phi_j(x) - \phi_j(x_j)} \tag{9.21}$$

where $\mu_j = \dfrac{\phi_j'(x_j)}{\left(\omega^{\phi_k}(x)\right)'\big|_{x=x_j}}$. Choosing different functions for $\phi_i(x)$ leads us to obtain

many new basic functions at different domains, some of which are summarized in Table 9.1.

TABLE 9.1
Properties of Choosing Different $\phi_i(x)$

$\phi_i(x)$	Property
x	Classical Lagrange functions will be produced.
x^δ	Where $\delta \in \mathbb{R}^+$, the fractional Lagrange functions will be produced and it is worth mentioning that suppose $0 < \delta \leq 1$, $\phi_i(x)$ is the function used by Sabermahani et al. (2018; 2019; 2020).
$\Phi(x)$	Where $\Phi(x)$ is a certain function, the generalized Lagrange functions, introduced by Delkhosh and Parand (2019), will be produced.
$\dfrac{x - L_j}{x + L_j}$ or $\dfrac{x}{x + L_j}$	Where L_j are the positive real values, the rational Lagrange functions will be produced on the semi-infinite domain $[0, \infty)$.
e^{ix}	The exponential Lagrange functions will be produced on the infinite domain $(-\infty, \infty)$.
$\sin(ix)$ or $\cos(ix)$	The Fourier Lagrange functions will be produced on the infinite domain $(-\infty, \infty)$.
$\begin{cases} e^{ix} & i = 0,\ldots,k \\ \sin(ix) & i = k+1, \ldots n \end{cases}$	The exponential Fourier Lagrange functions will be produced on the infinite domain $(-\infty, \infty)$.
$\begin{cases} x^\delta & ; i = 0 \\ x & ; i \neq 0 \end{cases}$	The produced space will be the same as the one proposed in Esmaeili and Shamsi (2011), Zayernouri and Karniadakis (2014).

DLFs have some useful properties discussed next.

Property 1. The Kronecker delta property will satisfy the DLFs in Eqn. (9.18), for instance, $L_j^{\phi_k}(x_i) = \delta_{i,j}$, for all i, j and it is worth mentioning that this property reduces computational costs.

Property 2. The property of $\sum_{j=0}^{n} L_j^{\phi_k}(x) = 1$ will satisfy the DLFs in Eqn. (9.18).

Property 3. By knowing that the functions of $\phi_i(x)$ are bounded, it can be concluded that the functions of $L_j^{\phi_k}(x)$ are also bounded.

9.3 METHODOLOGY

The definition of distributed-order fractional derivatives equations was introduced in previous section, and in this section, we are going to represent a method to solve fractional differential equations and fractional partial differential equations. For this method, we used the universal function-approximating property of feed-forward neural networks which is presented in Hornik et al. (1989).

The proposed method is based on training an orthogonal neural network in which fractional order of Lagrange functions is used as the essential activation functions of the first hidden layer. There are different and well-known algorithms that can be used for training the neural network, for instance, gradient descent, genetic algorithm, Adam optimizer algorithm, but in this chapter, the Levenberg–Marquardt algorithm is used as the optimizer for training the neural network. Moreover, we utilize the Lagrange polynomial roots as the training points. In the remainder of this section, we first introduce the neural network architecture used in this study, and the Levenberg-Marquardt is expressed, and at last a discussion about the time complexity of the proposed method is indicated.

9.3.1 NEURAL NETWORK ARCHITECTURE FOR DFODE

In this section, we introduce the architecture of the neural network; for achieving this target, first suppose a general form of an initial value DFODE over the interval $[a, b]$:

$$\int_a^b G_1\left(p, D^p u(t)\right)dp + G_2\left(t, u(t), D^{\alpha_i}u(t)\right) = F(t), \ t \in [0, T] \qquad (9.22)$$

where F, G_1, and G_2 can be either linear or nonlinear functions. If we consider Eqn. (9.22) as an initial value problem, it has the following conditions:

$$u^{(h)}(0) = u_h \qquad (9.23)$$

where $h = 0, 1, \cdots, \lfloor \max\{b, \alpha_i\} \rfloor$.

The first essential step for solving Eqn. (9.22) by considering Eqn. (9.23) is approximating the function $u(t)$ with the help of the DLF neural network, and the second

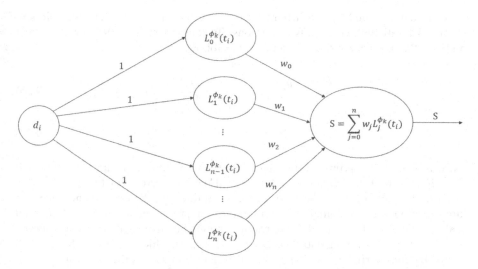

FIGURE 9.2   The topology of the DLF neural network proposed for solving DFODEs.

step would be establishing an appropriate cost function and then minimizing it, and finally, training the network in which $u(t)$ satisfies Eqn. (9.22) aided by Eqn. (9.23).

By considering what Chakraverty and Mall (2017) first mentioned, it can be concluded that we can approximate the solution of the mentioned problem by using a oneinput-layer DLF neural network, a DLF layer for the hidden layer role in which it can also be considered as a functional expansion block, and an output layer with a topology as shown in Figure 9.2, where $L_j^{\phi_k}(t)$s and w_js are presenting the DLFs and weights of the network, respectively.

Finding the exact solution or at least increasing the accuracy of the fractional derivatives is our main target. For achieving this purpose, for the activation function of the output neuron, we use identity function; therefore, by supposing the network which is drawn in Figure 9.2, the following equation will be the output of this fractional neural network:

$$S(t, W) = \sum_{j=0}^{n} w_j L_j^{\phi_k}(t) \tag{9.24}$$

where $W = (w_0, w_1, \ldots, w_n)$. The other required steps would be the implementation of the conditions and the network training to achieve the most accurate and efficient approximation.

Finding the solution procedure in the mentioned method is not complicated. Primarily, we had a drawback and it was initial or boundary condition; therefore, we employed the Ritz method to eliminate this problem in our calculations and thereby achieving the following. Suppose the approximate solution is

$$u(t) \approx u_{app}(t, W) = A(t) + B(t)S(t, W) \tag{9.25}$$

Consider $A(t)$ and $B(t)$ as arbitrary functions that cause the calculated solution, $u_{app}(t, W)$, satisfy the problems' conditions. By considering Eqn. (9.22) with respect to Eqn. (9.23), we can write $A(t)$ and $B(t)$ as follows:

$$A(t) = u_0 + u_1 t + \frac{u_2}{2!}t^2 + \cdots + \frac{u_{\lfloor \max\{b, \, \alpha_i\} \rfloor}}{\lfloor \max\{b, \, \alpha_i\} \rfloor !} t^{\lfloor \max\{b, \, \alpha_i\} \rfloor} \tag{9.26}$$

$$B(t) = t^{\lfloor \max\{b, \, \alpha_i\} \rfloor + 1} \tag{9.27}$$

Next is to find the optimized training dataset, and by that we mean a network that is trained to find the demanded solution with the least error. For this purpose, we used DLFs roots because they are useful in minimizing the convergence rate and are scattered over the domain. Therefore, it can approximate a suitable solution not only over the neighbourhood of the training points but also over the whole interval. Now suppose the training dataset is $D = \{t_1, t_2, \ldots, t_r\}$ in which t_is are the shifted DLF roots. By considering $u_{app}(t, W)$ as the solution of Eqn. (9.22), the cost function will be as follows:

$$Cost(W) =$$

$$\frac{1}{2r}\sum_{k=1}^{r}\left[\int_a^b G_1\left(p, D^p u_{app}(t_k, W)\right)dp + G_2\left(t_k, u_{app}(t_k, W), D^{\alpha_i}u_{app}(t_k, W)\right) - F(t_k)\right]^2$$

$$\tag{9.28}$$

and by using the mentioned numerical integration,

$$Cost(W) =$$

$$\frac{1}{2r}\sum_{k=1}^{r}\left[\sum_{j=0}^{M}\omega_j^* G_1\left(p_j, D^{p_j} u_{app}(t_k, W)\right) + G_2\left(t_k, u_{app}(t_k, W), D^{\alpha_i}u_{app}(t_k, W)\right) - F(t_k)\right]^2$$

$$\tag{9.29}$$

in which M is the Gaussian integration nodes. For having a good approximation with the least error, we should minimize the cost function. Therefore, the mentioned problem will convert to the following problem:

$$argmin_W \frac{1}{2r}$$

$$\sum_{k=1}^{r}\left[\sum_{j=0}^{M}\omega_j^* G_1\left(p_j, D^{p_j} u_{app}(t_k, W)\right) + G_2\left(t_k, u_{app}(t_k, W), D^{\alpha_i}u_{app}(t_k, W)\right) - F(t_k)\right]^2$$

$$\tag{9.30}$$

The procedure of minimizing the aforementioned optimization problem by the Levenberg–Marchard algorithm is presented in Hadian-Rasanan et al. (2020) in detail.

9.3.2 NEURAL NETWORK ARCHITECTURE FOR DFPDE

The neural network developed in the last part can be extended to solve DFPDEs. Now consider the general form of DFPDE which is introduced in Section 9.1:

$$\int_0^1 G\big(p, D^p u(x,t)\big)dp = \frac{\partial^2 u(x,t)}{\partial x^2} + F(x,t), (x,t) \in [x_0, x_1] \times [0, T] \quad (9.31)$$

with initial and Dirichlet boundary conditions mentioned in Eqns. (9.10) and (9.11). First, $u(x,t)$ is approximated by a DLF neural network and a cost function is obtained which should be minimized. In fact, the proposed neural network for solving DFPDE is an extension of the neural network that is used for solving DFODE. At the first layer, there are inputs for spatial and temporal dimensions. The second layer is a hidden layer called functional expansion block and contains DLFs for both x and t. Finally, there is an output layer that represents the approximation function for $u(x,t)$. This neural network can be demonstrated is Figure 9.3.

In this figure, $L_p^{\phi_k}(x)$ and $L_q^{\psi_k}(t)$ are DLFs for spatial and temporal dimensions, respectively, and ω_{pq} express weights in the network. Thus, according to the architecture of the proposed neural network, the output of the neural network is obtained as

$$S(x,t,W) = \sum_{p=0}^{n}\sum_{q=0}^{m} w_{pq} L_p^{\phi_k}(x) L_q^{\psi_k}(t) \quad (9.32)$$

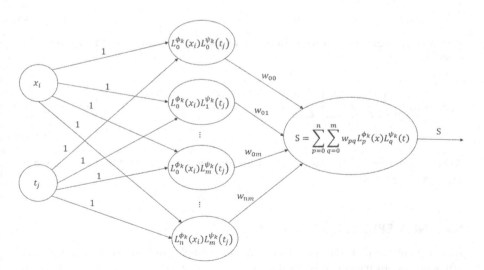

FIGURE 9.3 The topology of the DLF neural network proposed for solving DFPDEs.

where $W = (w_{00}, w_{01}, \ldots, w_{nn-1}, w_{nn})$. Now we should impose the initial and boundary conditions appropriately and then construct the cost function. By applying the initial and boundary conditions to the approximation, we have:

$$u(x,t) \approx u_{app}(x,t,W) = A(x,t) + xt(x-1)S(x,t,W) \qquad (9.33)$$

in which $A(x,t)$ is a function of x and t and is defined as follows:

$$A(x,t) = (1-x)g_0(t) + xg_1(t) + (T-t)\left[f(t) - \left((T-t)f(0) + tf(T)\right)\right] \quad (9.34)$$

As mentioned previously, DLFs roots are considered as the training data to achieve a suitable convergence rate. Therefore, we consider the training dataset in two-dimension as $D = \{(x_1,t_1), (x_1,t_2), \ldots, (x_r,t_{l-1}), (x_r,t_l)\}$ where x_is and y_is are the roots of shifted DLF. By imposing the approximation function into the DFPDE Eqn. (9.31), the cost function is obtained as follows:

$$Cost(W) = \frac{1}{2(r+l)} \sum_{i=1}^{r} \sum_{j=1}^{l} \left[\int_0^1 G\left(p,\, D^p u_{app}(x_i,t_j)\right) dp - \frac{\partial^2 u_{app}(x_i,t_j)}{\partial x^2} - F(x_i,t_j) \right]^2$$

$$(9.35)$$

Like in the previous section, the numerical integration with Gaussian integration nodes is used and we have:

$$Cost(W) = \frac{1}{2(r+l)} \sum_{i=1}^{r} \sum_{j=1}^{l} \left[\sum_{k=0}^{M} \omega_k^* G\left(p,\, D^p u_{app}(x_i,t_j)\right) dp - \frac{\partial^2 u_{app}(x_i,t_j)}{\partial x^2} - F(x_i,t_j) \right]^2$$

$$(9.36)$$

Now the problem is converted into the following non-constrained minimization problem:

$$argmin_W \frac{1}{2(r+l)} \sum_{i=1}^{r} \sum_{j=1}^{l} \left[\sum_{k=0}^{M} \omega_k^* G\left(p,\, D^p u_{app}(x_i,t_j)\right) - \frac{\partial^2 u_{app}(x_i,t_j)}{\partial x^2} - F(x_i,t_j) \right]^2$$

$$(9.37)$$

As mentioned before, the procedure of minimizing is performed by the Levenberg–Marchard algorithm.

9.4   NUMERICAL EXPERIMENTS

In this section, we present various numerical examples to indicate the accuracy and efficiency of our introduced neural network; here, we solve three DFODEs and three DFPDEs and evaluate our results in different aspects. By considering the interval

$\Omega = [a, b]$ as our intended domain, the following norm which is denoted by $\|e\|_2$ is utilized for comparing the exact solution and the approximation:

- For DFODE:

$$\|e\|_2 = \sqrt{\int_0^T \left(u(t) - u_{app}(t, W)\right)^2 dt} \qquad (9.38)$$

- For DFPDE:

$$\|e\|_2 = \sqrt{\int_0^T \int_0^1 \left(u(x,t) - u_{app}(x,t, W)\right)^2 dxdt} \qquad (9.39)$$

where u and u_{app} are the exact and approximated solutions, respectively. It is worth mentioning that our obtained result is way more accurate than the other proposed methods and they were all calculated in Maple software on a 3.5 GHz Intel Core i7 CPU machine with 8 Gbyte of RAM.

9.4.1 TEST PROBLEM 1

As the first example consider the following linear distributed-order fractional equation:

$$\int_{0.2}^{1.5} \Gamma(3 - p) \, D^p u(t) dp = 2 \frac{t^{1.8} - t^{0.5}}{\ln t}, \, t \in [0, 100] \qquad (9.40)$$

with the following initial conditions:

$$u(0) = u'(0) = 0 \qquad (9.41)$$

where $u(t) = t^2$ is the exact solution and $u_{app}(t, W) = t^2 \sum_{j=0}^{n} w_j L_j^{\phi_k}(t)$ is the approximated solution in which the basis function $L_j^{\phi_k}(t)$ is constructed by using $\phi_k(t) = \dfrac{t}{k}$ (i.e., see Eqn. (9.18)). By applying the proposed neural network, we obtain the following cost function:

$$Cost(W) = \frac{1}{2r} \sum_{i=1}^{r} \left[\sum_{j=0}^{M} \omega_j^* \Gamma(3 - p_j) D^{p_j} u_{app}(t_i, W) - 2 \frac{t_i^{1.8} - t_i^{0.5}}{\ln t_i} \right]^2 \qquad (9.42)$$

in which ω_j^*s are the weight of the numerical integration. Furthermore, by minimizing Eqn. (9.42) and applying Levenberg–Marquardt as the optimizer, we will achieve

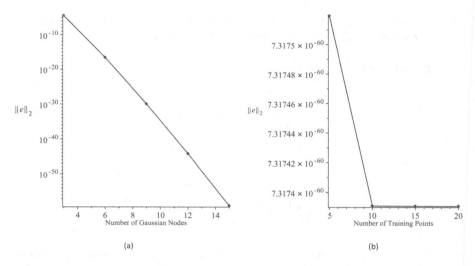

FIGURE 9.4   The convergence graph of test problem 1 for diverse Gaussian nodes and training points: (a) Figure of the error for different Gaussian nodes (b) Figure of the error for different training points.

the solution. Yuttanan and Razzaghi (2019) solved this problem with the help of two Legendre wavelets basis and achieved the exact solution. We also obtained the exact solution with two Lagrange basis, which implies with increasing numerical integration accuracy, we can get an exactness about 10^{-60}, which is beyond machine accuracy and technically the answer is exact. The convergence graph for different Gaussian nodes and diverse training points is illustrated in Figure 9.4a and b.

As can be seen in Figure 9.4a, by increasing the number of Gaussian integration points, the $\|e\|_2$ decreases exponentially, and similarly Figure 9.4b indicates that by increasing the number of training points, the approximated solution converges to the exact solution very rapidly.

9.4.2 Test Problem 2

Suppose the following nonlinear DFODEs, which is also solved in Xu et al. (2019):

$$\int_0^1 \Gamma(4.5-p)\, D^p u(t)\, dp = \sin(u(t)) + \Gamma(4.5)\frac{t^{2.5}(t-1)}{\ln t} - \sin(t^{3.5}),\ t \in [0, 1] \qquad (9.43)$$

by considering the initial conditions as follows:

$$u(0) = 0 \qquad (9.44)$$

Here, $u(t) = t^{3.5}$ is the exact solution and $u_{app}(t, W) = t\sum_{j=0}^n w_j L_j^{\phi_k}(t)$ is the approximated solution in which we have considered $\phi_k(t) = t^{0.5}$ to construct the basis

TABLE 9.2

Comparison between Obtained $\|e\|_2$ of Proposed Method and Method of Xu et al. (2019) for Test Problem 2

M	Method of (Xu et al., 2019) with N = 6	Presented Method with N = 6	Method of (Xu et al., 2019) with N = 14	Presented Method with N = 14
2	7.70E-6	1.29E-9	6.74E-8	1.29E-9
3	7.66E-6	5.97E-14	2.02E-8	5.96E-14
4	7.66E-6	2.73E-18	2.00E-8	2.72E-18
5	7.66E-6	1.26E-22	2.00E0	1.62E-22

functions for this problem (i.e., see Eqn. (9.18)). Now, similar to the previous examples, we will achieve Eqn. (9.45) as our cost function:

$$Cost(W) = \frac{1}{2r}\sum_{i=1}^{r}\left[\sum_{j=0}^{M}\omega_j^*\Gamma(4.5 - p_j)D^{p_j}u_{app}(t_i, W) - \sin(u_{app}(t_i, W)) \right.$$
$$\left. + \Gamma(4.5)\frac{t_i^{2.5}(t_i - 1)}{\ln t_i} + \sin(t_i^{3.5})\right]^2 \tag{9.45}$$

First we minimize it and, thereafter, with the help of Levenberg–Marquradt as our optimizer, we will obtain the required solution. This example is studied by Xu et al. (2019) and in Table 9.2, we compare his results with ours.

The convergence graph of example 2 is indicated in Figure 9.5a and b for various Gaussian nodes and training points.

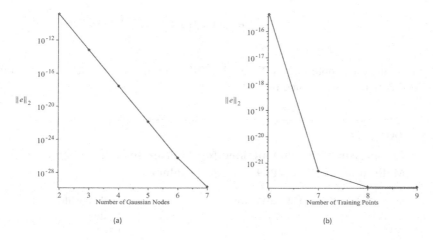

(a)

(b)

FIGURE 9.5 The convergence graph of test problem 2 for diverse Gaussian nodes and training points: (a) Figure of the error for different Gaussian nodes (b) Figure of the error for different training points.

By considering Figure 9.5a, it can be concluded that by increasing the Gaussian integration point the $\|e\|_2$ will be exponentially decreased and, similarly, as it can be seen in Figure 9.5b, the exact solution and the approximated solution are converging to each other by increasing the training points.

9.4.3 Test Problem 3

Consider the following equation which is also solved by Xu et al. (2019):

$$\int_0^1 e^p \Gamma(4.1-p) \, D^p u(t) dp = -u^3(t) - u(t) + \Gamma(4.1)\frac{t^{2.1}(e-t)}{1-\ln t} + t^{9.3} + t^{3.1} \quad (9.46)$$

where $t \in [0,2]$ and the initial condition is as follows:

$$u(0) = 0 \quad (9.47)$$

Note that the exact solution and the approximated one are $u(t) = t^{3.1}$ and $u_{app}(t, W) = t\sum_{j=0}^n w_j L_j^{\phi_k}(t)$, respectively. Equation (9.48) is the cost function we desired and by minimizing it and applying the Levenberg–Marquardt as the optimizer and assuming $\phi_k(t) = \dfrac{t^{3.1}}{k}$, the desired solution will be obtained.

$$Cost(W) = \frac{1}{2r} \sum_{i=1}^r \left[\sum_{j=0}^M \omega_j^* \Gamma(4.4 - p_j) D^{p_j} u_{app}(t_i, W) + u_{app}^3(t_i, W) \right.$$
$$\left. + u_{app}(t_i, W) - \Gamma(4.1)\frac{t_i^{2.1}(e-t_i)}{1-\ln t_i} - t_i^{9.3} - t_i^{3.1} \right]^2 \quad (9.48)$$

The best answer is obtained by Xu et al. (2019) and the best answer of the recommended method is mentioned in Table 9.3.

TABLE 9.3

Comparison between Obtained $\|e_2\|$ of Proposed Method and Method of Xu et al. (2019) for Test Problem 3

M	Method of (Xu et al., 2019) with $N = 18$	Presented Method with $N = 3$
2	2.36E-6	2.28E-8
3	6.03E-8	1.34E-12
4	5.13E-8	1.88E-16
5	5.13E-8	7.36E-17

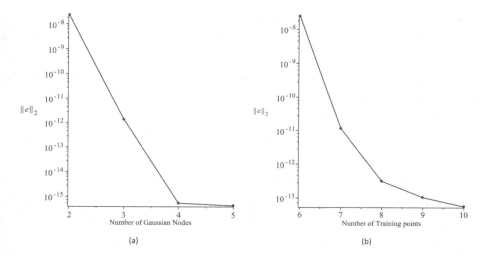

FIGURE 9.6 The convergence graph of test problem 3 for diverse Gaussian nodes and training points: (a) Figure of the error for different Gaussian nodes (b) Figure of the error for different training points.

The convergence graph of the third example for different Gaussian nodes and training points is demonstrated in Figure 9.6a and b.

By decreasing the Gaussian integration and training points, $\|e\|_2$ will decrease which is indicated in Figure 9.6a and b.

9.4.4 Test Problem 4

Now, for the last example, consider the following DFODE which is also solved in Katsikadelis (2014), Kharazmi et al. (2017), Mashayekhi and Razzaghi (2016), and Xu et al. (2019):

$$\frac{1}{120}\int_0^2 \Gamma(6-p)\, D^p u(t)\,dp = \Gamma(7.1)\frac{t^5-t^3}{\ln t},\, t\in[0,1] \qquad (9.49)$$

with the following initial condition

$$u(0)=u'(0)=0 \qquad (9.50)$$

Here, $u(t)=t^5$ is the exact solution and $u_{app}(t,\,W)=t^2\sum_{j=0}^n w_j L_j^{\phi_k}(t)$ is the approximated solution. By utilizing the mentioned method, the following function as the cost function is achieved:

$$Cost(W)=\frac{1}{2r}\sum_{i=1}^r\left[\frac{1}{120}\sum_{j=0}^M \omega_j^*\Gamma(6-p_j)D^{p_j}u_{app}(t_i,\,W)-\Gamma(7.1)\frac{t_i^5-t_i^3}{\ln t}\right]^2 \qquad (9.51)$$

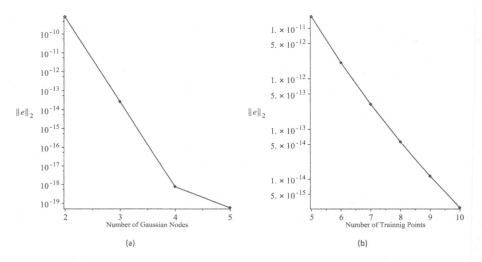

FIGURE 9.7 The convergence graph of test problem 4 for diverse Gaussian nodes and training points: (a) Figure of the error for different Gaussian nodes (b) Figure of the error for different training points.

and by minimizing it and applying the Levenberg–Marquardt algorithm, we will achieve the solution we wanted. It is notable that the best approximation obtained in Xu et al. (2019) with $N = 9$ and $M = 4$ was 6.75E-011, but with our mentioned method and with $N = 3$, $M = 4$, $\phi_k(t) = \dfrac{t^{1.5}}{k}$ and 20-training points, we achieved 7.5601E-019. The convergence graph of different Gaussian nodes and training points is indicated in Figure 9.7a and b.

As it is indicated in Figure 9.7a and b, by increasing the integration and training points, respectively, the approximated solution will converge to the exact solution or equivalently, $\|e\|_2$ will decrease.

9.4.5 TEST PROBLEM 5

Now, consider

$$
\int_0^1 \Gamma(2.5 - p) D^p u(x,t)\,dp = \frac{\partial^2 u(x,t)}{\partial x^2} +
$$

$$
\frac{\sqrt{t}\,(x-1)^2 \left(3\sqrt{\pi}\,(t-1)(x-1)^2 x^2 - 8t\left(5x(3x-2)+1\right)\ln(t)\right)}{4\ln(t)}
\tag{9.52}
$$

with the following initial and boundary conditions (Pourbabaee & Saadatmandi, 2019):

$$
u(x,0) = 0
\tag{9.53}
$$

$$
u(0,t) = 0,\; u(1,t) = 0
\tag{9.54}
$$

The exact solution is $u(x,t)=t^{\frac{3}{2}}x^2(x-1)^4$. Also, the approximate function for this example is $u_{app}(x,t,W)=xt(x-1)\sum_{p=0}^{n}\sum_{q=0}^{m}\omega_{pq}L_p^{\phi_k}(x)L_q^{\psi_k}(t)$, where $\phi_k(x)=\frac{x}{k}$ and $\psi_k(t)=t^{\frac{1}{2}}$. The following cost function is obtained by applying the proposed approach on the aforementioned problem:

$$Cost(W)=\frac{1}{2(r+l)}\sum_{i=1}^{r}\sum_{j=1}^{l}\left[\sum_{k=0}^{M}\omega_k^*\Gamma(2.5-p)D^p u_{app}(x_i,t_j)-\frac{\partial^2 u_{app}(x_i,t_j)}{\partial x^2}\right.$$

$$\left.-\frac{\sqrt{t_j}(x_i-1)^2\left(3\sqrt{\pi}(t_j-1)(x_i-1)^2 x_i^2-8t_j\left(5x_i(3x_i-2)+1\right)\ln(t_j)\right)}{4\ln(t_j)}\right]^2$$

$$(9.55)$$

The Levenberg–Marquardt algorithm is used to minimize the cost function. The proposed method can find an accurate solution with 10^{-28} the absolute error which can be considered as an exact solution. Figure 9.8a and b shows the convergence of the approximated solution according to the Gaussian nodes and number of training points. According to this figure, by increasing the number of training points and integration nodes, $\|e\|_2$ decreases quickly and the approximated solution converges to the exact one.

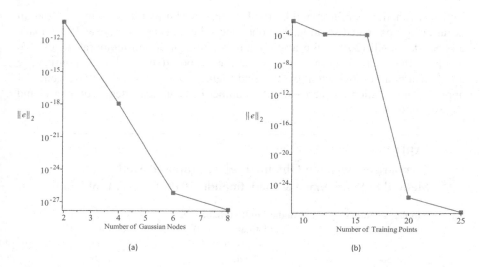

(a)   (b)

FIGURE 9.8   The convergence graph of test problem 5 for diverse Gaussian nodes and training points: (a) Figure of the error for different Gaussian nodes (b) Figure of the error for different training points.

9.4.6 Test Problem 6

In this example, consider the following equation (Pourbabaee & Saadatmandi, 2019):

$$\int_0^1 \Gamma(3-p)D^p u(x,t)dp = \frac{\partial^2 u(x,t)}{\partial x^2} + \frac{2t(1-x)(t-1)\cos(x)}{\ln(t)} + 2t^2 \sin(x) - t^2(1-x)\cos(x)$$

$$(9.56)$$

the initial and boundary conditions are:

$$u(x,0) = 0 \tag{9.57}$$

$$u(0,t) = t^2, \, u(1,t) = 0 \tag{9.58}$$

The exact solution is $u(x,t) = t^2(1-x)\cos(x)$. We define the approximate function as $u_{app} = (1-x)t^2 + xt(x-1)\sum_{p=0}^{n}\sum_{q=0}^{m}\omega_{pq}L_p^{\phi_k}(x)L_q^{\psi_k}(t)$, in which $\phi_k(x) = \dfrac{x}{k!}$ and $\psi_k(t) = t$. Using the proposed neural network, the following cost function is obtained:

$$Cost(W) = \frac{1}{2(r+l)}\sum_{i=1}^{r}\sum_{j=1}^{l}\left[\sum_{k=0}^{M}\omega_k^* \Gamma(3-p)D^p u_{app}(x_i,t_j) - \frac{\partial^2 u_{app}(x_i,t_j)}{\partial x^2}\right.$$

$$(9.59)$$

$$\left. - \frac{2t_j(1-x_i)(t-1)\cos(x_i)}{\ln(t_j)} - 2t_j^2 \sin(x_i) + t_j^2(1-x_i)\cos(x_i)\right]^2$$

The cost function is minimized by the Levenberg–Marquardt algorithm to find an accurate approximated solution. The obtained solution is as accurate as Pourbabaee and Saadatmandi (2019) with a lower number of spatial and temporal nodes. We compare the approximation solutions of the proposed method with Pourbabaee and Saadatmandi (2019) in Table 9.4, and Figure 9.9 expresses the convergence of approximate solution by increasing the number of Gaussian integration points and training points.

TABLE 9.4

Comparison between Obtained $\|e\|_2$ of Proposed Method and Method of Pourbabaee & Saadatmandi (2019) for Test Problem 6

Number of Nodes in the Domain	Method of (Pourbabaee & Saadatmandi, 2019)	Presented Method
16	1.81E-6	1.20E-10
36	2.40E-9	3.12E-11
49	1.54E-4	3.17E-12
64	2.19E-11	3.19E-12

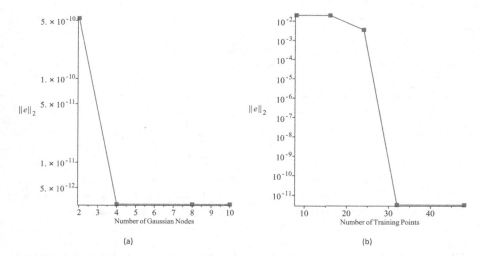

FIGURE 9.9 The convergence graph of test problem 6 for diverse Gaussian nodes and training points: (a) Figure of the error for different Gaussian nodes (b) Figure of the error for different training points.

9.4.7 TEST PROBLEM 7

As the last example, consider the following DFPDE (Pourbabaee & Saadatmandi, 2019):

$$\int_0^1 \Gamma(4-p) D^p u(x,t) dp = \frac{\partial^2 u(x,t)}{\partial x^2} + \frac{xt^2\left(6t + t\ln(t) - 6\right)\sin(x)}{\ln(t)} + 2t^3 \cos(x) \quad (9.60)$$

subject to the following conditions:

$$u(x,0) = 0 \quad (9.61)$$

$$u(0,t) = 0, \ u(1,t) = t^3 \sin(1) \quad (9.62)$$

The exact solution is $u(x,t) = t^3 x \sin(x)$. In order to solve this equation, we define the cost function as:

$$Cost(W) = \frac{1}{2(r+l)} \sum_{i=1}^{r} \sum_{j=1}^{l} \left[\sum_{k=0}^{M} \omega_k^* \Gamma(3-p) D^p u_{app}\left(x_i, t_j\right) - \frac{\partial^2 u_{app}\left(x_i, t_j\right)}{\partial x^2} \right.$$

$$\left. - \frac{x_i t_j^2 \left(6t_j + t\ln\left(t_j\right) - 6\right)\sin(x_i)}{\ln\left(t_j\right)} - 2t_j^3 \cos(x_i) \right]^2 \quad (9.63)$$

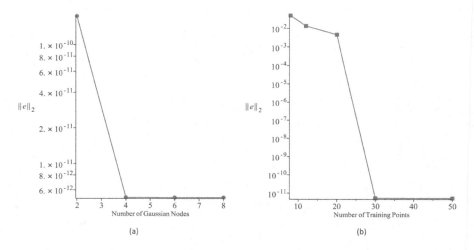

(a)   (b)

FIGURE 9.10   The convergence graph of test problem 7 for diverse Gaussian nodes and training points: (a) Figure of the error for different Gaussian nodes (b) Figure of the error for different training points.

TABLE 9.5

Comparison between Obtained $\|e\|_2$ of Proposed Method and Method of Pourbabaee & Saadatmandi (2019) for Test Problem 7

Number of Nodes on the Domain	Method of (Pourbabaee & Saadatmandi, 2019)	Presented Method
16	1.36E-3	1.20E-2
25	4.87E-5	5.23E-12
36	3.05E-6	5.24E-12
49	7.59E-8	5.19E-12

We consider $u_{app} = xt^3 \sin(1) + xt(x-1) \sum_{p=0}^{n} \sum_{q=0}^{m} \omega_{pq} L_p^{\phi_k}(x) L_q^{\psi_k}(t)$, in which $\phi_k = x$ and $\psi_k = t$. As in previous examples, the approximate solution is obtained by minimizing the cost function. The effects of the number of integration nodes and training points on $\|e\|_2$ are represented in Figure 9.10a and b. It is deduced that by increasing them, the approximated solution tends towards the exact solution. Moreover, the accuracy of the obtained solution by the proposed neural network is compared with the method in Pourbabaee and Saadatmandi (2019) in Table 9.5.

9.5 CONCLUSION AND DISCUSSIONS

In this chapter, by introducing the DLFs, we proposed a new neural network framework for solving distributed-order fractional dynamical systems. Additionally, the efficiency and accuracy of the presented method are considered by proposing seven numerical examples of both ordinary and partial fractional differential equations. According to the numerical results, the mentioned method has high accuracy; moreover, due to utilizing less number of neurons, this method is highly efficient. Since the examined model in this research is a distributed-order fractional derivative model and it is the general form of many other differential equations, it can be used for solving other differential equations. Despite many significant features of this algorithm, it has some limitations. One of the important restrictions is that there are no algorithms to find the ϕ_is which is chosen experimentally and by trial and error, and also proposing an intelligent algorithm for choosing ϕ_is accurately can be the subject of further researches.

REFERENCES

Ahmad, I., Raja, M.A.Z., Bilal, M., & Ashraf, F. (2017). Neural network methods to solve the Lane–Emden type equations arising in thermodynamic studies of the spherical gas cloud model. *Neural Computing and Applications*, 28(s1), 929–944. https://doi.org/10.1007/s00521-016-2400-y

Archambeau, C., Cornford, D., Opper, M., & Shawe-Taylor, J. (2007). Gaussian process approximations of stochastic differential equations. *Journal of Machine Learning Research: Workshop and Conference Proceedings*, 1, 1–16. http://eprints.aston.ac.uk/11096/

Asari, S.S., Amirfakhrian, M., & Chakraverty, S. (2019). Application of radial basis functions in solving fuzzy integral equations. *Neural Computing and Applications*, 31(10), 6373–6381. https://doi.org/10.1007/s00521-018-3459-4

Atanacković, T.M., Oparnica, L., & Pilipović, S. (2007). On a nonlinear distributed order fractional differential equation. *Journal of Mathematical Analysis and Applications*, 328(1), 590–608. https://doi.org/10.1016/j.jmaa.2006.05.038

Brown, S.D., & Heathcote, A. (2008). The simplest complete model of choice response time: Linear ballistic accumulation. *Cognitive Psychology*, 57(3), 153–178. https://doi.org/10.1016/j.cogpsych.2007.12.002

Brunton, S.L., Noack, B.R., & Koumoutsakos, P. (2020). Machine learning for fluid mechanics. *Annual Review of Fluid Mechanics*, 52, 477–508. https://doi.org/10.1146/annurev-fluid-010719-060214

Carlson, T., Goddard, E., Kaplan, D.M., Klein, C., & Ritchie, J. B. (2018). Ghosts in machine learning for cognitive neuroscience: Moving from data to theory. *NeuroImage*, 180(July 2017), 88–100. https://doi.org/10.1016/j.neuroimage.2017.08.019

Chakraverty, S., & Mall, S. (2014). Regression-based weight generation algorithm in neural network for solution of initial and boundary value problems. *Neural Computing and Applications*, 25(3–4), 585–594. https://doi.org/10.1007/s00521-013-1526-4

Chakraverty, S., & Mall, S. (2017). Artificial neural networks for engineers and scientists: Solving ordinary differential equations (1st ed.). CRC Press. https://doi.org/10.1201/9781315155265

Dehestani, H., Ordokhani, Y., & Razzaghi, M. (2019). A numerical technique for solving various kinds of fractional partial differential equations via Genocchi hybrid functions. *Revista de La Real Academia de Ciencias Exactas, Fisicas y Naturales – Serie A: Matematicas*, 113(4), 3297–3321. https://doi.org/10.1007/s13398-019-00694-5

Delkhosh, M., & Parand, K. (2019). Generalized pseudospectral method: Theory and applications. *Journal of Computational Science*, 34, 11–32. https://doi.org/10.1016/j.jocs.2019.04.007

Delkhosh, M., Parand, K., & Hadian-Rasanan, A.H. (2019). *A development of Lagrange interpolation, Part I: Theory*. http://arxiv.org/abs/1904.12145

Diethelm, K., & Ford, N.J. (2009). Numerical analysis for distributed-order differential equations. *Journal of Computational and Applied Mathematics*, *225*(1), 96–104. https://doi.org/10.1016/j.cam.2008.07.018

Ding, W., Patnaik, S., Sidhardh, S., & Semperlotti, F. (2021). Applications of distributed-order fractional operators: A review. *Entropy*, *23*(1), 1–42. https://doi.org/10.3390/e23010110

Dissanayake, M.W.M.G., & Phan-Thien, N. (1994). Neural-network-based approximations for solving partial differential equations. *Communications in Numerical Methods in Engineering*, *10*(3), 195–201. https://doi.org/10.1002/cnm.1640100303

Effati, S., & Buzhabadi, R. (2012). A neural network approach for solving Fredholm integral equations of the second kind. *Neural Computing and Applications*, *21*(5), 843–852. https://doi.org/10.1007/s00521-010-0489-y

Effati, S., & Pakdaman, M. (2010). Artificial neural network approach for solving fuzzy differential equations. *Information Sciences*, *180*(8), 1434–1457. https://doi.org/10.1016/j.ins.2009.12.016

Ejlali, N., & Hosseini, S.M. (2017). A pseudospectral method for fractional optimal control problems. *Journal of Optimization Theory and Applications*, *174*(1), 83–107. https://doi.org/10.1007/s10957-016-0936-8

Esmaeili, S., & Shamsi, M. (2011). A pseudo-spectral scheme for the approximate solution of a family of fractional differential equations. *Communications in Nonlinear Science and Numerical Simulation*, *16*(9), 3646–3654. https://doi.org/10.1016/j.cnsns.2010.12.008

Golbabai, A., Mammadov, M., & Seifollahi, S. (2009). Solving a system of nonlinear integral equations by an RBF network. *Computers and Mathematics with Applications*, *57*(10), 1651–1658. https://doi.org/10.1016/j.camwa.2009.03.038

Golbabai, A., & Seifollahi, S. (2006). Numerical solution of the second kind integral equations using radial basis function networks. *Applied Mathematics and Computation*, *174*(2), 877–883. https://doi.org/10.1016/j.amc.2005.05.034

Groza, G., & Razzaghi, M. (2019). Approximation of solutions of polynomial partial differential equations in two independent variables. *Journal of Computational and Applied Mathematics*, *346*, 205–223. https://doi.org/10.1016/j.cam.2018.07.011

Hadian Rasanan, A.H., Amani Rad, J., & Sewell, D. K. (2021). Are there jumps in evidence accumulation, and what, if anything, do they reflect psychologically ? An analysis of Lévy-Flights models of decision-making. *PsyArXiv*. https://doi.org/10.31234/osf.io/vy2mh

Hadian Rasanan, A.H., Bajalan, N., Parand, K., & Rad, J.A. (2020). Simulation of nonlinear fractional dynamics arising in the modeling of cognitive decision making using a new fractional neural network. *Mathematical Methods in the Applied Sciences*, *43*(3), 1437–1466. https://doi.org/10.1002/mma.5981

Hadian-Rasanan, A.H., Rahmati, D., Gorgin, S., & Parand, K. (2020). A single layer fractional orthogonal neural network for solving various types of Lane–Emden equation. *New Astronomy*, *75*(July 2019), 101307. https://doi.org/10.1016/j.newast.2019.101307

Hawkins, G.E., & Heathcote, A. (2021). Racing against the clock: Evidence-based versus time-based decisions. *Psychological Review*, *128*(2), 222–263. https://doi.org/10.1037/rev0000259

Hornik, K., Stinchcombe, M., & White, H. (1989). Multilayer feedforward networks are universal approximators. *Neural Networks*, *2*(5), 359–366. https://doi.org/10.1016/0893-6080(89)90020-8

Huang-Pollock, C., Ratcliff, R., McKoon, G., Shapiro, Z., Weigard, A., & Galloway-Long, H. (2017). Using the diffusion model to explain cognitive deficits in attention deficit hyperactivity disorder. *Journal of Abnormal Child Psychology*, *45*(1), 57–68. https://doi.org/10.1007/s10802-016-0151-y

Javad Sabouri K., Effati, S., & Pakdaman, M. (2017). A neural network approach for solving a class of fractional optimal control problems. *Neural Processing Letters, 45*(1), 59–74. https://doi.org/10.1007/s11063-016-9510-5

Jeswal, S.K., & Chakraverty, S. (2018). Solving transcendental equation using artificial neural network. *Applied Soft Computing Journal, 73*, 562–571. https://doi.org/10.1016/j.asoc.2018.09.003

Jianyu, L., Siwei, L., Yingjian, Q., & Yaping, H. (2003). Numerical solution of elliptic partial differential equation using radial basis function neural networks. *Neural Networks, 16*(5–6), 729–734. https://doi.org/10.1016/S0893-6080(03)00083-2

Katsikadelis, J.T. (2014). Numerical solution of distributed order fractional differential equations. *Journal of Computational Physics, 259*, 11–22. https://doi.org/10.1016/j.jcp.2013.11.013

Kharazmi, E., Zayernouri, M., & Karniadakis, G.E.M. (2017). Petrov-galerkin and spectral collocation methods for distributed order differential equations. *SIAM Journal on Scientific Computing, 39*(3), A1003–A1037. https://doi.org/10.1137/16M1073121

Khosravian-Arab, H., Dehghan, M., & Eslahchi, M.R. (2015). Fractional Sturm-Liouville boundary value problems in unbounded domains: Theory and applications. *Journal of Computational Physics, 299*, 526–560. https://doi.org/10.1016/j.jcp.2015.06.030

Khosravian-Arab, H., Dehghan, M., & Eslahchi, M.R. (2017). Fractional spectral and pseudo-spectral methods in unbounded domains: Theory and applications. *Journal of Computational Physics, 338*, 527–566. https://doi.org/10.1016/j.jcp.2017.02.060

Krajbich, I., Lu, D., Camerer, C., & Rangel, A. (June 2012). The attentional drift-diffusion model extends to simple purchasing decisions. *Frontiers in Psychology, 3*. https://doi.org/10.3389/fpsyg.2012.00193

Kremer, J., Stensbo-Smidt, K., Gieseke, F., Pedersen, K.S., & Igel, C. (2017). Big universe, big data: Machine learning and image analysis for astronomy. *IEEE Intelligent Systems, 32*(2), 16–22. https://doi.org/10.1109/MIS.2017.40

Lagaris, I.E., Likas, A., & Fotiadis, D.I. (1998). Artificial neural networks for solving ordinary and partial differential equations. *IEEE Transactions on Neural Networks, 9*(5), 987–1000. https://doi.org/10.1109/72.712178

Lee, H., & Kang, I.S. (1990). Neural algorithm for solving differential equations. *Journal of Computational Physics, 91*(1), 110–131. https://doi.org/10.1016/0021-9991(90)90007-N

Lehmann, M.P., Xu, H.A., Liakoni, V., Herzog, M.H., Gerstner, W., & Preuschoff, K. (2019). One-shot learning and behavioral eligibility traces in sequential decision making. *ELife, 8*, 1–32. https://doi.org/10.7554/eLife.47463

Mai-Duy, N., & Tran-Cong, T. (2001). Numerical solution of differential equations using multiquadric radial basis function networks. *Neural Networks, 14*(2), 185–199. https://doi.org/10.1016/S0893-6080(00)00095-2

Mainardi, F., Mura, A., Pagnini, G., & Gorenflo, R. (2008). Time-fractional diffusion of distributed order. *JVC/Journal of Vibration and Control, 14*(9–10), 1267–1290. https://doi.org/10.1177/1077546307087452

Magdziarz, M., & Teuerle, M. (2017). Fractional diffusion equation with distributed-order material derivative. Stochastic foundations. *Journal of Physics A: Mathematical and Theoretical, 50*(18), 184005. https://doi.org/10.1088/1751-8121/aa651e

Mall, S., & Chakraverty, S. (2014). Chebyshev Neural Network based model for solving Lane-Emden type equations. *Applied Mathematics and Computation, 247*, 100–114. https://doi.org/10.1016/j.amc.2014.08.085

Mall, S., & Chakraverty, S. (2015). Numerical solution of nonlinear singular initial value problems of Emden-Fowler type using Chebyshev Neural Network method. *Neurocomputing, 149*(PB), 975–982. https://doi.org/10.1016/j.neucom.2014.07.036

Mall, S., & Chakraverty, S. (2016a). Application of Legendre Neural Network for solving ordinary differential equations. *Applied Soft Computing Journal, 43*, 347–356. https://doi.org/10.1016/j.asoc.2015.10.069

Mall, S., & Chakraverty, S. (March 2016b). Hermite functional link neural network for solving the Van der Pol-Duffing Oscillator Equation. *Neural Computation, 28*, 1574–1598. https://doi.org/10.1162/NECO_a_00858

Mall, S., & Chakraverty, S. (2017). Single layer Chebyshev neural network model for solving elliptic partial differential equations. *Neural Processing Letters, 45*(3), 825–840. https://doi.org/10.1007/s11063-016-9551-9

Mashayekhi, S., & Razzaghi, M. (January 2016). Numerical solution of distributed order fractional differential equations by hybrid functions. *Journal of Computational Physics, 315*, 169–181. https://doi.org/10.1016/j.jcp.2016.01.041

Masood, Z., Majeed, K., Samar, R., & Raja, M.A.Z. (2017). Design of Mexican hat wavelet neural networks for solving Bratu type nonlinear systems. *Neurocomputing, 221*, 1–14. https://doi.org/10.1016/j.neucom.2016.08.079

Meade, A.J., & Fernandez, A.A. (1994a). Solution of nonlinear ordinary differential equations by feedforward neural networks. *Mathematical and Computer Modelling, 20*(9), 19–44. https://doi.org/10.1016/0895-7177(94)00160-X

Meade, A.J., & Fernandez, A.A. (1994b). The numerical solution of linear ordinary differential equations by feedforward neural networks. *Mathematical and Computer Modelling, 19*(12), 1–25. https://doi.org/10.1016/0895-7177(94)90095-7

Meerschaert, M.M., Nane, E., & Vellaisamy, P. (2011). Distributed-order fractional diffusions on bounded domains. *Journal of Mathematical Analysis and Applications, 379*(1), 216–228. https://doi.org/10.1016/j.jmaa.2010.12.056

Najafi, H.S., Sheikhani, A.R., & Ansari, A. (2011). Stability analysis of distributed order fractional differential equations. *Abstract and Applied Analysis, 2011*. https://doi.org/10.1155/2011/175323

Padash, A., Chechkin, A.V., Dybiec, B., Pavlyukevich, I., Shokri, B., & Metzler, R. (2019). First-passage properties of asymmetric Lévy flights. *Journal of Physics A: Mathematical and Theoretical, 52*(45). https://doi.org/10.1088/1751-8121/ab493e

Pakdaman, M., Ahmadian, A., Effati, S., Salahshour, S., & Baleanu, D. (2017). Solving differential equations of fractional order using an optimization technique based on training artificial neural network. *Applied Mathematics and Computation, 293*, 81–95. https://doi.org/10.1016/j.amc.2016.07.021

Parand, K., Aghaei, A.A., Jani, M., & Ghodsi, A. (2021). A new approach to the numerical solution of Fredholm integral equations using least squares-support vector regression. *Mathematics and Computers in Simulation, 180*, 114–128. https://doi.org/10.1016/j.matcom.2020.08.010

Pedersen, M.L., Frank, M.J., & Biele, G. (2017). The drift diffusion model as the choice rule in reinforcement learning. *Psychonomic Bulletin and Review, 24*(4), 1234–1251. https://doi.org/10.3758/s13423-016-1199-y

Pourbabaee, M., & Saadatmandi, A. (2019). A novel Legendre operational matrix for distributed order fractional differential equations. *Applied Mathematics and Computation, 361*, 215–231. https://doi.org/10.1016/j.amc.2019.05.030

Raissi, M., Perdikaris, P., & Karniadakis, G.E. (October 2019). Physics-informed neural networks: A deep learning framework for solving forward and inverse problems involving nonlinear partial differential equations. *Journal of Computational Physics, 378*, 686–707. https://doi.org/10.1016/j.jcp.2018.10.045

Raissi, M., Perdikaris, P., & Karniadakis, G.E. (2017). Machine learning of linear differential equations using Gaussian processes. *Journal of Computational Physics, 348*, 683–693. https://doi.org/10.1016/j.jcp.2017.07.050

Raissi, M., Perdikaris, P., & Karniadakis, G.E. (2018). Numerical Gaussian processes for time-dependent and nonlinear partial differential equations. *SIAM Journal on Scientific Computing, 40*(1), A172–A198. https://doi.org/10.1137/17M1120762

Raja, M.A.Z., & Ahmad, S.Ul I. (2014). Numerical treatment for solving one-dimensional Bratu problem using neural networks. *Neural Computing and Applications, 24*(3–4), 549–561. https://doi.org/10.1007/s00521-012-1261-2

Raja, M.A.Z., Khan, J.A., & Qureshi, I.M. (2011). Solution of fractional order system of Bagley-Torvik equation using evolutionary computational intelligence. *Mathematical Problems in Engineering, 2011.* https://doi.org/10.1155/2011/675075

Raja, M.A.Z., Manzar, M.A., & Samar, R. (2015). An efficient computational intelligence approach for solving fractional order Riccati equations using ANN and SQP. *Applied Mathematical Modelling, 39*(10–11), 3075–3093. https://doi.org/10.1016/j.apm.2014.11.024

Raja, M.A.Z., Mehmood, J., Sabir, Z., Nasab, A.K., & Manzar, M.A. (2019). Numerical solution of doubly singular nonlinear systems using neural networks-based integrated intelligent computing. *Neural Computing and Applications, 31*(3), 793–812. https://doi.org/10.1007/s00521-017-3110-9

Raja, M.A.Z., & Samar, R. (2014). Numerical treatment for nonlinear MHD Jeffery-Hamel problem using neural networks optimized with interior point algorithm. *Neurocomputing, 124*, 178–193. https://doi.org/10.1016/j.neucom.2013.07.013

Raja, M.A.Z., Samar, R., Manzar, M.A., & Shah, S.M. (2017). Design of unsupervised fractional neural network model optimized with interior point algorithm for solving Bagley–Torvik equation. *Mathematics and Computers in Simulation, 132*, 139–158. https://doi.org/10.1016/j.matcom.2016.08.002

Ratcliff, R. (1978). A theory of memory retrieval. *Psychological Review, 85*(2), 59–108. https://doi.org/10.1037/0033-295X.85.2.59

Ratcliff, R., & McKoon, G. (2008). The diffusion decision model: Theory and data for two-choice decision tasks. *Neural Computation, 20*(4), 873–922. https://doi.org/10.1162/neco.2008.12-06-420

Ratcliff, R., McKoon, G., & Gomez, P. (2004). A diffusion model account of the lexical decision task. *Psychological Review, 111*(1), 159–182. https://doi.org/10.1037/0033-295X.111.1.159

Ratcliff, R., & Smith, P.L. (2004). A comparison of sequential sampling models for two-choice reaction time. *Psychological Review, 111*(2), 333–367. https://doi.org/10.1037/0033-295X.111.2.333

Ratcliff, R., Smith, P.L., Brown, S.D., & McKoon, G. (2016). Diffusion decision model: Current issues and history. *Trends in Cognitive Sciences, 20*(4), 260–281. https://doi.org/10.1016/j.tics.2016.01.007

Rieskamp, J. (2008). The probabilistic nature of preferential choice. *Journal of Experimental Psychology: Learning, Memory, and Cognition, 34*(6), 1446. https://doi.org/10.1037/a0013646

Rieskamp, J., Busemeyer, J.R., & Mellers, B.A. (2006). Extending the bounds of rationality: Evidence and theories of preferential choice. *Journal of Economic Literature, 44*(3), 631–661. https://doi.org/10.1257/jel.44.3.631

Rostami, F., & Jafarian, A. (2018). A new artificial neural network structure for solving high-order linear fractional differential equations. *International Journal of Computer Mathematics, 95*(3), 528–539. https://doi.org/10.1080/00207160.2017.1291932

Sabermahani, S., Ordokhani, Y., & Yousefi, S.A. (2018). Numerical approach based on fractional-order Lagrange polynomials for solving a class of fractional differential equations. *Computational and Applied Mathematics, 37*(3), 3846–3868. https://doi.org/10.1007/s40314-017-0547-5

Sabermahani, S., Ordokhani, Y., & Yousefi, S.A. (2019). Fractional-order Lagrange polynomials: An application for solving delay fractional optimal control problems. *Transactions of the Institute of Measurement and Control, 41*(11), 2997–3009. https://doi.org/10.1177/0142331218819048

Sabermahani, S., Ordokhani, Y., & Yousefi, S.A. (2020). Fractional-order general Lagrange scaling functions and their applications. *BIT Numerical Mathematics, 60*(1), 101–128. https://doi.org/10.1007/s10543-019-00769-0

Sadeghi, M., Khosrowabadi, R., Bakouie, F., Mahdavi, H., Eslahchi, C., & Pouretemad, H. (March 2017). Screening of autism based on task-free fMRI using graph theoretical approach. *Psychiatry Research – Neuroimaging, 263*, 48–56. https://doi.org/10.1016/j.pscychresns.2017.02.004

Shinn, M., Lam, N.H., & Murray, J.D. (2020). A flexible framework for simulating and fitting generalized drift-diffusion models. *ELife, 9*, 1–27. https://doi.org/10.7554/ELIFE.56938

Smith, P.L. (2000). Stochastic dynamic models of response time and accuracy: A foundational primer. *Journal of Mathematical Psychology, 44*(3), 408–463. https://doi.org/10.1006/jmps.1999.1260

Smith, P.L., & Ratcliff, R. (2021). Modeling evidence accumulation decision processes using integral equations: Urgency-gating and collapsing boundaries. *Psychological Review.* https://doi.org/10.1037/rev0000301

Sun, H., Chen, Y., & Chen, W. (2011). Random-order fractional differential equation models. *Signal Processing, 91*(3), 525–530. https://doi.org/10.1016/j.sigpro.2010.01.027

Usher, M., & McClelland, J.L. (2001). The time course of perceptual choice: The leaky, competing accumulator model. *Psychological Review, 108*(3), 550–592. https://doi.org/10.1037/0033-295X.108.3.550

van Ravenzwaaij, D., Brown, S.D., Marley, A.A.J., & Heathcote, A. (2019). Accumulating advantages: A new conceptualization of rapid multiple choice. *Psychological Review, 127*(2), 186–215. https://doi.org/10.1037/rev0000166

Voskuilen, C., Ratcliff, R., & Smith, P.L. (2016). Comparing fixed and collapsing boundary versions of the diffusion model. *Journal of Mathematical Psychology, 73*, 59–79. https://doi.org/10.1016/j.jmp.2016.04.008

Voss, A., Lerche, V., Mertens, U., & Voss, J. (2019). Sequential sampling models with variable boundaries and non-normal noise: A comparison of six models. *Psychonomic Bulletin and Review, 26*(3), 813–832. https://doi.org/10.3758/s13423-018-1560-4

Voss, A., Rothermund, K., & Voss, J. (2004). Interpreting the parameters of the diffusion model: An empirical validation. *Memory and Cognition, 32*(7), 1206–1220. https://doi.org/10.3758/BF03196893

Voss, A., & Voss, J. (2008). A fast numerical algorithm for the estimation of diffusion model parameters. *Journal of Mathematical Psychology, 52*(1), 1–9. https://doi.org/10.1016/j.jmp.2007.09.005

Wagenmakers, E.J., Van Der Maas, H.L.J. & Grasman, R.P.P.P. An EZ-diffusion model for response time and accuracy. *Psychonomic Bulletin & Review 14*(1), 3–22 (2007). https://doi.org/10.3758/BF03194023

Wieschen, E.M., Voss, A., & Radev, S. (2020). Jumping to Conclusion? A Lévy flight model of decision making. *The Quantitative Methods for Psychology, 16*(2), 120–132. https://doi.org/10.20982/tqmp.16.2.p120

Xu, H., Jiang, X., & Yu, B. (2017). Numerical analysis of the space fractional Navier–Stokes equations. *Applied Mathematics Letters, 69*, 94–100. https://doi.org/10.1016/j.aml.2017.02.006

Xu, Y., Zhang, Y., & Zhao, J. (March 2019). Error analysis of the Legendre-Gauss collocation methods for the nonlinear distributed-order fractional differential equation. *Applied Numerical Mathematics, 142*, 122–138. https://doi.org/10.1016/j.apnum.2019.03.005

Yuttanan, B., & Razzaghi, M. (2019). Legendre wavelets approach for numerical solutions of distributed order fractional differential equations. *Applied Mathematical Modelling, 70*, 350–364. https://doi.org/10.1016/j.apm.2019.01.013

Zayernouri, M., & Karniadakis, G.E. (2014). Fractional spectral collocation method. *SIAM Journal on Scientific Computing, 36*(1), A40–A62. https://doi.org/10.1137/130933216

Zúñiga-Aguilar, C.J., Coronel-Escamilla, A., Gómez-Aguilar, J.F., Alvarado-Martínez, V.M., & Romero-Ugalde, H.M. (2018). New numerical approximation for solving fractional delay differential equations of variable order using artificial neural networks. *European Physical Journal Plus, 133*(2). https://doi.org/10.1140/epjp/i2018-11917-0

Zúñiga-Aguilar, C.J., Romero-Ugalde, H.M., Gómez-Aguilar, J.F., Escobar-Jiménez, R.F., & Valtierra-Rodríguez, M. (2017). Solving fractional differential equations of variable-order involving operators with Mittag-Leffler kernel using artificial neural networks. *Chaos, Solitons and Fractals, 103*, 382–403. https://doi.org/10.1016/j.chaos.2017.06.030

10 Dynamics of Slender Single-Link Flexible Robotic Manipulator Based on Timoshenko Beam Theory

Priya Rao, S. Chakraverty, and Debanik Roy

CONTENTS

10.1 INTRODUCTION

The successive progeny of flexible manipulators has established its mantle through great applications in real world where compliance comes as a boon, notwithstanding the technical challenges in the realization of hardware. This charismatic domain has opened up a wide area for research in robotics. Subtle, precise, and high-performance applicable research in robotics demands for customized manipulator with sufficient compliance, optimum stability low-tare weight, and low rate of energy consumption. Several pieces of research have been conducted in the past to study the flexible

DOI: 10.1201/9781003328032-10

robotic system (FRS) in totality, where the goal was to keep the tip position of the flexible manipulator stable in the face of joint friction and payload variations [1] or rapid evaluation of damping and frequency of vibration [2]. The use of a particularly robust control strategy has been taken for flexible manipulators, which overcome both linear and non-linear frictions. The control technique of FRS in such case consists of two-layered feedback loops: an inner loop for controlling the motor position and an outside loop for controlling the tip position.

Significant research has been done for slewing links in the past few decades. Bellezza et al. [3] has used the novel technique of 'assumed modes' in order to model slewing flexible link, using the traditional pinned or clamped eigenfunctions. The exact eigenfunctions for the slewing link has been presented in their study. A slewing link, modelled as a Timoshenko beam, was examined by Heppler et al. [4] in a rotating frame. In their study, equations of motion has been constructed for two cases: pseudo-clamped and pseudo-pinned, which differ in the non-inertial frame of reference used. Boundary conditions, frequency equations, orthogonality relations, and transformations between the two situations are discussed by the authors to a considerable extent.

Ower and Vegte [5] studied the planar motion of a manipulator made up of two flexible links and two rotational joints has been modelled using a Lagrangian dynamics technique. Transfer function matrix has been used in the study to represent the linearized equations. In addition, a strategy based on classical principles is used to develop a multivariable control system. Zhao et al. [6] investigated the boundary disturbance observer-based control for a vibrating single-link flexible manipulator system with external disturbances. Kalyoncu investigated the dynamic reaction and mathematical modelling of a flexible robot manipulator with rotating-prismatic joint [7]. The study by Kobayashi et al. [8] includes dynamic model of the flexible robotic arm which is derived using a new description of deformation on the flexible shaft and Hamilton's principle, as the bending process of a flexible beam was not accurately expressed by traditional modelling methods. The dynamic behaviour of the flexible beam has been described using a non-linear dynamic model, and the dynamic equations of the flexible robotic arm were also presented. Some non-linear forces and non-linear stiffness are included in the derived equations formed.

The study by Tomei and Tornambe [9] describes a method for generating approximate dynamic models that use a system of ordinary differential equations whose order is related to the desired order of approximation for a given robot. The exact model is described by non-linear partial differential equations, which reflect the distributed nature of link flexibility. These models are created by taking a Lagrangian approach and expanding the generalised co-ordinates unearthing the specific geometry of the beam that makes up the robot in a minimal number of terms. The work by [10] has used Lagrangian approach to derive the dynamic model of the structure. Links are modelled as Euler–Bernoulli beams with proper clamped-mass boundary conditions. The assumed modes method was adopted in order to obtain a finite-dimensional model.

Roy [11] successfully modelled and demonstrated the hardware interfacing of magnetic grippers, undergoing perpetual vibration, for use in the unstructured robotic workspace. The basic modelling of the unstructured robotic environment of [11] accelerated the focal point to the vibration signature of robotic systems having alternative designs. For example, the real-time dynamics as well as source function identification of inherent vibration of flexible manipulators have been described in a novel way in [12] so as to assess its impact on the associated dynamics of a typical multi-degrees-of-freedom FRS. In addition, the research focuses on the modelling and theoretical analysis of a novel rheological rule base, with a focus on the zone-based relative dependency of the finite-numbered sensor units in battling the inherent vibration in the flexible robot. A new proposal has also made employing a stochastic model for analysing the decision threshold band, signalling the activation of the FRS gripper. With the base work of [12], the author substantiated the theory further in developing sensor model-based dynamic analysis of the in-situ vibration of FRS and its impact on the control system hardware [13, 14].

Huang [15] developed the frequency equations pertaining to the normal modes of free flexural vibrations of finite beams including the effect of shear and rotatory inertia for various cases of simple beams. A comparatively sophisticated analytical modelling technique is described by di castri [16] that accounts for elastic complicating effects as well as lumped inertial loads, effectively symbolizing the dynamics of the manipulator's joints. Milford and Asokanthan [17] studied the said vibration issues by matching the boundary equations at the elbow of a general two-link flexible manipulator. The exact eigenfrequencies corresponding to arbitrary elbow angles are obtained by solving the partial differential equation formulation numerically.

In this chapter, a comparison of representative approaches for model-based and model-free control of flexible-link robots is presented. Modelling for a single-link system has been considered with the dynamics presented in [16], which may be compared with [18] where the study has been carried out for mass-loaded clamped-free beam or can be compared with [19] for non-zero tip mass. Governing differential equations and the boundary conditions esentially leads to an eigenvalue problem. The analytical treatment for solving the eigenvalue problem is, by and large, shape-independent as long as the computing plane is not altered. The solution for the eigenvalue problem in vibrating plates was investigated by Chakraverty [20], which can be extended to the solution of the same in beams too.

Traditionally, boundary conditions in a flexible robotic system are provided by taking the relations between forces and moments in each joint. Through substitution of the boundary conditions into the solution of the dynamic equation of the FRS system, we get a fresh set of equations, which leads to a homogeneous system of linear equations, resulting to an eigenvalue problem. Applications of the eigenvalue problem has been investigated by many researchers in the past few years [21, 22]. The buckling behaviour of an Euler–Bernoulli nanobeam has been investigated by Jena et al. [23] where the shifted Chebyshev polynomials-based Rayleigh–Ritz method has been used to calculate critical buckling loads for all classical boundary conditions, such as Pined–Pined (P-P), Clamped–Pined (C-P), Clamped–Clamped (C-C), and Clamped–Free (C-F). For different cross-sections of beam, shear coefficient has been derived by Cowper [24] that was verified with values obtained by other authors. A new formula for the shear coefficient eventually comes out of the derivation thereof.

10.2 PRELIMINARIES

The traditional and widely used hardware of flexible manipulator is built with links having circular cross-section. Figure 10.1 schematically illustrates a single flexible link with circular cross-section and end (tip) mass.

With this concept of flexible link in mind, a few important basics of beam theory and terms related to single-link manipulator are given for a better understanding of readers in this section.

10.2.1 JOINTS [25]

The joint is the conglomeration of two movable members that cause relative motion between the adjacent links of a robotic manipulator. We will investigate two predominant types of joints that are used in flexible robots.

10.2.1.1 Revolute Joint

A revolute joint (also called pin joint or hinge joint) is a one-degree-of-freedom kinematic pair used frequently as inflexible manipulators. The joint constrains the motion of two bodies to pure rotation along a common axis; it does not allow translation or sliding linear motion. The parlances of a typical revolute joint of a flexible manipulator are illustrated in Figure 10.2. While the basic physics of the rotational motion of two mating components is explained schematically in Figure 10.2a, the external envelope for the prototyping of the revolute joint is conceptualized in Figure 10.2b. The technical details of the mechanical assembly of the revolute joint are highlighted schematically in Figure 10.2c.

10.2.1.2 Prismatic Joint

A prismatic joint provides a linear sliding movement between two bodies and is often called translatory joint, due to its characteristic features. A prismatic joint can be formed with a polygonal cross-section to resist rotation. A prismatic pair is also called a sliding pair. Figure 10.3 illustrates the fundamental motion generation for a prismatic/sliding/translatory joint in flexible robot.

Irrespective of the type of joint, the crucial-most aspect of the design of flexible manipulator is the placement of joints with respect to either global origin or the

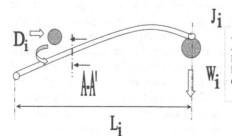

Index: L_i: Length of the Link of Flexible Manipulator; D_i: Area of Cross-section of the Flexible Link; $\{J_{i-1}, J_i\}$: Revolute Joints of the Flexible manipulator; W_i: End-Mass (Payload) of the Manipulator; (X,Y,Z): Cartesian Co-ordinate Frame; $A-A'$: Section plane

FIGURE 10.1 Schematic disposition of the flexible link with circular cross-section and end mass.

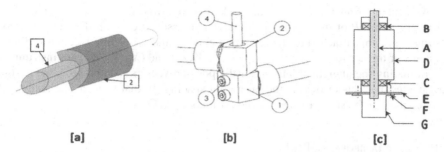

[a] **[b]** **[c]**

Legends: **1:** Lower Housing; **2:** Upper Housing; **3:** Access for Bearings; **4:** Joint Shaft; **A:** Pin; **B:** Micro-Bearing (Upper Rung); **C:** Micro-Bearing (Lower Rung); **D:** Joint Housing; **E:** Adapter Plate; **F:** Fixing Screws; **G:** Extension Plate

FIGURE 10.2 Parlances of revolute joint of flexible manipulator: (a) basic physics, (b) prototyping (external envelope), (c) technical details of the assembly.

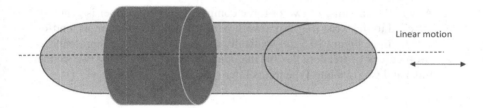

FIGURE 10.3 Prismatic joint.

(local) origin of the preceeding joint. The ensemble kinematic analysis of a typical multi-degrees-of-freedom flexible manipulator will depend largely on the locational syntax of the joints. This locational paradigm is essentially time-dependent and it serves as the fundamental scientific seeding for the twist of the joint(s) of the under run-time condition of the flexible manipulator. Subsequently, this run-time dynamic twist of the joint(s) gives rise to random in-situ vibration of the flexible robotic system. Figure 10.4 schematically illustrates a representative disposition of two adjacent revolute joints, post twisting (under in-situ vibration/trembling).

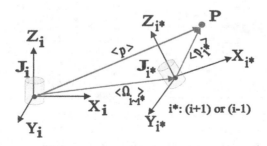

FIGURE 10.4 Representative disposition of two adjacent revolute joints under twist.

As evident from Figure 10.4, the task was to define the position vector of an arbitrary point, P of the flexible manipulator in association with the Cartesian co-ordinate systems attached to the revolute joints, J_i' and J_{i*}'. While local co-ordinate systems of these two adjacent joints, viz. (X_i, Y_i, Z_i) and (X_{i*}, Y_{i*}, Z_{i*}) are instrumental in finding the analytical lemma for the position vectors $\langle p \rangle$ and $\langle p^* \rangle$, the design parameter that gets critical is the intra-joint spacing. In that respect, the analytical model needs to be strengthened for the estimation of $\langle \Omega_{i-1*} \rangle$.

10.2.2   MANIPULATOR [26]

A robot manipulator is a mechanical connection that can reach, grip, and handle objects in a similar way to a human arm. The physical structure of the manipulator can be divided into two segments:

1. One or more 'arm-and-body' sub-system, which usually consists of N joints connected by M long slender links $(M \leq N)$;
2. A 'wrist', consisting of two or three compact joints. A gripper is generally attached to the wrist to grasp a work part or a tool, to function as 'end-of-arm tooling'. Figure 10.5 schematically shows the layout of a representative two-degrees-of-freedom flexible manipulator system, involving two slender links and two revolute-type joints inter-spaced.

10.2.3   CANTILEVER BEAM [20]

A cantilever beam is one that has one free end and one end that is supported by a fixed end. Figure 10.6 illustrates typical configuration of a cantilever beam-type single-link flexible manipulator with an end-effector (two-jaw gripper) attached at its free end. It is important to note here that all analytical model of flexible manipulator must incorporate 'pesudo-spring' in order to simulate the random vibration of the system. This pesudo-spring will take care of the tare weight of the manipulator as well as the 'payload' at the gripper. The spring-based vibration model of the flexible

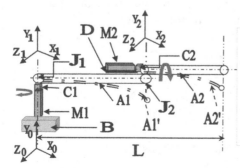

Index: **A1**: First Link of the Flexible Manipulator; **A2**: Second Link of the Flexible Manipulator; **B**: Base of the Flexible Manipulator; **{A1', A2'}**: Deflected Postures of the Links; **J₁**: First Revolute Joint; **J₂**: Second Revolute Joint; **{C1, C2}**: Couplers for the Revolute Joints; **{M1, M2}**: Servomotors for the Joint Actuations; **D**: Seat for the Motor 'M2'; **L**: Total Horizontal Span of the Flexible Manipulator; **(X₀,Y₀,Z₀)**: Global / Base Co-ordinate System; **(X₁,Y₁,Z₁)** & **(X₂,Y₂,Z₂)**: Local / Joint Co-ordinate Systems

FIGURE 10.5   Schematic layout of a representative two-degrees-of-freedom flexible manipulator.

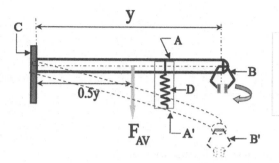

Index: **A**: Flexible Link (Stable Posture); **A'**: Flexible Link (Deflected Posture); **{B, B'}**: Stable & Deflected Postures of the Gripper; **C**: Fixed Support of the *'Cantilever Beam'* (Flexible Link); **D**: Peuso-Spring; **y**: Length of the Flexible Link; F$_{AV}$: Tare-weight of the Flexible Link

FIGURE 10.6 Conceptual schematic of the spring-based structural model for flexible manipulator.

manipulator is a sort of foundation for the evaluation of natural frequency of vibration and subsequent modal analysis for the flexible manipulator.

10.2.4 BENDING MOMENT [27]

The material that is oriented in the normal axis of the relevant cross-section to the bending moment undergoes bending stress as a result of the bending moment.

10.2.5 BOUNDARY CONDITIONS FOR BEAM [20]

1. Clamped Boundary Conditions:
    Both displacement and slope are considered to be zero, i.e.,

    at $x = 0$; $v(x) = 0$; $\dfrac{dv}{dx} = 0$

    at $x = l$; $v(x) = 0$; $\dfrac{dv}{dx} = 0.$

2. Simply Supported Boundary Conditions:
    Here displacement and bending moment must be zero, i.e.,

    at $x = 0$; $v(x) = 0$ and $\dfrac{d^2v}{dx^2} = 0$

    at $x = l$; $v(x) = 0$ and $\dfrac{d^2v}{dx^2} = 0.$

10.3 THEORETICAL ANALYSIS

The flexible manipulator under study is constrained in horizontal plane. Link designed of appropriate material with prime properties that guarantees the manufacturing material to be homogeneous, isotropic, and linearly elastic. We have the following parameters of the flexible link for modelling, namely length l, cross-sectional area A (with uniform cross-sections), volumetric density ρ, Young's modulus E, cross-sectional moment of inertia I, shear modulus G, and shear correction factor K.

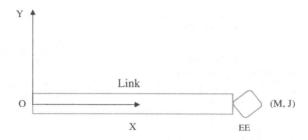

FIGURE 10.7   Model of a single-link flexible manipulator with payload at its tip [16].

The link has a solid body of sufficiently long size with mass and moment of inertia M and J, respectively, at the tip representing a payload at tip.

A single-link flexible manipulator is shown in Figure 10.7 [16], for the purpose of mathematical analysis. The configuration has been chosen such that at point O, the manipulator is fixed.

The co-ordinates for the generic point P, with regard to the corresponding reference system (O,X,Y,Z) of the link can be represented by $[x+u(x,t),y(x,t),0]^T$, where the two functions $u(x,t)$ and $y(x,t)$ reflect the axial and transverse displacements endured by the considered point, respectively.

10.3.1 MATHEMATICAL MODELLING

The bending moment $M(x,t)$ and the shear force $T(x,t)$ in the beam are related to the displacement $y(x,t)$ and the rotation $\phi(x,t)$. Applying Timoshenko beam theory, the stress displacement relations are given by [16]:

$$M(x,t) = EI\frac{\partial \phi}{\partial x}$$

$$T(x,t) = KGA\left[\frac{\partial y}{\partial x} - \phi(x,t)\right] \tag{10.1}$$

$$N(x,t) = EA\frac{\partial u(x,t)}{\partial x}$$

where $u(x,t)$ is the axial displacement function, $N(x,t)$ is axial force of the link. The dynamic equilibrium equations of stress for a link subjected to a transverse distributed load $q(x,t)$ and an axial distributed load $p(x,t)$ are given by [16]

$$\frac{\partial M(x,t)}{\partial x} + T(x,t) = \rho I\frac{\partial^2 \phi}{\partial t^2}$$

$$\frac{\partial T(x,t)}{\partial x} - \rho A\frac{\partial^2 y}{\partial t^2} = -q(x,t) \tag{10.2}$$

$$\frac{\partial N(x,t)}{\partial x} - \rho A\frac{\partial^2 \phi}{\partial t^2} = -p(x,t)$$

We get a system of partial differential equations for $u(x,t)$, $y(x,t)$, and $\phi(x,t)$ characterising the dynamics of the single link when we substitute expressions (10.1) in the latter Eqn. (10.2) given in [16]:

$$EI\frac{\partial^2\phi(x,t)}{\partial x^2} + KGA\left(\frac{\partial y(x,t)}{\partial x} - \phi(x,t)\right) = \rho I\frac{\partial^2\phi(x,t)}{\partial t^2} \tag{10.3}$$

$$KGA\left(\frac{\partial\phi(x,t)}{\partial x} - \frac{\partial^2 y(x,t)}{\partial x^2}\right) + \phi A\frac{\partial^2 y(x,t)}{\partial t^2} = q(x,t) \tag{10.4}$$

$$EA\frac{\partial^2 u(x,t)}{\partial x^2} - \rho A\frac{\partial^2 u(x,t)}{\partial t^2} = -p(x,t) \tag{10.5}$$

For the link of the manipulator, there will be a system of partial differential equations describing the deformations of the link of the manipulator. It will be feasible to solve the differential eigenvalue problem of the system associated with a generic configuration, the eigenvalues and eigenfunctions of which will represent the needed modal data, by articulating the exact boundary conditions in the case of free vibration.

10.3.2 MODAL ANALYSIS

The system's free vibration problem is now investigated; in this situation, the distributed forces $q(x,t)$ and $p(x,t)$ are equal to zero. In this regard, the dynamics of the single-link system can be re-established on decoupling Eqns. (10.3) and (10.4) for ϕ and y, [16, 28] wherein we have the following differential equations:

$$\frac{EI}{\rho A}\frac{\partial^4\phi}{\partial x^4} - \frac{I}{A}\left(1+\frac{E}{KG}\right)\frac{\partial^4\phi}{\partial x^2\partial t^2} + \frac{\partial^2\phi}{\partial t^2} + \frac{\rho I}{KGA}\frac{\partial^4\phi}{\partial t^4} = 0 \tag{10.6}$$

$$\frac{EI}{\rho A}\frac{\partial^4 y}{\partial x^4} - \frac{I}{A}\left(1+\frac{E}{KG}\right)\frac{\partial^4 y}{\partial x^2\partial t^2} + \frac{\partial^2 y}{\partial t^4} + \frac{\rho I}{KGA}\frac{\partial^4 y}{\partial t^4} = 0 \tag{10.7}$$

$$EA\frac{\partial^2 u}{\partial x^2} - \rho A\frac{\partial^2 u}{\partial t^2} = 0 \tag{10.8}$$

Assuming variables separation in the way, having [16]

$$\phi(x,t) = \Phi(x)e^{j\lambda t}$$
$$y(x,t) = Y(x)e^{j\lambda t} \tag{10.9}$$
$$u(x,t) = U(x)e^{j\lambda t}$$

Equations (10.6), (10.7), and (10.8) can be represented as ordinary differential equations in the following form as [18] with natural frequency λ.

$$EI\Phi''''(x) + \lambda^2\rho I\left(1+\frac{E}{KG}\right)\Phi''(x) + \lambda^2\left(\lambda^2\frac{\rho^2 I}{KG} - \rho A\right)\Phi(x) = 0 \tag{10.10}$$

$$EIY''''(x) + \lambda^2 \rho I \left(1 + \frac{E}{KG}\right) V''(x) + \lambda^2 \left(\lambda^2 \frac{\rho^2 I}{KG} - \rho A\right) Y(x) = 0 \quad (10.11)$$

$$EAU''(x) + \lambda^2 \rho A U(x) = 0 \tag{10.12}$$

$\Phi(x)$, $Y(x)$, and $U(x)$ are unknown functions with unknown parameter λ. The boundary conditions are written as relations between forces and moments in correspondence to joint section for a given configuration and the free end, given as [16].
    For $O(x = 0)$ section:

$$\Phi(x) = 0 \tag{10.13}$$

$$Y(x) = 0 \tag{10.14}$$

$$U(x) = 0 \tag{10.15}$$

For EE section (end effector):

$$EI\Phi'(x) - \lambda^2 J\Phi(x)\big|_{x=l} = 0 \tag{10.16}$$

$$KGA[Y'(x) - \Phi(x)] - \lambda^2 MY(x)\big|_{x=l} = 0 \tag{10.17}$$

$$EAU'(x) - \lambda^2 MU(x)\big|_{x=l} = 0 \tag{10.18}$$

10.4 FORMATION OF EIGENVALUE PROBLEM FOR SINGLE-LINK MANIPULATOR

For single-link manipulator with $\lambda < \sqrt{\dfrac{KGA}{\rho I}}$ solution to Eqns. (10.10), (10.11), and (10.12) on solving the characteristic equations for the above, we have solution as [16]:

$$\Phi(x) = \overline{C_1} \cosh(\omega_{11}x) + \overline{C_2} \sinh(\omega_{11}x) + \overline{C_3} \cos(\omega_{12}x) + \overline{C_4} \sin(\omega_{12}x) \quad (10.19)$$

$$Y(x) = C_1 \cosh(\omega_{11}x) + C_2 \sinh(\omega_{11}x) + C_3 \cos(\omega_{12}x) + C_4 \sin(\omega_{12}) \quad (10.20)$$

$$U(x) = D_1 \cos\left(\frac{\lambda x}{a}\right) + D_2 \sin\left(\frac{\lambda x}{a}\right) \tag{10.21}$$

where $a = \sqrt{\dfrac{E}{\rho}}$ is the longitudinal-wave speed. The coefficients $\overline{C_k}$ and C_k ($k = 1,\ldots,4$) are connected through Eqn. (10.4); in fact, we may get the following relations by substituting solutions in Eqn. (10.9) for $y(x,t)$ and $\phi(x,t)$ and using expressions (10.19) and (10.20) as [16].

$$\overline{C_1} = \xi_1 C_2 \tag{10.22}$$

$$\overline{C_2} = \xi_1 C_1 \tag{10.23}$$

$$\overline{C_3} = \xi_2 C_4 \tag{10.24}$$

$$\overline{C_4} = -\xi_2 C_3 \tag{10.25}$$

where,

$$\xi_1 = \frac{1}{\omega_{11}} \left(\omega_{11}^2 + \frac{\lambda^2 \rho}{KG} \right) \tag{10.26}$$

$$\xi_2 = \frac{1}{\omega_{11}} \left(\omega_{11}^2 - \frac{\lambda^2 \rho}{KG} \right) \tag{10.27}$$

where $c = [C_1, C_2, C_3, C_4, D_1, D_2]^T$. Obtaining the equations in matrix form on substituting Eqns. (10.19)–(10.21) into Eqns. (10.13)–(10.18), as [16]:

$$M(\lambda)c = 0 \tag{10.28}$$

$$M(\lambda) = \begin{bmatrix} 0 & t_{11} & 0 & t_{12} & 0 & 0 \\ 1 & 0 & 1 & 0 & 0 & 0 \\ 0 & 0 & 0 & 0 & 1 & 0 \\ r_{41} & r_{42} & r_{43} & r_{44} & 0 & 0 \\ r_{51} & r_{52} & r_{53} & r_{54} & 0 & 0 \\ 0 & 0 & 0 & 0 & r_{65} & r_{66} \end{bmatrix} \tag{10.29}$$

where

$$r_{41} = -Jt_{11}sinh(l\omega_{11})\lambda^2 + EI\omega_{11}t_{11}cosh(l\omega_{11})$$
$$r_{42} = -Jt_{11}cosh(l\omega_{11})\lambda^2 + EI\omega_{11}t_{11}sinh(l\omega_{11})$$
$$r_{43} = Jt_{12}sin(l\omega_{12})\lambda^2 - EI\omega_{12}t_{12}cos(l\omega_{12})$$
$$r_{44} = -Jt_{12}cos(l\omega_{12})\lambda^2 - EI\omega_{12}t_{12}sin(l\omega_{12})$$
$$r_{51} = -Mcosh(l\omega_{11})\lambda^2 + GKA(\omega_{11}sinh(L_1\omega_{11}) - t_{11}sinh(l\omega_{11})) \tag{10.30}$$
$$r_{52} = -Msinh(l\omega_{11})\lambda^2 + GKA(\omega_{11}cosh(l\omega_{11}) - t_{11}cosh(l\omega_{11}))$$
$$r_{53} = -Mcos(l\omega_{12})\lambda^2 - GKA(\omega_{12}sin(l\omega_{12}) - t_{12}sin(l\omega_{12}))$$
$$r_{54} = -Msin(l\omega_{12})\lambda^2 + GKA(\omega_{12}cos(l\omega_{12}) - t_{12}cos(l\omega_{12}))$$
$$r_{65} = -M\lambda^2cos((l\lambda)/a) - (EA\lambda sin((l\lambda)/a))/a$$
$$r_{66} = (EA\lambda cos((l\lambda)/a))/a - M\lambda^2sin((l\lambda)/a)$$

$M(\lambda)$ reflects the system characteristic matrix associated with the multi-link manipulator's nominal setup. This is a matrix the entries of which are functions of natural frequency λ.

10.5 NUMERICAL RESULTS AND DISCUSSIONS

Horizontal vibration of an undamped Euler–Bernoulli beam with tip mass attached at free end along with clamped-free boundary conditions has been studied by [19]. The governing differential equation of motion of homogeneous Euler–Bernoulli beam for undamped free vibration is obtained as [19]

$$EI\frac{\partial^4 y(x,t)}{\partial x^4} + m\frac{\partial^2 y(x,t)}{\partial t^2} = 0 \tag{10.31}$$

where $y(x,t)$ is transverse displacement due to bending, EI is bending stiffness, and m is mass per unit length of beam. The boundary conditions for clamped-free beam with tip mass are expressed as [19]:

$$y(0,t) \quad = 0,$$

$$\left.\frac{\partial y(x,t)}{\partial x}\right|_{x=0} = 0 \tag{10.32}$$

$$\left[EI\frac{\partial^2 y(x,t)}{\partial x^2} + J\frac{\partial^3 y(x,t)}{\partial t^2 \partial x}\right]_{x=l} = 0$$

$$\left[EI\frac{\partial^3 y(x,t)}{\partial x^3} - M\frac{\partial^2 y(x,t)}{\partial t^2}\right]_{x=l} = 0 \tag{10.33}$$

where M and J are tip mass and moment of inertia of tip mass, respectively. Assuming variable separation [19, 29] by

$$y(x,t) = \phi(x)n(t) \tag{10.34}$$

Finally, Eqn. (10.31) reduces to [19]

$$\frac{EI}{m}\frac{1}{\phi(x)}\frac{d^4\phi(x)}{dx^4} = -\frac{1}{n(t)}\frac{d^2 n(t)}{dt^2} \tag{10.35}$$

The standard argument of separation of variables states that both sides of Eqn. (10.35) must be equal to a constant χ (say). Accordingly, the solution to Eqn. (10.35) can be obtained as per [19]

$$\phi(x) = A\cos\alpha x + B\cosh\alpha x + C\sin\alpha x + D\sinh\alpha x \tag{10.36}$$

TABLE 10.1

Parameters for Single-Link Manipulator [16]

Parameters	Symbols	Units	Values
Length	l	m	0.5
Thickness	h	m	0.00061
Width	b	m	0.051
Mass of link	m	kg	0.121
Stiffness	EI	N m^2	0.203
Tip load mass	M	kg	0.61
Tip load inertia	J	kg m^2	$1.5 * 10^{-3}$

$$n(t) = E \cos \lambda t + F \sin \lambda t \tag{10.37}$$

where $\alpha = \dfrac{\omega}{l}$, A, B, C, D, E, F are constants with $\lambda^2 = \chi$ and

$$\omega^4 = \lambda^2 \frac{ml^4}{EI} \tag{10.38}$$

By substituting Eqns. (10.36) and (10.37) into Eqns. (10.32) and (10.33), the differential eigenvalue problem has been formed as $M(\omega)c = 0$, where $c = [A, C]$ (as in [21]):

$$\begin{bmatrix} \cos\omega + \cosh\omega - \dfrac{\omega^3 J}{ml^3}(\sin\omega + \sinh\omega) & \sin\omega + \sinh\omega + \dfrac{\omega^3 J}{ml^3}(\cos\omega - \cosh\omega) \\ \sin\omega - \sinh\omega + \dfrac{\omega M}{ml}(\cos\omega - \cosh\omega) & -\cos\omega - \cosh\omega - \dfrac{\omega M}{ml}(\sin\omega - \sinh\omega) \end{bmatrix} \begin{bmatrix} A \\ C \end{bmatrix} = \begin{bmatrix} 0 \\ 0 \end{bmatrix} \tag{10.39}$$

For non-trivial solution of Eqn. (10.39), necessary condition is $det(M) = 0$. Solving Eqn. (10.39), the parameters taken for the calculations are from Table 10.1 [16].

Frequency has been calculated by solving Eqns. (10.28) and (10.39) using Newton Raphson's method with tip mass zero. Comparing both the obtained values of λ,

TABLE 10.2

Frequency Comparison for Zero Tip Mass

Frequencies	λ by [16] (in Hz)	λ by [19] (in Hz)	Absolute Error
1	2.0500795032	2.0500748243	4.6789×10^{-6}
2	12.847539111	12.847639556	1.00445×10^{-4}
3	35.973075019	35.973775913	7.00894×10^{-4}
4	70.491705382	70.494244376	2.538978×10^{-3}
5	116.5253395	116.53203989	6.70039×10^{-3}

natural frequency for undamped Euler–Bernoulli cantilever beam and single-link manipulator, we get the following result as shown in Table 10.2.

For mass-loaded clamped-free beam, from [18] the equations of motion for transverse deflection and rotation are

$$\frac{d^4 y}{dx^4} + b(r+s)\frac{d^2 y}{dx^2} - b(1-brs)y = 0 \tag{10.40}$$

$$\frac{d^4 \psi}{dx^4} + b(r+s)\frac{d^2 \psi}{dx^2} - b(1-brs)\psi = 0 \tag{10.41}$$

where x is non-dimensional length, $b = (1/EI)(\eta A/g)l^4 \lambda^2$, $r = I/Al^2$, $s = EI/KAGl^2$, $\bar{m} = m/M$, $\sigma = k/l$ are also dimensionless variables defined by [15, 18]. The boundary conditions as in [18] are given by

$$y = 0, \ \psi = 0 \ \text{at} \ x = 0 \tag{10.42}$$

$$\psi - \left(\frac{1}{l}\right)\frac{dy}{dx} + \bar{m}bs\frac{y}{l} = 0 \ \text{at} \ x = 1 \tag{10.43}$$

$$\left(\frac{1}{l}\right)\frac{d\psi}{dx} - b(1-brs)\psi = 0 \ \text{at} \ x = 1 \tag{10.44}$$

Solution to Eqns. (10.40) and (10.41) is [18],

$$y = A_1 \cosh \sqrt{b}\alpha x + A_2 \sinh \sqrt{b}\alpha x + A_3 \cos \sqrt{b}\beta x + A_4 \sin \sqrt{b}\beta x \tag{10.45}$$

$$\psi = A_1' \sinh \sqrt{b}\alpha x + A_2' \cosh \sqrt{b}\alpha x + A_3' \sin \sqrt{b}\beta x + A_4' \cos \sqrt{b}\beta x \tag{10.46}$$

where $\alpha = \dfrac{\sqrt{-(r+s)+\sqrt{(r-s)^2 + \dfrac{4}{b}}}}{\sqrt{2}}$, $\beta = \dfrac{\sqrt{-(r+s)-\sqrt{(r-s)^2 + \dfrac{4}{b}}}}{\sqrt{2}}$. Substituting

Eqns. (10.45) and (10.46) into Eqns. (10.42)–(10.44) yields the homogeneous system [18]:

$$\begin{bmatrix} 1 & 0 & 1 & 0 \\ 0 & (\alpha^2+s)/\alpha & 0 & (\beta^2-s)/\beta \\ r_1 & r_2 & r_3 & r_4 \\ r_1' & r_2' & r_3' & r_4' \end{bmatrix} \begin{bmatrix} A_1 \\ A_2 \\ A_3 \\ A_4 \end{bmatrix} = \begin{bmatrix} 0 \\ 0 \\ 0 \\ 0 \end{bmatrix} \tag{10.47}$$

where,

$$r_1 = \frac{\sqrt{b}}{\alpha} \sinh \sqrt{b}\alpha + \bar{m}b \cosh \sqrt{b}\alpha$$

$$r_2 = \frac{\sqrt{b}}{\alpha} \cosh \sqrt{b}\alpha + \bar{m}b \sinh \sqrt{b}\alpha$$

$$r_3 = \frac{\sqrt{b}}{\beta} \sin \sqrt{b}\beta + \bar{m}b \cos \sqrt{b}\beta$$

$$r_4 = -\frac{\sqrt{b}}{\beta} \cos \sqrt{b}\beta + \bar{m}b \sin \sqrt{b}\beta$$

$$r_1' = \frac{(\alpha^2 + s)}{\alpha} \left(\sqrt{b}\alpha \cosh \sqrt{b}\alpha - \frac{1}{2}\bar{m}\sigma b \sinh \sqrt{b}\alpha \right)$$

$$r_2' = \frac{(\alpha^2 + s)}{\alpha} \left(\sqrt{b}\alpha \sinh \sqrt{b}\alpha - \frac{1}{2}\bar{m}\sigma b \cosh \sqrt{b}\alpha \right)$$

$$r_3' = -\frac{(\beta^2 - s)}{\beta} \left(\sqrt{b}\beta \cos \sqrt{b}\beta - \frac{1}{2}\bar{m}\sigma b \sin \sqrt{b}\beta \right)$$

$$r_4' = \frac{(\beta^2 - s)}{\beta} \left(-\sqrt{b}\beta \sin \sqrt{b}\beta - \frac{1}{2}\bar{m}\sigma b \cos \sqrt{b}\beta \right)$$

(10.48)

Frequency has been calculated by solving Eqns. (10.28) and (10.47) using Newton Raphson's method with non-zero tip mass. Comparing both the obtained values of λ, natural frequency for the model [18] and single-link manipulator with Timoshenko beam theory, we get the following result as shown in Table 10.3. Solving Eqns. (10.47) and (10.28), the parameters taken for the calculations are from Table 10.1 [16].

Table 10.2 shows the comparison between obtained frequencies of single-link manipulator using Timoshenko beam theory with frequencies of cantilever beam using Euler–Bernoulli with zero tip mass [19]. Table 10.3 shows the comparison between obtained frequency for single-link manipulator using Timoshenko beam theory and frequency for cantilever beam with non-zero tip mass [18].

TABLE 10.3
Frequency Comparison for Non-Zero Tip Mass

Frequencies	λ by [16] (in Hz)	λ by [18] (in Hz)	Absolute Error
1	0.43504211718	0.43503412194	7.99524×10^{-6}
2	4.7543985962	4.7544037894	5.1932×10^{-6}
3	14.346262292	14.346226587	3.5705×10^{-5}
4	36.572002716	36.571800612	2.02104×10^{-4}
5	70.84895912	70.848952998	5.7086×10^{-4}

10.6 CONCLUSIONS

In this chapter, comparative study between single-link manipulator and cantilever beam with tip mass has been done. Eigenvalue problem corresponding to the configuration of the single-link manipulator, with boundary conditions, has been explained using Timoshenko beam theory. From the comparison of frequency values, as presented in Tables 10.2 and 10.3, we can see that single-link flexible manipulator shows good agreement with cantilever beam with tip mass. Besides, the single-link flexible manipulator shows a good agreement with cantilever beam with tip mass using classical beam theory in case there is no presence of tip mass.

REFERENCES

[1] Feliu, V., Rattan, K.S. and Brown, H.B. 1990. Adaptive control of a single-link flexible manipulator. *IEEE Control Systems Magazine*, 10(2), pp. 29–33.

[2] Trapero-Arenas, J.R., Mboup, M., Pereira-Gonzalez, E. and Feliu, V. 2008, June. On-line frequency and damping estimation in a single-link flexible manipulator based on algebraic identification. In 2008, 16th Mediterranean Conference on Control and Automation (pp. 338–343). IEEE.

[3] Bellezza, F., Lanari, L. and Ulivi, G. 1990, May. Exact modeling of the flexible slewing link. In Proceedings., IEEE International Conference on Robotics and Automation (pp. 734–739). IEEE.

[4] White, M.W.D. and Heppler, G.R. 1995, June. A Timoshenko model of a flexible slewing link. In Proceedings of 1995 American Control Conference-ACC'95 (Vol. 4, pp. 2815–2819). IEEE.

[5] Ower, J. and De Vegte, J. 1987. Classical control design for a flexible manipulator: Modeling and control system design. *IEEE Journal on Robotics and Automation*, 3(5), pp. 485–489.

[6] Zhao, Z., He, X. and Ahn, C.K. 2019. Boundary disturbance observer-based control of a vibrating single-link flexible manipulator. *IEEE Transactions on Systems, Man, and Cybernetics: Systems*, 51(4), pp. 2382–2390.

[7] Kalyoncu, M. 2008. Mathematical modelling and dynamic response of a multi-straight-line path tracing flexible robot manipulator with rotating-prismatic joint. *Applied Mathematical Modelling*, 32(6), pp. 1087–1098.

[8] Yin, H., Kobayashi, Y., Hoshino, Y. and Emaru, T. 2011. Modeling and vibration analysis of flexible robotic arm under fast motion in consideration of nonlinearity. *Journal of System Design and Dynamics*, 5(5), pp. 909–924.

[9] Tomei, P. and Tornambe, A. 1988. Approximate modeling of robots having elastic links. *IEEE Transactions on Systems, Man, and Cybernetics*, 18(5), pp. 831–840.

[10] De Luca, A. and Siciliano, B. 1991. Closed-form dynamic model of planar multi-link lightweight robots. *IEEE Transactions on Systems, Man, and Cybernetics*, 21(4), pp. 826–839.

[11] Roy, D. 2015. Development of novel magnetic grippers for use in unstructured robotic workspace. *Robotics and Computer-Integrated Manufacturing*, 35, pp. 16–41.

[12] Roy, D. 2020. Control of inherent vibration of flexible robotic systems and associated dynamics. In *Recent Trends in Wave Mechanics and Vibrations*, Chakraverty, S. and Biswas, P. (eds.) (pp. 201–222). Singapore: Springer.

[13] Roy, D. 2020. Design, modeling and indigenous firmware of patient assistance flexible robotic system-type i: Beta version. *Advances in Robotics & Mechanical Engineering*, 2, pp. 148–159.

[14] Debanik, R. 2019. Towards the control of inherent vibration of flexible robotic systems and associated dynamics: New proposition and model. *International Journal of Robotics Research, Applications and Automation*, 1(1), pp. 6–17.

[15] Huang, T.C. 1961. The effect of rotatory inertia and of shear deformation on the frequency and normal mode equations of uniform beams with simple end conditions. *Journal of Applied Mechanics*, 028(4), p. 579.

[16] di Castri, C. and Messina, A. 2011. Vibration analysis of multilink manipulators based on timoshenko beam theory. *Journal of Robotics*, 2011, https://doi.org/10.1155/2011/890258

[17] Milford, R.I. and Asokanthan, S.F. 1999. Configuration dependent eigenfrequencies for a two-link flexible manipulator: Experimental verification. *Journal of Sound and Vibration*, 222(2), pp. 191–207.

[18] Bruch Jr., J.C. and Mitchell, T.P. 1987. Vibrations of a mass-loaded clamped-free Timoshenko beam. *Journal of Sound and Vibration*, 114(2), pp. 341–345.

[19] Erturk, A. and Inman, D.J. 2011. *Piezoelectric Energy Harvesting*. New York: John Wiley and Sons.

[20] Chakraverty, S. 2008. *Vibration of Plates*. CRC Press.

[21] Jeswal, S.K. and Chakraverty, S. 2021. Fuzzy eigenvalue problems of structural dynamics using ANN. In *New Paradigms in Computational Modeling and its Applications* (pp. 145–161). Academic Press.

[22] Chakraverty, S. and Mahato, N.R. 2018. Nonlinear interval eigenvalue problems for damped spring-mass system. *Engineering Computations*, 35(6), pp. 2272–2286.

[23] Jena, S.K., Chakraverty, S. and Tornabene, F. 2019. Buckling behavior of nanobeams placed in electromagnetic field using shifted Chebyshev polynomials-based Rayleigh-Ritz method. *Nanomaterials*, 9(9), p. 1326.

[24] Cowper, G.R. 1966. The shear coefficient in Timoshenko's beam theory. *Journal of Applied Mechanics*, 33(2), pp. 335–340.

[25] Britannica, E. 2019. The Editors of Encyclopaedia Britannica. de la Enciclopedia Británica.

[26] Hollerbach, John M. 2003. Robotics, computer simulations for. Edited by Robert A. Meyers. In *Encyclopedia of Physical Science and Technology* (third edition) (pp. 275–281). Academic Press.

[27] Goharian, A., Kadir, M.R.A. and Abdullah, M.R., 2017. Trauma Plating Systems: Biomechanical, Material, Biological, and Clinical Aspects. Amsterdam, The Netherlands: Elsevier Science.

[28] Graff, K.F., 1991. Wave Motion in Elastic Solids. New York: Dover Publications Inc.

[29] Greenberg, M.D. 1998. Book notices1. *Journal of Optimization Theory and Applications*, 98(2), pp. 503–504.

11 Non-Probabilistic Solution of Imprecisely Defined Structural Problem with Beams and Trusses Using Interval Finite Element Method

Sukanta Nayak and Shravani V Shetgaonkar

CONTENTS

11.1 INTRODUCTION

Finite element method (FEM) have been popularized due to the importance of applicability and easy use to investigate many engineering and science problems. There are plenty of excellent manuscripts available for its first-hand use. Few of these are viz. (Bhavikatti [2005]; Narasaiah [2008]; Seshu [2003]). These manuscripts provide a comprehensive knowledge to understand FEM and its insight. Structural engineering problem is one of the research areas in which extensive use of FEM is found. Due to the easy application and illustration of the FEM, it is

DOI: 10.1201/9781003328032-11

used for complicated structures that are discretized to finite number of elements. After discretization of the domain, elements are assembled for complete study of the structure. Then boundary conditions are provided with the assembled element and the field variables are estimated. The approximation in finite elements are performed with nodal values of the structure that we take. As such, the continuous system is converted to discrete finite element problem with the help of unknown nodal values. Algebraic equations are obtained for the problem and hence the values of field variable inside the elements are recovered by using the nodes. Two important features of FEM makes it different from other numerical techniques. First is piece-wise approximation on elements give good precision. That is, if one increase the number of discretized element then desired precision can be achieved. Second is the locality of approximation provide a sparse system of equations. These two features greatly help to solve partial differential equation (PDE) by converting it into algebraic equations. In this respect, Bhavikatti (2005) and Gerald and Wheatley (1985) presents a variety of application-oriented problems with simple understanding. The mentioned theory are based on the deterministic or exact (crisp) values of involved system parameters, coefficients, and conditions. Instead of the deterministic, we may have imprecise or vague values depending on the nature of the problem. The impreciseness or vagueness comes into play due to the occurrence of errors in the experiments, observations, and partial information we have. As a result, this can make a critical impact on safety of the system. So, proper modeling of the same and deployment in the system give overall idea of uncertain outcome with approximate prediction. Generally, the uncertainties are dealt with probabilistic, fuzzy, and interval set theory. Random variables are assumed in probabilistic approach, whereas epistemic variables are considered in fuzzy set theory approach. The common issue for both is that one need to know the distribution of uncertainty in advance. But in case of intervals, one may start with uncertainties without prior knowledge of the distribution. Hence, many researchers have taken the interval theory to investigate various uncertain-based system and contributed a lot. Few of the important research works are reported in (Behera and Chakraverty [2013], [2020]; Herzberger [1983]; Nayak [2020]; Kabir and Papadopoulos [2018]; Nayak and Chakraverty [2013]; Nirmala et al. [2011]; Moore et al. [2009]; Nikishkov [2004]). Now, it is worth mentioning that the interval set theory and FEM both are responsible for the effective modeling and design. So, the combination of both may be major to illustrate real-life problems. Moore et al. (2009) include fundamental principles with the interval theory and its applications. Nayak and Chakraverty (2018) included structural problems such as spring mass, fin, bar, beam truss, frame, and brick and modeled them using uncertain environment. They have taken interval finite element method (IFEM) to solve the problem with MATLAB codes supplied for the same. Further, the concept of interval equations was used by Behera and Chakraverty (2020) to solve static structural problem subjected to various interval external forces. However, there is no definite method of solving system of interval equations for uncertainty problems. Concepts of existence and uniqueness of the solutions are still to be verified. In this regards, Nayak (2020) has given a one variable mapping technique for intervals with finite element analysis. The advantage of the same is that it transforms any arbitrary interval $[a,b]$ to $[0,1]$ and $[-1,1]$, which provides a confidence of choosing the same with problem type. In this chapter, the same is used to solve structural engineering problems, viz. beam and truss problems. Then the sensitiveness of the uncertain parameters are analyzed and compared with the obtained results with special cases.

11.2   PRELIMINARIES OF FINITE ELEMENT METHOD (FEM)

Finite element analysis (FEA) is a simulation-based method used to analyze the change (deformation, stress, heat transfer, etc.) that vary over time in a system. This is used to understand certain physical phenomenon like structural deformation, thermal transfer, fluid behavior, and wave propagation. The processes that occur are modeled mathematically using partial differential equations (PDEs). These PDEs are solved using numerical techniques computationally to determine the end result which could be total deformation, stress, strain, etc.

One may say that a computational FEM technique can be used to find out the approximate solution of boundary value problems. It is a widely used method for numerically solving differential equations arising in engineering and mathematical modeling which finally results in a system of algebraic equations to solve unknown nodal quantity like displacement.

Motivation of FEM: The computational efficient FEM transforms the original component into a simplified model. The structure is divided into a finite number of elements that are connected with common points called nodes. To estimate the total structure and behavior of each field variables, all the elements are assembled subjected to the boundary conditions.

In other words, meshing is performed in FEA to discretize or divide the object into finite number of elements. This is done so that the deformation that occurs in every single element can be calculated which would give a clear result of the deformations at every single point in the object. Hence, it is safe to assume that when the number of nodes and elements are increased for meshing, it would increase the accuracy of the result while requiring high-computational power to solve it.

Steps involved in the FEA

1. Select suitable field variables and the elements.

   Field represents the domain of physical structure. Field variables are the dependents which are operated by the governing differential equation of the system.

2. Discretize the continua.

   A node is the common point of the elements where degrees of freedom are defined. The set of elements and node is known as mesh, whereas the process of obtaining is called meshing. A sample model is given in Figures 11.1 and 11.3.

3. Select shape or interpolation or approximating functions.

   The polynomial chosen to interpolate the field variables over the element are interpolation functions.

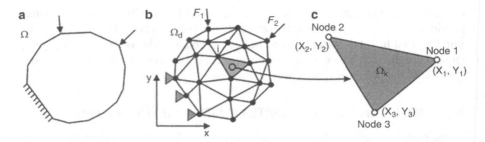

FIGURE 11.1 A model finite element discretization of a two-dimensional domain with triangular elements.

4. Find the element properties.

The elements can be lines (1-D trusses or 1-D beams), areas (2-D or 3-D plates and membranes), or solids (bricks or tetrahedral) which establishe a mathematical relation that tells the connection between two consecutive nodes in terms of degree of freedom. Also, it relates how the deflections create stresses.

5. Assembling of element to get global stiffness.

The global stiffness matrix (GSM) is obtained by assembling all the element stiffness matrices (ESM) to get a system of equations which describes the behavior of the system.

Approximate acceptable solutions (with reasonable accuracy) for the unknown quantities only at discrete or finite number of points in the component is carried out in the following two stages.

- Formulation of the mathematical model, with respect to the physical behavior of the component.
- Obtain numerical solution to the simplified mathematical model. Generally, the three methods to do so are
  1. Weighted residual methods,
  2. Variational method, and
  3. Principle of minimum potential energy.

11.2.1 Weighted Residual Methods

Here, an approximate solution, in the form $y = \sum_{i=1}^{n} N_i C_i$ where C_i are the unknown coefficients or weights (constants) and N_i are functions of the independent variable satisfying the given kinematic boundary conditions, is used in the differential equation. Difference between the two sides of the equation with known terms, on one side (usually functions of the applied loads), and unknown terms, on the other side (functions of constants C_i), is called the residual, R. This residual value may vary from point to point in the component, depending on the particular approximate solution. Three such popular methods are stated next.

11.2.1.1   Galerkin Method

In this method, solution is obtained by equating the integral of the product of function N_i and residual R over the entire component to zero, for each N_i. Thus, the n constants in the approximate solution are evaluated from the n conditions:

$$\int N_i R \, dx = 0 \ \text{for} \ i = 1, 2, \ldots, n \tag{11.1}$$

11.2.1.2   Collocation Method

The residual is equated to zero at n select points of the component where displacement value is not specified, and where n is the number of unknown coefficients in the assumed displacement field:

$$R(\{C\}, x_i) = 0 \ \ \text{for} \ i = 1, 2, \ldots, n \tag{11.2}$$

Collocation method applied on some selected surfaces or volumes is called sub-domain collocation method.

$$\int R(\{C\}, x).dS_j = 0 \ \ \text{for} \ j = 1, 2, \ldots, m \tag{11.3}$$

$$\int R(\{C\}, x).dV_k = 0 \ \ \text{for} \ k = 1, 2, \ldots, m \tag{11.4}$$

11.2.1.3   Least Squares Method

In this method, integral of the residual over the entire component is minimized:

$$\frac{\partial I}{\partial C_i} = 0 \ \ \text{for} \ i = 1, 2, \ldots, n \tag{11.5}$$

where

$$I = \int [R(\{a\}, x)]^2 \tag{11.6}$$

From the aforementioned discussion, we know the basic requirement and steps of FEM. We shall now proceed towards interval analysis.

An interval (denoted by $\sim$ over the parameter) is an orderd pair of values. The left end and right end values are the left end and right end bound of uncertainty. Generally, the interval is represented as $[a, b]$, where left end point is a and right end point is b. The values a and b follow the relation $a \leq b$.

11.3   INTERVAL ARITHMETIC

An interval $\tilde{x}$ is the set of real or complex numbers given by

$$\tilde{x} = [\underline{x}, \overline{x}] = \{x \in \mathbb{R} \ \text{or} \ \mathbb{C} : \underline{x} \leq x \leq \overline{x}\}.$$

Then, the **Radius** of the interval is given by $x_\Delta = \dfrac{\overline{x} - \underline{x}}{2}$.

Midpoint of the interval is given by $x_m = \dfrac{\underline{x} + \overline{x}}{2}$.

Let $\tilde{x} = [\underline{x}, \overline{x}]$ and $\tilde{y} = [\underline{y}, \overline{y}]$, then the two intervals are **equal**, i.e.

$$\tilde{x} = \tilde{y} \text{ iff } \underline{x} = \underline{y} \text{ and } \overline{x} = \overline{y} \text{ or } x_m = y_m \text{ and } x_\Delta = y_\Delta.$$

Addition of two intervals is given by $\tilde{x} + \tilde{y} = [\underline{x} + \underline{y}, \overline{x} + \overline{y}]$.
Subtraction of two intervals is given by $\tilde{x} - \tilde{y} = [\underline{x} - \overline{y}, \overline{x} - \underline{y}]$.
Product of two intervals is given by

$$\tilde{x} \cdot \tilde{y} = [min\, S, max\, S] \text{ where } S = \left\{\underline{x} \cdot \underline{y}, \overline{x} \cdot \overline{y}, \underline{x} \cdot \overline{y}, \overline{x} \cdot \underline{y}\right\}$$

Division of one interval by another interval is given by

$$\frac{\tilde{x}}{\tilde{y}} = [\underline{x}, \overline{x}] \cdot \left[min\left(\frac{1}{\overline{y}}, \frac{1}{\underline{y}}\right), max\left(\frac{1}{\overline{y}}, \frac{1}{\underline{y}}\right)\right], \text{ if } 0 \notin \tilde{y}.$$

Scalar multiplication of interval by $k \in \mathbb{R}$, $k \cdot \tilde{x} = \begin{cases} [k\overline{x}, k\underline{x}] & k < 0, \\ [k\underline{x}, k\overline{x}] & k \geq 0 \end{cases}$

From the mentioned definitions, it is clear that computation of interval uncertainties is a challenging task and may lead to weak solution. Hence, there is a need of interval transformation to single variable function which reduces the complexity of computation.

11.3.1   INTERVAL TRANSFORMATION

It is observed that the classical interval arithmetic becomes tedious to apply when a large number of intervals need to be computed. Hence, one should go with the alternative approach which preserves the classical arithmetic concept as well as is computationally effective. Therefore, here we may use the modified interval arithmetic (Nayak [2020]).

Using the idea presented in (Nayak [2020]), one may define an arbitrary interval $\tilde{x} = [\underline{x}, \overline{x}]$ as $\tilde{x} = [\underline{x}, \overline{x}] = [x_m - x_\Delta, x_m + x_\Delta]$, where members of the interval are shown in the form $x_m - x_\Delta + 2\alpha x_\Delta = x(\alpha), \alpha \in [0,1]$.

In similar fashion, an interval $J = [x_1, x_2]$ can be written in the following ordered pair form of two linear functions, $[p_1(t), p_2(t)], t \in [0,1]$. Here, $p_1(t)$ and $p_2(t)$ possess the following properties:

i. $\lim\limits_{t \to 0} p_1(t) = \lim\limits_{t \to 0} p_2(t) = m$ where $m = \dfrac{x_1 + x_2}{2}$

ii. $\dfrac{d}{dt}(p_1(t))$ and $\dfrac{d}{dt}(p_2(t))$ preserve the inverse properties under addition.

$$(p \circ q)(x) = p((q))(x), \forall x \in J_1.$$

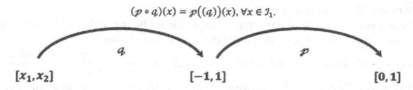

$[x_1, x_2]$ $\qquad$ $[-1, 1]$ $\qquad$ $[0, 1]$

FIGURE 11.2   Transformations of arbitrary intervals into standard intervals.

Further, it is worth mentioning that the linear functions $p_1(t)$ and $p_2(t)$ can be defined as $p_1(t) = m - (m - x_1)t$ and $p_2(t) = m + (x_2 - m)t$.

Consider three intervals $J_1 = [x_1, x_2], J_2 = [-1, 1]$, and $J_3 = [0, 1]$. We can define functions $q : J_1 \to J_2$ and $p : J_2 \to J_3$ shown in Figure 11.2. We define the composition function as:

$$p \circ q : J_1 \to J_3 \text{ as } (p \circ q)(x) = p((q))(x), \forall x \in J_1$$

Here, $p(y) = \dfrac{y+1}{2}, y \in [-1, 1]$ and $q(x) = \dfrac{2}{w}(x - m), x \in [x_1, x_2]$, where $w = x_2 - x_1$ is the width of the interval.

Based on the aforementioned definition and modified arithmetic, the following two results are concluded. (1) If two linear functions $p(y)$ and $q(x)$ are surjective to each other, then the composition of these two linear functions is also surjective. (2) If two linear functions $p(y)$ and $q(x)$ are injective to each other, then the composition of these two linear functions is also injective.

From the above statements it can be seen that any interval $[z_1, z_2]$ can be mapped to $[-1, 1]$ and then $[0, 1]$ or directly. This gives us the confidence to choose convenient mapping and the representation of interval arithmetic depending on the problem requirement. For example, if we need to apply intervals whose values run from negative to positive sense then the mapping q is useful. Next, the same is used to manage imprecise or vague uncertainties with FEM to develop modified interval FEM (IFEM).

11.4 STRUCTURAL PROPERTIES

{Force Vector} = {Nodal Vector}{Stiffness Matrix}

Stiffness is the relation between the force applied and displacement. Elements stiffness is used as a building block for more complex systems. GSM describes the behavior of the complete system; we achieve this by assembling individual stiffness matrices by enforcing boundary conditions.

Displacement is the length of the shortest distance from the initial to the final point.

Force is the push or pull upon an object resulting from the objects interaction with another object. It is given by

Force = Mass × Acceleration = Stress × Area = Youngs Modulus × Strain × Area

Stress is the force acting on the unit area of a material.

Strain is the amount of deformation experienced by the body in the direction of the force applied, divided by the initial dimension of the body.

Youngs Modulus (E) is a measure of elasticity, $E = \dfrac{\text{Stress acting on a substance}}{\text{Strain Produced}}$

Moment of inertia (I) is a measure of the resistance of a body to angular acceleration about a given axis that is equal to the sum of the products of each element of mass in the body and the square of the element's distance from the axis.

Bending moment is the reaction induced in a structural element when an external force or moment is applied to the element, causing the element to bend.

Shearing forces are unaligned forces pushing one part of a body in one specific direction, and another part of the body in the opposite direction.

Load deflection/moment curvature relation of a beam is given by

$$M = EI \frac{\partial^2 y}{\partial y^2}, \text{ where } M \text{ is Bending moment, } y \text{ is deflection.}$$

Now we shall apply these concepts to simple structural elements like beam and truss to solve a numerical example.

11.5   STRUCTURAL FINITE ELEMENT MODEL FOR BEAM

A beam in a space frame is generally subjected to axial load, torsion load, and bending loads in two planes, due to the combined effect of loads acting at different locations of the space frame and in different directions. In beam theory, transverse displacement/lateral deflection v and rotational displacement θ due to load and moment applied are treated as independent variables. Here the cross-section, which is perpendicular to the neutral axis, remains unchanged after bending; hence, there occurs a small deflection. So, we assume that the rotation is same as that of the slope of deflection ($\frac{dv}{dx} = \theta$). The load and moment reactions obtained are then used to calculate the deflections and bending stresses by strength of materials approach.

Take a two-noded beam element (Figure 11.3), then there occurs two vertical displacements and two rotations. Degree of freedom is limited to vertical displacements and rotations in one plane and has the same physical and mechanical properties in all directions. Hence, in this case, at each node, unknowns are the displacement and slope that can be represented as

$$\{\tilde{\delta}\} = \begin{Bmatrix} \tilde{v}_1 \\ \tilde{\theta}_1 \\ \tilde{v}_2 \\ \tilde{\theta}_2 \end{Bmatrix} \text{ where, } \tilde{\theta}_1 = \frac{\partial \tilde{v}_1}{\partial x} \; \tilde{\theta}_2 = \frac{\partial \tilde{v}_2}{\partial x}$$

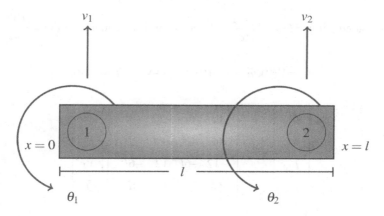

FIGURE 11.3   A beam element.

Since there are four nodal values, we select polynomial with four constants.

$$\tilde{v}(x) = \tilde{a}_1 + \tilde{a}_2 x + \tilde{a}_3 x^2 + \tilde{a}_4 x^3 \tag{11.7}$$

$$\tilde{\theta} = \frac{\partial \tilde{v}}{\partial x} = \tilde{a}_2 + 2\tilde{a}_3 x + 3\tilde{a}_4 x^2 \tag{11.8}$$

$$x = 0 : \tilde{v}_1 = \tilde{a}_1, \tilde{\theta}_1 = \tilde{a}_2$$
$$x = l : \tilde{v}_2 = \tilde{v}_1 + \tilde{\theta}_1 l + \tilde{a}_3 l^2 + \tilde{a}_4 l^3, \tilde{\theta}_2 = \tilde{\theta}_1 + 2\tilde{a}_3 l + 3\tilde{a}_4 l^2$$

By solving them, we get

$$\tilde{a}_3 = \frac{3(\tilde{v}_2 - \tilde{v}_1)}{l^2} - \frac{2\tilde{\theta}_1 + \tilde{\theta}_2}{l} \text{ and } \tilde{a}_4 = \frac{2(\tilde{v}_1 - \tilde{v}_2)}{l^3} - \frac{\tilde{\theta}_1 + \tilde{\theta}_2}{l^2}$$

Substituting $\tilde{a}_1, \tilde{a}_2, \tilde{a}_3, \tilde{a}_4$ in Eqns. (11.7) and (11.8)

$$\tilde{v}(x) = \tilde{v}_1 + \tilde{\theta}_1 x + \left(\frac{3(\tilde{v}_2 - \tilde{v}_1)}{l^2} - \frac{2\tilde{\theta}_1 + \tilde{\theta}_2}{l} \right) x^2 + \left(\frac{2(\tilde{v}_1 - \tilde{v}_2)}{l^3} - \frac{\tilde{\theta}_1 + \tilde{\theta}_2}{l^2} \right) x^3$$

We know that $\tilde{v} = N_1 \tilde{v}_1 + N_2 \tilde{\theta}_1 + N_3 \tilde{v}_2 + N_4 \tilde{\theta}_2$.
Substituting we get shape functions

$$N_1 = 1 - \frac{3x^2}{l^2} + \frac{2x^3}{l^2}, \ N_2 = x - \frac{2x^2}{l} + \frac{x^2}{l^2}, \ N_3 = \frac{3x^2}{l^2} - \frac{2x^3}{l^3}, \ N_4 = \frac{-x^2}{l} + \frac{x^3}{l^2}$$

Moment equation $\tilde{m}(x) = \dfrac{d^2\tilde{v}(x)}{dx^2}\tilde{E}\tilde{I}$ and shear force equation $\tilde{V}(x) = \dfrac{d^3\tilde{v}(x)}{dx^3}$

$$\tilde{F}_{1y} = \tilde{V}(0), \tilde{m}_1 = -\tilde{m}(0), \tilde{F}_{2y} = -\tilde{V}(l), \tilde{m}_1 = -\tilde{m}(l)$$

$$\begin{Bmatrix} \tilde{F}_{1y} \\ \tilde{m}_1 \\ \tilde{F}_{2y} \\ \tilde{m}_2 \end{Bmatrix} = \frac{\tilde{E}\tilde{I}}{l^3} \begin{bmatrix} 12 & 6l & -12 & 6l \\ 6l & 4l^2 & -6l & 2l^2 \\ -12 & -6l & 12 & -6l^2 \\ 6l & 2l^2 & 6l & 4l^2 \end{bmatrix} \begin{Bmatrix} \tilde{v}_1 \\ \tilde{\theta}_1 \\ \tilde{v}_2 \\ \tilde{\theta}_2 \end{Bmatrix}$$

Let p be distributed load per unit length, then by potential energy of a beam element, which is equal to strain energy minus the work done by the external forces acting on the element, the equivalent load on the element will be:

$$\{F\}^T = \left[\frac{pl}{2} \quad \frac{pl^2}{12} \quad \frac{pl}{2} \quad \frac{pl^2}{12} \right]$$

Numerical example of an application problem

We need to analyze the beam shown in Figure 11.4 by finite element method and determine the end reactions. Also, determine the deflections at mid-spans with 5% error of $E = 2 \times 10^5 \text{ N/mm}^2$ and $I = 5 \times 10^6 \text{ mm}^4$.

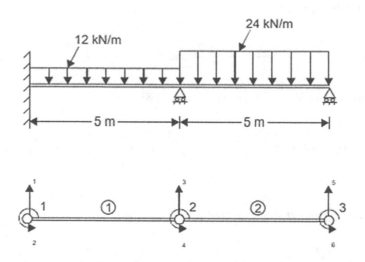

FIGURE 11.4 Model diagram of a beam.

(We use N and m units throughout.) Substituting values for load p and length l, we get load on the corresponding element

$$F = \begin{Bmatrix} \dfrac{pl}{2} \\ \dfrac{pl^2}{12} \\ \dfrac{pl}{2} \\ \dfrac{pl^2}{12} \end{Bmatrix}, \quad F_1 = \begin{Bmatrix} -30 \\ -25 \\ -30 \\ 25 \end{Bmatrix} \quad F_2 = \begin{Bmatrix} -60 \\ -50 \\ -60 \\ 50 \end{Bmatrix} \Rightarrow F = \begin{Bmatrix} -30 \\ -25 \\ -90 \\ -25 \\ -60 \\ 50 \end{Bmatrix}$$

Here, uncertain parameters (E and I) are taken as interval numbers, and we solve the problem considering three cases and notice variation in displacement value at nodes 4 and 6.

Case 1: E has an interval and I has a definite value.

$$E = [1.9 \times 10^8, 2.1 \times 10^8] \text{N/m}^2, I = 5 \times 10^{-6} \text{ m}^4$$

The nodal displacement vector is $\{\tilde{\delta}\}^T = [\tilde{\delta}_1 \ \tilde{\delta}_2 \ \tilde{\delta}_3 \ \tilde{\delta}_4 \ \tilde{\delta}_5 \ \tilde{\delta}_6]$

$$k_1 = k_2 = \dfrac{\tilde{E}\tilde{I}}{l^3} \begin{bmatrix} 12 & 6l & -12 & 6l \\ 6l & 4l^2 & -6l & 2l^2 \\ -12 & -6l & 12 & -6l^2 \\ 6l & 2l^2 & 6l & 4l^2 \end{bmatrix} = \dfrac{[950,1050]}{5^3} \begin{bmatrix} 12 & 30 & -12 & 30 \\ 30 & 100 & -30 & 50 \\ -12 & -30 & 12 & -30 \\ 30 & 50 & -30 & 100 \end{bmatrix}$$

$$= \begin{bmatrix} [91.2,100.8] & [228,252] & [-100.8,-91.2] & [228,252] \\ [228,252] & [760,840] & [-252,-228] & [380,420] \\ [-100.8,-91.2] & [-252,-228] & [91.2,100.8] & [-252,-228] \\ [228,252] & [380,420] & [-252,-228] & [760,840] \end{bmatrix}$$

Global stiffness matrix

$$K = \begin{bmatrix} [91.2,100.8] & [228,252] & [-100.8,-91.2] & [228,252] & [0,0] & [0,0] \\ [228,252] & [760,840] & [-252,-228] & [380,420] & [0,0] & [0,0] \\ [-100.8,-91.2] & [-252,-228] & [182.4,201.6] & [0,0] & [-100.8,-91.2] & [228,252] \\ [228,252] & [380,420] & [0,0] & [1520,1680] & [-252,-228] & [380,420] \\ [0,0] & [0,0] & [-100.8,-91.2] & [-252,-228] & [91.2,100.8] & [-252,-228] \\ [0,0] & [0,0] & [228,252] & [380,420] & [-252,-228] & [760,840] \end{bmatrix}$$

In the given problem, the boundary conditions are $\delta_1 = \delta_2 = \delta_3 = \delta_5 = 0$. Imposing them by elimination method, we get

$$\{\delta\} = [K]^{-1}\{F\}$$

$$\begin{bmatrix} \tilde{\delta}_4 \\ \tilde{\delta}_6 \end{bmatrix} = \begin{bmatrix} [6.8 \times 10^{-4}, 7.5 \times 10^{-4}] & [-3.75 \times 10^{-4}, -3.4 \times 10^{-4}] \\ [-3.75 \times 10^{-4}, -3.4 \times 10^{-4}] & [13.6 \times 10^{-4}, 15 \times 10^{-4}] \end{bmatrix} \begin{bmatrix} -25 \\ 50 \end{bmatrix}$$

$$= \begin{bmatrix} [-0.03759, -0.03401] \\ [0.07653, 0.08458] \end{bmatrix}$$

Similarly,

Case 2: Taking E as definite and I as interval number,

$$E = 2 \times 10^8 \, \text{N/m}^2, \ I = [4.9 \times 10^{-6}, 5.1 \times 10^{-6}] \text{m}^4$$

$$k_1 = k_2 = \begin{bmatrix} [94.08, 97.92] & [235.2, 244.8] & [-97.92, -94.08] & [235.2, 244.8] \\ [235.2, 244.8] & [784, 816] & [-244.8, -235.2] & [392, 408] \\ [-97.92, -94.08] & [-244.8, -235.2] & [94.08, 97.92] & [-244.8, -235.2] \\ [235.2, 244.8] & [392, 408] & [-244.8, -235.2] & [784, 816] \end{bmatrix}$$

Assembling we get,

$$K = \begin{bmatrix} [94.08, 97.92] & [235.2, 244.8] & [-97.92, -94.08] \\ [235.2, 244.8] & [784, 816] & [-244.8, -235.2] \\ [-97.92, -94.08] & [-244.8, -235.2] & [188.2, 195.8] \\ [235.2, 244.8] & [392, 408] & [0, 0] \\ [0, 0] & [0, 0] & [-97.92, -94.08] \\ [0, 0] & [0, 0] & [235.2, 244.8] \end{bmatrix}$$

$$\times \begin{bmatrix} [235.2, 244.8] & [0, 0] & [0, 0] \\ [392, 408] & [0, 0] & [0, 0] \\ [0, 0] & [94.08, 97.92] & [-244.8, -235.2] \\ [1568, 1632] & [-244.8, -235.2] & [392, 408] \\ [-244.8, -235.2] & [94.08, 97.92] & [-244.8, -235.2] \\ [392, 408] & [-244.8, -235.2] & [784, 816] \end{bmatrix}$$

$$\begin{bmatrix} \tilde{\delta}_4 \\ \tilde{\delta}_6 \end{bmatrix} = \begin{bmatrix} [7 \times 10^{-4}, 7.28 \times 10^{-4}] & [-3.64 \times 10^{-4}, -3.5 \times 10^{-4}] \\ [-3.64 \times 10^{-4}, -3.5 \times 10^{-4}] & [14.0 \times 10^{-4}, 14.57 \times 10^{-4}] \end{bmatrix} \begin{bmatrix} -25 \\ 50 \end{bmatrix}$$

$$= \begin{bmatrix} [-0.035, -0.0364] \\ [0.07878, 0.08199] \end{bmatrix}$$

Case 3: E and I both are taken as intervals.

$$E = [1.9 \times 10^8, 2.1 \times 10^8] \text{N/m}^2, I = [4.9 \times 10^{-6}, 5.1 \times 10^{-6}] \text{m}^4$$

$$k_1 = k_2 = \begin{bmatrix} [89.376, 102.816] & [223.44, 257.04] & [-102.816, -89.376] & [223.44, 257.04] \\ [223.44, 257.04] & [744.8, 856.8] & [-257.04, -223.44] & [372.4, 428.4] \\ [-102.816, -89.376] & [-257.04, -223.44] & [89.376, 102.816] & [-257.04, -223.44] \\ [223.44, 257.04] & [372.4, 428.4] & [-257.04, -223.44] & [744.8, 856.8] \end{bmatrix}$$

$$K = \begin{bmatrix} [89.376, 102.816] & [223.44, 257.04] & [-102.816, -89.376] \\ [223.44, 257.04] & [744.8, 856.8] & [-257.04, -223.44] \\ [-102.816, -89.376] & [-257.04, -223.44] & [178.8, 192.2] \\ [223.44, 257.04] & [372.4, 428.4] & [-33.6, 0] \\ [0, 0] & [0, 0] & [-102.816, -89.376] \\ [0, 0] & [0, 0] & [223.44, 257.04] \end{bmatrix}$$

$$\times \begin{bmatrix} [223.44, 257.04] & [0, 0] & [0, 0] \\ [372.4, 428.4] & [0, 0] & [0, 0] \\ [-33.6, 0] & [-102.816, -89.376] & [223.44, 257.04] \\ [1489.6, 1601.6] & [-257.04, -223.44] & [372.4, 428.4] \\ [-257.04, -223.44] & [89.376, 102.816] & [-257.04, -223.44] \\ [372.4, 428.4] & [-257.04, -223.44] & [744.8, 856.8] \end{bmatrix}$$

We get displacement at nodes 4 and 6 as,

$$\begin{bmatrix} \tilde{\delta}_4 \\ \tilde{\delta}_6 \end{bmatrix} = \begin{bmatrix} [6.67 \times 10^{-4}, 7.67 \times 10^{-4}] & [-3.83 \times 10^{-4}, -3.33 \times 10^{-4}] \\ [-3.83 \times 10^{-4}, -3.33 \times 10^{-4}] & [13.3 \times 10^{-4}, 15.3 \times 10^{-4}] \end{bmatrix} \begin{bmatrix} -25 \\ 50 \end{bmatrix}$$

$$= \begin{bmatrix} [-0.03835, -0.0383] \\ [0.0703, 0.08607] \end{bmatrix}$$

The reactions at supports are nothing but end equilibrium forces. Hence, $\{\tilde{R}\} = [\tilde{K}]_e [\tilde{\delta}]_e - \{\tilde{F}\}_e$

$$
\begin{Bmatrix} \tilde{R}_1 \\ \tilde{R}_2 \\ \tilde{R}_3 \\ \tilde{R}_4 \end{Bmatrix} = \frac{\tilde{E}\tilde{I}}{l^3} \begin{bmatrix} 12 & 6l & -12 & 6l \\ 6l & 4l^2 & -6l & 2l^2 \\ -12 & -6l & 12 & -6l^2 \\ 6l & 2l^2 & 6l & 4l^2 \end{bmatrix} \begin{Bmatrix} \tilde{\delta}_1 \\ \tilde{\delta}_2 \\ \tilde{\delta}_3 \\ \tilde{\delta}_4 \end{Bmatrix} - \begin{Bmatrix} \dfrac{pl}{2} \\[4pt] \dfrac{pl^2}{12} \\[4pt] \dfrac{pl}{2} \\[4pt] \dfrac{pl^2}{12} \end{Bmatrix}
$$

$$
\begin{Bmatrix} \tilde{R}_1 \\ \tilde{R}_2 \\ \tilde{R}_3 \\ \tilde{R}_4 \end{Bmatrix} = [K] \begin{bmatrix} 0 \\ 0 \\ 0 \\ [-0.03835,-0.0383] \end{bmatrix} - \begin{Bmatrix} -30 \\ -25 \\ -30 \\ 25 \end{Bmatrix} = \begin{bmatrix} [21.398, 26.062] \\ [8.5066, 10.737] \\ [38.558, 39.896] \\ [-57.9868, -53.526] \end{bmatrix}
$$

$$
\begin{Bmatrix} \tilde{R}_3 \\ \tilde{R}_4 \\ \tilde{R}_5 \\ \tilde{R}_6 \end{Bmatrix} = [K] \begin{bmatrix} 0 \\ [-0.03835,-0.0383] \\ 0 \\ [0.0703, 0.08607] \end{bmatrix} - \begin{Bmatrix} -60 \\ -50 \\ -60 \\ 50 \end{Bmatrix} = \begin{bmatrix} [58.3722, 69.38] \\ [47.655, 54.072] \\ [47.7275, 52.848] \\ [-11.903, -6.571] \end{bmatrix}
$$

11.6 STRUCTURAL FINITE ELEMENT MODEL FOR TRUSS

A truss is an assemblage of long, slender structural elements that are connected at their ends. Usually, a truss is an assembly of beams or other elements that create a rigid structure. The truss elements are assumed to be joined together by pins or other such connections that allow free rotation around the joints. The elements are organized so that the assemblage as a whole behaves as a single object. Trusses find substantial use in modern construction, for instance, towers, bridges, and scaffolding. In addition to their practical importance as useful structures, truss elements have a dimensional simplicity that helps to extend further the concepts of mechanics. A planar truss is one where all members and nodes lie within a two-dimensional plane, while a space truss has members and nodes that extend into three dimensions.

Plane truss is a two-dimensional, simplest, and commonly used structural element. In the analysis, all joints are assumed pin connected and all loads act at joints only, i.e., the loads can only be applied at the two ends. The forces are subjected axially in truss element; hence, due to the application of forces, deformation happens in the axial direction. These assumptions result into no bending of any member, and

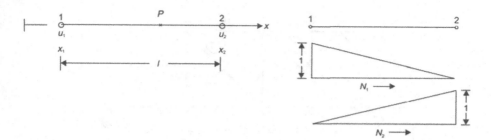

FIGURE 11.5   Truss element with two nodes.

trusses cannot sustain shear and moment. All members are subjected to only direct stresses, tensile or compressive.

To find shape functions for two-noded truss element

Figure 11.5 shows the typical truss element. In this case, nodal unknowns are displacements u_1 and u_2 along x axis. For this element, we have to select polynomial with only two constants to represent displacement at any point in the elements. Hence we select

$$\tilde{u} = \tilde{a}_1 + \tilde{a}_2 x$$

$$\tilde{u} = [1 \quad x] \left\{ \begin{array}{c} \tilde{a}_1 \\ \tilde{a}_2 \end{array} \right\}$$

Since $\tilde{u} = \tilde{u}_1$ at node 1 and equal to $\tilde{u}_2$ at node 2, we have

$$\{\delta\} = \left\{ \begin{array}{c} \tilde{u}_1 \\ \tilde{u}_2 \end{array} \right\} = \left[\begin{array}{cc} 1 & x_1 \\ 1 & x_2 \end{array} \right] \left\{ \begin{array}{c} \tilde{a}_1 \\ \tilde{a}_2 \end{array} \right\}$$

$$\left\{ \begin{array}{c} \tilde{a}_1 \\ \tilde{a}_2 \end{array} \right\} = \frac{1}{x_2 - x_1} \left[\begin{array}{cc} x_2 & -1 \\ -x_1 & 1 \end{array} \right] \left\{ \begin{array}{c} \tilde{u}_1 \\ \tilde{u}_2 \end{array} \right\} = \frac{1}{l} \left[\begin{array}{cc} x_2 & -1 \\ -x_1 & 1 \end{array} \right] \left\{ \begin{array}{c} \tilde{u}_1 \\ \tilde{u}_2 \end{array} \right\}$$

$$\therefore \tilde{u} = [1 \quad x] \frac{1}{l} \left[\begin{array}{cc} x_2 & -1 \\ -x_1 & 1 \end{array} \right] \left\{ \begin{array}{c} \tilde{u}_1 \\ \tilde{u}_2 \end{array} \right\} = \left[\begin{array}{cc} \dfrac{x_2 - x}{l} & \dfrac{x - x_1}{l} \end{array} \right] \left\{ \begin{array}{c} \tilde{u}_1 \\ \tilde{u}_2 \end{array} \right\}$$

$$\tilde{u} = [N_1 \quad N_2] \left\{ \begin{array}{c} \tilde{u}_1 \\ \tilde{u}_2 \end{array} \right\} = N_1 \tilde{u}_1 + N_2 \tilde{u}_2$$

Thus, the shape function is $[N] = [N_1 \quad N_2] = \left[\dfrac{x_2 - x}{l} \quad \dfrac{x - x_1}{l} \right]$

Analysis of two-dimensional truss (plane truss) with 15 elements

Consider a truss structure consisting of 15 bars shown in Figure 11.6. Truss is subjected to horizontal and vertical loads at nodes 3 and 5, respectively.

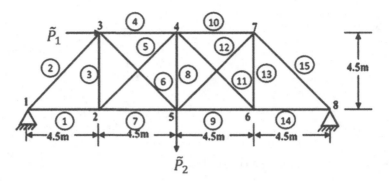

FIGURE 11.6   Truss with 15 elements.

Deterministic (crisp) material and geometric properties are considered, whereas the external forces at the nodes are uncertain in terms of interval. To investigate the variable static responses, we consider three different cases with nodal force varying the uncertainty width.

$$P_1 = 150\,\text{N},\; P_2 = 250\,\text{N}$$

Cross-sectional area for $A_1, A_2, A_3, A_{13}, A_{14}, A_{15}$ is $10 \times 10^5\,\text{m}^2$ and for others is $6 \times 10^5\,\text{m}^2$.

Modulus of the elasticity is $2 \times 10^{11}\,\text{N/m}^2$.

Step 1: In partitioned elements and field variables, the joint displacements are taken as field variables. There is no bending and hence it is ensured that only displacement continuity (C^0 continuity) occur with no slope continuity (C^1 continuity). As such, two-noded bar elements are considered to be the same. As the structural elements are exerted axial external forces, the axial displacement takes place at the nodes. As a result, nodal vector of the bar element can be represented as:

$$\left\{ \tilde{\delta}' \right\} = \begin{Bmatrix} \tilde{\delta}_1' \\ \tilde{\delta}_2' \end{Bmatrix}$$

where $\tilde{\delta}_1'$ and $\tilde{\delta}_1'$ take place axially. Here, one point may be noted that the axial direction varies with the elements depending on the orientation. For example, in $x - y$ coordinate system, two different displacements occur at two end nodes of the element. Therefore, nodal vector becomes $\left\{ \tilde{\delta}' \right\} = [\tilde{\delta}_1\ \tilde{\delta}_2\ \tilde{\delta}_3\ \tilde{\delta}_4]$

From Figure 11.7, it can be seen that

$$\tilde{\delta}_1' = \tilde{\delta}_1 cos(\theta) + \tilde{\delta}_2 sin(\theta)$$
$$\tilde{\delta}_2' = \tilde{\delta}_3 cos(\theta) + \tilde{\delta}_4 sin(\theta)$$

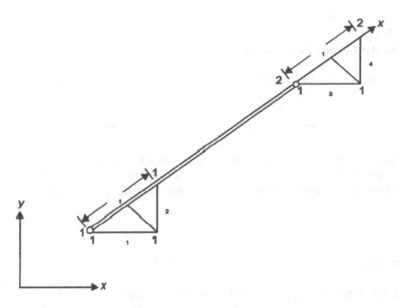

FIGURE 11.7 Typical element and its nodal displacement.

Taking l and m as direction cosines, that is, $l = cos(\theta)$, $m = sin(\theta)$, we get:

$$\tilde{\delta}_1' = \tilde{\delta}_1 l + \tilde{\delta}_2 m$$
$$\tilde{\delta}_2' = \tilde{\delta}_3 l + \tilde{\delta}_4 m$$

$$\{\tilde{\delta}'\} = \left\{ \begin{array}{c} \tilde{\delta}_1' \\ \tilde{\delta}_2' \end{array} \right\} = \left[\begin{array}{cccc} l & m & 0 & 0 \\ 0 & 0 & l & m \end{array} \right] \left\{ \begin{array}{c} \tilde{\delta}_1 \\ \tilde{\delta}_2 \\ \tilde{\delta}_3 \\ \tilde{\delta}_4 \end{array} \right\}$$

$$\{\tilde{\delta}'\} = [L]\{\tilde{\delta}\}$$

$$[L] = \left[\begin{array}{cccc} l & m & 0 & 0 \\ 0 & 0 & l & m \end{array} \right],$$

$[L]$ is called transformation (or rotation) matrix.

Step 2: Discretization of the domain

Here, each member can be considered as an element and 15 truss structure (Table 11.1) may possesses

a. two top chord members,
b. four bottom chord members,
c. three vertical members, and
d. six diagonal members.

TABLE 11.1

The Nodal Connectivity Details

Element No.	1	2	3	4	5	6	7	8	9	10	11	12	13	14	15
Element Node 1	1	1	2	3	2	3	2	4	5	4	4	5	6	6	7
Element Node 2	2	3	3	4	4	5	5	5	6	7	6	7	7	8	8

There are two degrees of freedom at each node along x- and y directions.

Therefore, we can say that the total number of degrees of freedom are nothing but the product of number of nodes with the degree of freedom of each node, that is,

$$8 \times 2 = 16,$$

the displacement vector is $\{\delta\}^T = [\tilde{\delta}_1 \ \tilde{\delta}_2 \ \tilde{\delta}_3 \ldots \tilde{\delta}_{16}]$

Step 3: Interpolating polynomials here, for truss element

$$\{u\} = [N_1 \ N_2]\{\tilde{\delta}'\}, \text{ where}$$

$$[N] = [N_1 \ N_2] = \left[\frac{x_2' - x'}{l} \ \frac{x' - x_1'}{l} \right],$$

Step 4: Element properties
Element stiffness matrix $[K_e]$

$$[K_e] = \frac{A_e E_e}{l_e} \begin{bmatrix} l^2 & lm & -l^2 & -lm \\ lm & m^2 & -lm & -m^2 \\ -l^2 & -lm & l^2 & lm \\ -lm & -m^2 & lm & m^2 \end{bmatrix}$$

where, cross-sectional area for the element is A_e, modulus of the elasticity for the element is E_e, and length of the element is l_e.

The element-wise stiffness matrices are given next.

$$[K_1] = [K_{14}] = \begin{bmatrix} 4.44 \times 10^6 & 0 & -4.44 \times 10^6 & 0 \\ 0 & 0 & 0 & 0 \\ -4.44 \times 10^6 & 0 & 4.44 \times 10^6 & 0 \\ 0 & 0 & 0 & 0 \end{bmatrix}$$

$$[K_2] = \begin{bmatrix} 1.5713 \times 10^6 & 1.5713 \times 10^6 & -1.5713 \times 10^6 & -1.5713 \times 10^6 \\ 1.5713 \times 10^6 & 1.5713 \times 10^6 & -1.5713 \times 10^6 & -1.5713 \times 10^6 \\ -1.5713 \times 10^6 & -1.5713 \times 10^6 & 1.5713 \times 10^6 & 1.5713 \times 10^6 \\ -1.5713 \times 10^6 & -1.5713 \times 10^6 & 1.5713 \times 10^6 & 1.5713 \times 10^6 \end{bmatrix}$$

$$[K_3] = [K_{13}] = \begin{bmatrix} 0 & 0 & 0 & 0 \\ 0 & 4.44 \times 10^6 & 0 & -4.44 \times 10^6 \\ 0 & 0 & 0 & 0 \\ 0 & -4.44 \times 10^6 & 0 & 4.44 \times 10^6 \end{bmatrix}$$

$$[K_4] = [K_7] = [K_9] = [K_{10}] = \begin{bmatrix} 2.6667 \times 10^6 & 0 & -2.6667 \times 10^6 & 0 \\ 0 & 0 & 0 & 0 \\ -2.6667 \times 10^6 & 0 & 2.6667 \times 10^6 & 0 \\ 0 & 0 & 0 & 0 \end{bmatrix}$$

$$[K_5] = \begin{bmatrix} 9.4281 \times 10^5 & 9.4281 \times 10^5 & -9.4281 \times 10^5 & -9.4281 \times 10^5 \\ 9.4281 \times 10^5 & 9.4281 \times 10^5 & -9.4281 \times 10^5 & -9.4281 \times 10^5 \\ -9.4281 \times 10^5 & -9.4281 \times 10^5 & 9.4281 \times 10^5 & 9.4281 \times 10^5 \\ -9.4281 \times 10^5 & -9.4281 \times 10^5 & 9.4281 \times 10^5 & 9.4281 \times 10^5 \end{bmatrix}$$

$$[K_6] = [K_{11}] = \begin{bmatrix} 9.4281 \times 10^5 & -9.4281 \times 10^5 & -9.4281 \times 10^5 & 9.4281 \times 10^5 \\ -9.4281 \times 10^5 & 9.4281 \times 10^5 & 9.4281 \times 10^5 & -9.4281 \times 10^5 \\ -9.4281 \times 10^5 & 9.4281 \times 10^5 & 9.4281 \times 10^5 & -9.4281 \times 10^5 \\ 9.4281 \times 10^5 & -9.4281 \times 10^5 & -9.4281 \times 10^5 & 9.4281 \times 10^5 \end{bmatrix}$$

$$[K_8] = \begin{bmatrix} 0 & 0 & 0 & 0 \\ 0 & 2.6667 \times 10^6 & 0 & -2.6667 \times 10^6 \\ 0 & 0 & 0 & 0 \\ 0 & -2.6667 \times 10^6 & 0 & 2.6667 \times 10^6 \end{bmatrix}$$

$$[K_{15}] = \begin{bmatrix} 1.5713 \times 10^6 & -1.5713 \times 10^6 & -1.5713 \times 10^6 & 1.5713 \times 10^6 \\ -1.5713 \times 10^6 & 1.5713 \times 10^6 & 1.5713 \times 10^6 & -1.5713 \times 10^6 \\ -1.5713 \times 10^6 & 1.5713 \times 10^6 & 1.5713 \times 10^6 & -1.5713 \times 10^6 \\ 1.5713 \times 10^6 & -1.5713 \times 10^6 & -1.5713 \times 10^6 & 1.5713 \times 10^6 \end{bmatrix}$$

Step 5: Global properties

1. Loads:
   As the loads are acting on the joints (nodes), it is simple to assemble all the force vectors. A null load vector F having a size of degrees of freedom is initialized and then load values are incorporated.

$$F_5 = P_1, \; F_9 = P_2$$

2. In a similar manner as mentioned in the previous point, a null matrix is constructed for the global stiffness matrix with required size. Then all the elemental stiffness values are posted against the specific null values.

Global stiffness matrix (GSM) $[K] =$

$$
\begin{bmatrix}
6.0158 & 1.5713 & -4.4444 & 0 & -1.5713 & -1.5713 & 0 & 0 & 0 & 0 \\
1.5713 & 1.5713 & 0 & 0 & -1.5713 & -1.5713 & 0 & 0 & 0 & 0 \\
-4.4444 & 0 & 8.0539 & 0.9428 & 0 & 0 & -0.9428 & -0.9428 & -2.6667 & 0 \\
0 & 0 & 0.9428 & 5.3873 & 0 & -4.4444 & -0.9428 & -0.9428 & 0 & 0 \\
-1.5713 & -1.5713 & 0 & 0 & 5.1808 & 0.6285 & -2.6667 & 0 & -0.9428 & 0.9428 \\
-1.5713 & -1.5713 & 0 & -4.4444 & 0.6285 & 6.9586 & 0 & 0 & 0.9428 & -0.9428 \\
0 & 0 & -0.9428 & -0.9428 & -2.6667 & 0 & 7.2190 & 0 & 0 & 0 \\
0 & 0 & -0.9428 & -0.9428 & 0 & 0 & 0 & 4.5523 & 0 & -2.6667 \\
0 & 0 & -2.6667 & 0 & -0.9428 & 0.9428 & 0 & 0 & 7.2190 & 0 \\
0 & 0 & 0 & 0 & 0.9428 & -0.9428 & 0 & -2.6667 & 0 & 4.5523 \\
0 & 0 & 0 & 0 & 0 & 0 & -0.9428 & 0.9428 & -2.6667 & 0 \\
0 & 0 & 0 & 0 & 0 & 0 & 0.9428 & -0.9428 & 0 & 0 \\
0 & 0 & 0 & 0 & 0 & 0 & -2.6667 & 0 & -0.9428 & -0.9428 \\
0 & 0 & 0 & 0 & 0 & 0 & 0 & -0.9428 & -0.9428 & 0 \\
0 & 0 & 0 & 0 & 0 & 0 & 0 & 0 & 0 & 0 \\
0 & 0 & 0 & 0 & 0 & 0 & 0 & 0 & 0 & 0
\end{bmatrix}
$$

$$
\times
\begin{bmatrix}
0 & 0 & 0 & 0 & 0 & 0 \\
0 & 0 & 0 & 0 & 0 & 0 \\
0 & 0 & 0 & 0 & 0 & 0 \\
0 & 0 & 0 & 0 & 0 & 0 \\
0 & 0 & 0 & 0 & 0 & 0 \\
0 & 0 & 0 & 0 & 0 & 0 \\
-0.9428 & 0.9428 & -2.6667 & 0 & 0 & 0 \\
0.9428 & -0.9428 & 0 & 0 & 0 & 0 \\
-2.6667 & 0 & -0.9428 & -0.9428 & 0 & 0 \\
0 & 0 & -0.9428 & -0.9428 & 0 & 0 \\
8.0539 & -0.9428 & 0 & 0 & -4.4444 & 0 \\
-0.9428 & 5.3873 & 0 & -4.4444 & 0 & 0 \\
0 & 0 & 5.1808 & -0.6285 & -1.5713 & 1.5713 \\
-4.4444 & -0.6285 & 6.9586 & 1.5713 & -1.5713 & \\
-4.4444 & 0 & -1.5713 & 1.5713 & 6.0158 & -1.5713 \\
0 & 0 & 1.5713 & -1.5713 & -1.5713 & 1.5713
\end{bmatrix}
$$

TABLE 11.2

Percentage of Error and Interval Considered

Error (%)	Interval Considered
3.3	$\tilde{P}_1(N) = [145,155], \tilde{P}_2(N) = [245,255]$
6.6	$\tilde{P}_1(N) = [140,160], \tilde{P}_2(N) = [240,260]$
10	$\tilde{P}_1(N) = [135,165], \tilde{P}_2(N) = [235,265]$

Boundary conditions

Since nodes 1 and 8 are fixed, their displacement value is zero.

$$\delta_1 = \delta_2 = \delta_{15} = \delta_{16} = 0.$$

Solving by using the formula $\{\tilde{\delta}\} = [K]^{-1}[\tilde{F}]$

by applying the proposed method, the obtained interval displacements are shown in Table 11.2 and a comparison is given in Table 11.3.

It may be noted from the above numerical results that the interval size gradually increases with the increase in uncertainity value of the load taken.

TABLE 11.3

Comparison of Uncertain Deflections with Special Case (mid-point)

Node	P1 = [145, 155], P2 = [245, 255]	P1 = [140, 160], P2 = [240, 260]	P1 = [135, 165], P2 = [235, 265]	P1 = 150, P2 = 250
δ_1	[0,0]	[0,0]	[0,0]	0
δ_2	[0,0]	[0,0]	[0,0]	0
δ_3	[−0.0015, −0.0014]	[−0.0016, −0.0013]	[−0.0017, −0.0013]	−0.0015
δ_4	[−0.1979, −0.1894]	[−0.2022, −0.1851]	[−0.2065, −0.1809]	−0.1937
δ_5	[0.1228, 0.1293]	[0.1195, 0.1326]	[0.1163, 0.1358]	0.1260
δ_6	[−0.1858, −0.1777]	[−0.1899, −0.1736]	[−0.1939, −0.1695]	−0.1817
δ_7	[0.0233, 0.0249]	[0.0225, 0.0257]	[0.0217, 0.0265]	0.0241
δ_8	[−0.2815, −0.2695]	[−0.2875, −0.2635]	[−0.2935, −0.2575]	−0.2755
δ_9	[0.0154, 0.0165]	[0.0149, 0.0170]	[0.0144, 0.0176]	0.0160
δ_{10}	[−0.3355, −0.3212]	[−0.3426, −0.3141]	[−0.3497, −0.3069]	−0.3283
δ_{11}	[0.0179, 0.0188]	[0.0174, 0.0193]	[0.0169, 0.0198]	0.0183
δ_{12}	[−0.1921, −0.1840]	[−0.1962, −0.1799]	[−0.2003, −0.1758]	−0.1880
δ_{13}	[−0.0661, −0.0636]	[−0.0673, −0.0624]	[−0.0685, −0.0612]	−0.0649
δ_{14}	[−0.1719, −0.1647]	[−0.1755, −0.1611]	[−0.1791, −0.1574]	−0.1683
δ_{15}	[0,0]	[0,0]	[0,0]	0
δ_{16}	[0,0]	[0,0]	[0,0]	0

11.7 CONCLUSION

This chapter demonstrates a transformation-based technique for intervals, as the presence of errors in the structural problem is tough to handle; hence, we convert them in terms of intervals by finding out the minimum and maximum possibility range for their uncertainties and taking them as the interval bounds. The investigation presents the IFEM in the vibration of two-dimensional beam and truss elements. This is applied in a known problem of beam and 15-bar truss structure to have the efficiency of the proposed method. As discussed earlier, the concepts of interval numbers have been used here while solving the numerical problems of the system of linear equations. In this chapter, one may easily verify that the approximate solution lies in the calculated interval space. Further, the variation of uncertain parameters and its consequence on the system can be studied. This may be analyzed by observing the width of uncertainty. If the width is big then the uncertain parameter is more sensitive, whereas if the width is small then the assumed uncertain parameter is less sensitive. This provides a measurement tool for better interpretation of interval results which are very often encountered in real-life applications where the material properties may not be obtained in terms of crisp but a vague values in term of intervals. The idea may easily be extended to other structural problems with various complicating effects. Although this require more complex form of interval computation to handle the corresponding problem.

REFERENCES

G. Alefeld and J. Herzberger. 1983. *Introduction to Interval Computation.* Germany: Academic Press.

D. Behera and S. Chakraverty. 2013. Solution to fuzzy system of linear equations with crisp coefficients. *Fuzzy Information and Engineering,* 5(2), 205–219.

D. Behera and S. Chakraverty. 2020. Solving the nondeterministic static governing equations of structures subjected to various forces under fuzzy and interval uncertainty. *International Journal of Approximate Reasoning,* 116, 43–61.

S.S. Bhavikatti. 2005. *Finite Element Analysis.* New Delhi: New Age International (P) Ltd.

C.F. Gerald and P.O. Wheatley. 1985. *Applied Numerical Analysis,* 7th ed. New Delhi: Addison-Wesley Pub. Co.

S. Kabir and Y. Papadopoulos. 2018. A review of applications of fuzzy sets to safety and reliability engineering. *International Journal of Approximate Reasoning,* 100, 29–55.

R.E. Moore, R.B. Kearfott, and M.J. Cloud. 2009. *Introduction to Interval Analysis.* Philadelphia, PA: SIAM.

G.L. Narasaiah. 2008. *Finite Element Analysis.* Hyderabad: PBS Publications.

S. Nayak. 2020. Uncertain quantification of field variables involved in transient convection diffusion problems for imprecisely defined parameters. *International Communications in Heat and Mass Transfer,* 119, 104894.

S. Nayak and S. Chakraverty. 2013. Non-probabilistic approach to investigate uncertain conjugate heat transfer in an imprecisely defined plate. *International Journal of Heat and Mass Transfer,* 67, 445–454.

S. Nayak and S. Chakraverty. 2018. *Interval Finite Element Method with MATLAB.* San Diego, CA: Academic Press, Elsevier Inc.

G.P. Nikishkov. 2004. Introduction to the Finite Element Method, University of Aizu, Aizu-Wakamatsu, 965–8580, Lecture Notes, Japan.

T. Nirmala, D. Datta, H.S. Kushwaha, and K. Ganesan. 2011. Analytical solution of one-dimensional advection-diffusion equation with interval parameters. *Applied Mathematical Sciences*, 5(13), 607–624.

P. Seshu. *Textbook of Finite Element Analysis*. 2003. New Delhi: PHI Learning Private Limited.

12 Linear Eigenvalue Problems in Dynamic Structure with Uncertainty
An Expectation-Based Approach

Mrutyunjaya Sahoo and S. Chakraverty

CONTENTS

DOI: 10.1201/9781003328032-12

12.1 INTRODUCTION

The governing differential equations of dynamic analysis from various science and engineering problems with different material and geometrical properties lead to linear eigenvalue problems (LEPs) such as the generalized eigenvalue problem (GEP) and standard eigenvalue problem (SEP). Linear eigenvalue problems (GEP or SEP) have a wide range of applications in science and engineering, such as structural mechanics, communication systems, designing bridges, designing car stereo systems, electrical engineering, and mechanical engineering. For instance, in structural dynamic problems, the linear eigenvalue problem is quite important.

The following differential equation governs the mathematical models of structural systems:

$$M\,\ddot{s}(t) + D\dot{s}(t) + Ss(t) = f(t), \tag{12.1}$$

where, M, D, and S are the mass, damping, and stiffness matrices, respectively. Further, $s(t)\,\&\,f(t)$ are the respective displacement vector and external load vector.

Under static conditions, the controlling differential Eqn. (12.1) transforms into the linear system

$$Ss = f \tag{12.2}$$

Furthermore, the controlling differential equation can be treated as eigenvalue problems in a dynamic case. For an undamped system, Eqn. (12.1) may be reduced to:

$$M\,\ddot{s}(t) + Ss(t) = f(t), \tag{12.3}$$

This could result in a generalized eigenvalue problem as:

$$Sx = \lambda Mx. \tag{12.4}$$

where, λ is the eigenvalue known as the natural frequency of the structural system, and x is the corresponding eigenvector, which are known as vibration characteristics.

All variables and parameters are traditionally considered as crisp to make computations simple and easy. However, when carrying out the experimental task, one may experience ambiguity or ambiguous information due to changes in the environment or observational errors. Such uncertainty may be handled using a probabilistic approach when a huge amount of experimental data is available. However, in some circumstances, the sample size available is insufficient to use the probabilistic approach. As a result, intervals and/or fuzzy numbers may be utilized to deal with uncertain and imprecise parameters.

There are several well-known methods for dealing with SEP and GEP with precise or crisp parameters. These kinds of problems have been discussed in [1–4]. However, there is some possibility of inaccuracy in the involved parameters. These uncertainties may be modeled using fuzzy or interval concepts. As a result, few researchers have investigated such issues with fuzzy/interval environments. In this regard, Hladik et al. [5] suggested a filtering strategy for enhancing an outer estimate of the eigenvalue set of an interval matrix, as well as attempting to construct sharp bounds for real eigenvalues of interval matrices [6]. Qui et al. [7] proposed an approximation approach for determining the lower- and upper-bound solutions of eigenvalues of the real non-symmetric interval matrix of IEPs. Different approaches to deal with interval standard eigenvalue problems (ISEP) and interval general-ized eigenvalue problems (IGEP) were presented by Rohn [8], Moore et al. [9], and Alefeld and Herzberger [10]. Further, Leng et al. [11] and Leng and He [12] computed real eigenvalue bounds of the real interval matrices.

In structural dynamics, an efficient method was suggested by Xia and Friswell [13] to obtain the solution of FGEPs. In addition, the parameter identification of a multi-story frame structure with imprecise dynamic data, which may lead to fuzzy generalized eigenvalue problem (FGEP), was studied by Chakraverty and Behera [14]. Mahato and Chakraverty [15] demonstrated the filtering approach for real eigenvalue bounds of both IGEP and FGEP. Static and dynamic analysis of structural systems have been studied by Chakraverty and Behera [16] in uncertain environ-ment. Jeswal and Chakraverty [17] established an ANN-based approach for dealing with fuzzy eigenvalue problems. To address linear eigenvalue problems with uncer-tainty, Mohapatra and Chakraverty [18] used type-2 fuzzy numbers.

This work deals with the fuzzy linear eigenvalue problems, which are based on the double parametric form of fuzzy numbers and their expectations. Here we have discussed the concept of fuzzy variables to deal with the expectation of fuzzy numbers. Further, in this investigation, TFN and TrFN have been used to handle the titled problem.

The remainder of this work is organized in the following manner. In Section 12.2, we consider some background information on fuzzy numbers. Section 12.3 delves into fuzzy variables and their expectations. The fuzzy linear eigenvalue problems and proposed methodology have been discussed in Section 12.4. In Section 12.5, the pro-posed approach is explained with two numerical and one application problem related to spring-mass systems. The improper results are obtained in one of the examples, which is written in a proper form. Finally, in Section 12.6, there are some closing observations.

12.2 PRELIMINARIES

A few fundamental basics of the fuzzy theory are presented in this part for readers' better comprehension.

12.2.1 Fuzzy Set [19]

A fuzzy set $\overline{F}$ may be defined as a collection of a set of order pairs, i.e.,

$$\overline{F} = \left\{ \left(x, \mu_{\overline{F}}(x) \right) : x \in X, \mu_{\overline{F}}(x) \in [0,1] \right\}, \tag{12.5}$$

where, $\mu_{\overline{F}}(x)$ is known as the membership function of $\overline{F}$ and X be the universal set.

12.2.2 Fuzzy Number [19]

A fuzzy number is a special type of fuzzy set $\overline{F}$, which is normalized and convex having a piecewise continuous membership function.

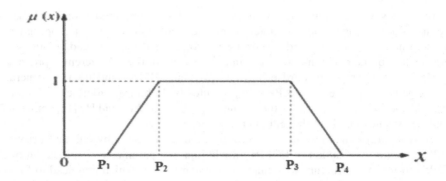

FIGURE 12.1   Membership function of triangular fuzzy number.

12.2.3 α-CUT OF A FUZZY NUMBER [19]

Given a fuzzy set $\bar{F}$ in X and any real number $\alpha \in [0,1]$, then the α-cut of $\bar{F}$, denoted by $\bar{F}(\alpha)$ is the crisp set and may be defined as:

$$\bar{F}(\alpha) = \{x \in X \mid \mu_{\bar{F}}(x) \geq \alpha\}. \tag{12.6}$$

12.2.4   TRIANGULAR FUZZY NUMBER (TFN) [19]

A TFN $\bar{P}$ may be interpreted as a triplet $\bar{P} = (p_1, p_2, p_3)$, $p_1 < p_2 < p_3$. It may be considered as a type of triangular-shaped fuzzy number as shown in Figure 12.1. The membership function $\mu_{\bar{P}}(x)$ of a TFN $\bar{P}$ may be described as follows:

$$\mu_{\bar{P}}(x) = \begin{cases} 0 & , x < p_1 \\[2mm] \dfrac{x - p_1}{p_2 - p_1} & , p_1 \leq x \leq p_2 \\[3mm] \dfrac{p_3 - x}{p_3 - p_2} & , p_2 \leq x \leq p_3 \\[2mm] 0 & , x > p_3 \end{cases} \tag{12.7}$$

12.2.4.1 α-Cut Representation of a TFN [20]

Considering a TFN $\bar{P} = (p_1, p_2, p_3)$. Then it may be parameterized into a fuzzy interval form by using the α-cut approach as follows:

$$\bar{P}_\alpha = \left[\bar{P}_\alpha^l, \bar{P}_\alpha^r\right] = \left[p_1 + \alpha(p_2 - p_1), p_3 - \alpha(p_3 - p_2)\right], \alpha \in [0,1] \tag{12.8}$$

12.2.4.2 (α, β)-Cut Representation of a TFN [20]

By using (α, β)-cut technique, Eqn. (12.8) may be written in the parametric form as:

$$\bar{P}_{(\alpha,\beta)} = \beta\left[\{p_3 - \alpha(p_3 - p_2)\} - \{p_1 + \alpha(p_2 - p_1)\}\right] + \{p_1 + \alpha(p_2 - p_1)\} \tag{12.9}$$

Notes:

1. Letting $\beta = 0$, we have the lower value of α-cut of TFN, i.e., $p_1 + \alpha(p_2 - p_1)$.
   a. By substituting $\alpha = 0$, in $p_1 + \alpha(p_2 - p_1)$, we may get p_1, i.e., the left spread of a TFN.
   b. By substituting $\alpha = 1$, in $p_1 + \alpha(p_2 - p_1)$, we may get p_2, i.e., the center of a TFN.
2. Letting $\beta = 1$, we have the upper value of α-cut of TFN i.e., $p_3 - \alpha(p_3 - p_2)$.
   a. By substituting $\alpha = 0$, in $p_3 - \alpha(p_3 - p_2)$, we may get p_3, i.e., the right spread of a TFN.
   b. By substituting $\alpha = 1$, in $p_3 - \alpha(p_3 - p_2)$, we may get p_2, i.e., the center of a TFN.

12.2.5 TRAPEZOIDAL FUZZY NUMBER (TrFN) [19]

A TrFN $\tilde{A}$ may be interpreted as $\bar{P} = (p_1, p_2, p_3, p_4)_{\text{TrFN}}$ which has the membership function (Figure 12.2) $\mu_{\bar{P}}(x)$ as follows:

$$\mu_{\bar{P}}(x) = \begin{cases} 0 & , x < p_1 \\ \dfrac{x - p_1}{p_2 - p_1} & , p_1 \le x \le p_2 \\ 1 & p_2 \le x \le p_3 \\ \dfrac{p_3 - x}{p_3 - p_2} & , p_3 \le x \le p_4 \\ 0 & , x > p_4 \end{cases} \tag{12.10}$$

12.2.5.1 α-Cut Representation of a TrFN [20]

Consider a TrFN $\bar{P} = (p_1, p_2, p_3, p_4)$. Then by using the α-cut approach, it may be parameterized into a fuzzy interval form as:

$$\bar{P}_\alpha = \left[\bar{P}_\alpha^l, \bar{P}_\alpha^r\right] = \left[p_1 + \alpha(p_2 - p_1), p_3 - \alpha(p_3 - p_2)\right], \alpha \in [0,1] \tag{12.11}$$

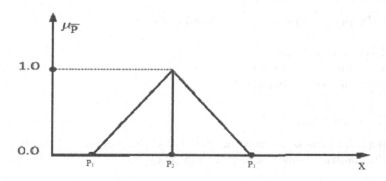

FIGURE 12.2 Membership function of trapezoidal fuzzy number.

12.2.5.2 (α, β)-Cut Representation of a TrFN [20]

Equation (12.11) may be written in the parametric form by using (α, β)-cut technique, such as:

$$\bar{P}_{(\alpha, \beta)} = \beta \Big[\big\{ p_4 - \alpha (p_4 - p_3) \big\} - \big\{ p_1 + \alpha (p_2 - p_1) \big\} \Big] + \big\{ p_1 + \alpha (p_2 - p_1) \big\} \qquad (12.12)$$

Notes:

1. Letting $\beta = 0$, we have the lower value of α-cut of TrFN, i.e., $p_1 + \alpha (p_2 - p_1)$.
   c. By substituting $\alpha = 0$, in $p_1 + \alpha (p_2 - p_1)$, we may get p_1, i.e., the left spread of a TrFN.
   d. By substituting $\alpha = 1$, in $p_1 + \alpha (p_2 - p_1)$, we may get p_2, i.e., the spread right to the left spread of TrFN.
2. Letting $\beta = 1$, we have the upper value of α-cut of TrFN, i.e., $p_4 - \alpha (p_4 - p_3)$.
   e. By substituting $\alpha = 0$, in $p_4 - \alpha (p_4 - p_3)$, we may get p_4, i.e., the right spread of a TrFN.
   f. By substituting $\alpha = 1$, in $p_4 + \alpha (p_4 - p_3)$, we may get p_3, i.e., the spread left to the right spread of a TrFN.

12.3   FUZZY VARIABLES

In 1978, Zadeh [21, 22] proposed the possibility theory as a method for studying ambiguous situations. The uncertain theory has become the fundamental structure for studying possibilistics. Kaufmann [23] coined the phrase "fuzzy variable", which has since been adopted by others. Nahmias [24] and Wang [25], respectively, extended the concept of a fuzzy variable to sample space and pattern space.

Probability space is related to crisp variables, whereas possibility space is associated with fuzzy variables. A mapping from a possibility space to a set of real numbers is referred to as a fuzzy variable. Liu and Liu [26] introduced a different version of credibility theory with a weighted mean based on the principles of possibility measure and necessity measure to investigate the behavior of fuzzy occurrence.

12.3.1   POSSIBILITY SPACE [21]

A triplet $\big(\kappa, p(\kappa), Pos \big)$ is said to be a possibility space, where

- κ is a non-empty set,
- $p(\kappa)$ is the power set of κ,
- Pos is the possibility measure.

12.3.2   DEFINITION 1 [21]

Let γ is a fuzzy variable specified on a possibility space, then the possibility measure is used to construct its membership function $\mu(x)$, i.e.,

$$\mu(x) = Pos \big\{ \omega \in \kappa \mid x(\omega) = x \big\}, x \in \Re \qquad (12.13)$$

12.3.3   Definition 2 [21, 22]

Let μ is the membership function for a fuzzy variable γ, and r^* is a real number. Then for the fuzzy event $\{\gamma \leq r\}$, the possibility and necessity may be defined as follows:

$$Pos\{\gamma \leq r^*\} = \sup_{x \leq r^*} \mu(x), Nec\{\gamma \leq r^*\} = 1 - \sup_{x \leq r^*} \mu(x) \qquad (12.14)$$

Let B is any set in $p(\kappa)$, then for a possibility space $(\kappa, p(\kappa), Pos)$, the credibility of B may be defined as:

$$Cr(B) = \frac{1}{2}\left(Pos\{B\} + Nec\{B\}\right) \qquad (12.15)$$

By using Eqn. (12.14), Eqn. (12.15) may be written as:

$$Cr\left(\gamma \leq r^*\right) = \frac{1}{2}\left[\sup_{x \leq r^*} \mu(x) + 1 - \sup_{x \leq r^*} \mu(x)\right] \qquad (12.16)$$

12.3.3.1   Credibility Measure of a TFN [26]

Suppose that $\gamma \sim \bar{P}(p_1, p_2, p_3)$ be a TFN having membership function given in Eqn. (12.7). Then the credibility distribution of γ is given by:

$$\vartheta(x) = \begin{cases} 0 & , x \leq p_1 \\[2mm] \dfrac{x - p_1}{2(p_2 - p_1)} & , p_1 < x \leq p_2 \\[3mm] \dfrac{x + p_3 - 2p_2}{2(p_3 - p_2)} & , p_2 < x \leq p_3 \\[3mm] 1 & , x > p_3 \end{cases} \qquad (12.17)$$

12.3.3.2   Credibility Measure of a TrFN [26]

Suppose that $\gamma \sim \bar{P}(p_1, p_2, p_3, p_4)$ be a TrFN having membership function given in Eqn. (12.10). Then the credibility distribution of γ is given by:

$$\vartheta(x) = \begin{cases} 0 & , x \leq p_1 \\[2mm] \dfrac{x - p_1}{2(p_2 - p_1)} & , p_1 < x \leq p_2 \\[3mm] \dfrac{1}{2} & , p_2 \leq x \leq p_3 \\[3mm] \dfrac{1}{2} - \dfrac{1}{2}\left(\dfrac{x - p_4}{2(p_3 - p_4)}\right) & , p_3 < x \leq p_4 \\[3mm] \dfrac{1}{2} & , x \geq p_3 \end{cases} \qquad (12.18)$$

12.3.4   Expectation of Fuzzy Variable [27]

For a fuzzy variable γ, the expected value of γ may be defined as:

$$E[\gamma] = \int_0^\infty Cr\{\gamma \geq r^*\} dr^* - \int_{-\infty}^0 Cr\{\gamma \leq r^*\} dr^* \qquad (12.19)$$

where one of the two integrals is finite.

12.3.4.1   Expectation of a TFN [27]

On solving Eqn. (12.19) and using Eqn. (12.16), we have the following for expectation of TFN as:

$$E[\gamma] = \frac{p_1 + 2p_2 + p_3}{4} \qquad (12.20)$$

12.3.4.2   Expectation of a TrFN [27]

We have the following result for expectation of a TrFN as

$$E[\gamma] = \frac{p_1 + p_2 + p_3 + p_4}{4} \qquad (12.21)$$

12.4   FUZZY LINEAR EIGENVALUE PROBLEMS (FLEP)

Here we have discussed some fundamental concepts in linear eigenvalue problems, such as crisp linear eigenvalue problems (CLEP). Based on their structure, CLEP may be classified into two parts.

    I. Crisp Generalized Eigenvalue Problem (CGEP)
    II. Crisp Standard Eigenvalue Problem (CSEP)

The general forms CGEP and CSEP may be defined as follows:

$$Kx = \lambda Nx \qquad (12.22)$$

and,

$$Px = \lambda x \qquad (12.23)$$

where, K, N and P are square matrices of the same order, i.e., $m \times m$(say). It is very clear from the previous two equations that for $N = I_{m \times m}$ ($I_{m \times m}$ is an $m \times m$ Identity matrix of order m), Eqn. (12.23) is a special case of Eqn. (12.22).

    Similarly, when all the parameters of GEP and SEP are in the form of fuzzy numbers, these may be referred to as FLEP and may be classified as follows:

    I. Fuzzy Generalized Eigenvalue Problem (FGEP)
    II. Fuzzy Standard Eigenvalue Problem (FSEP)

12.4.1 Fuzzy Generalized Eigenvalue Problem (FGEP)

Let $\bar{K}$ & $\bar{N}$ be two fuzzy square matrices of same order (say $m \times m$). Then FGEP may be defined as:

$$\bar{K}\bar{x} = \bar{\lambda}\bar{N}\bar{x} \qquad (12.24)$$

Here, the fuzzy eigenvalue is $\bar{\lambda}$ and the corresponding fuzzy eigenvector is $\bar{x}$.

Let the fuzzy matrix $\bar{K}$ & $\bar{N}$ be as follows:

$$\bar{K} = \begin{bmatrix} \bar{k}_{11} & \bar{k}_{12} & . & . & . & \bar{k}_{1m} \\ \bar{k}_{21} & \bar{k}_{22} & . & . & . & \bar{k}_{2m} \\ . & . & . & . & . & . \\ . & . & . & . & . & . \\ . & . & . & . & . & . \\ \bar{k}_{m1} & \bar{k}_{m2} & . & . & . & \bar{k}_{mm} \end{bmatrix} \text{ and } \bar{N} = \begin{bmatrix} \bar{n}_{11} & \bar{n}_{12} & . & . & . & \bar{n}_{1m} \\ \bar{n}_{21} & \bar{n}_{22} & . & . & . & \bar{n}_{2m} \\ . & . & . & . & . & . \\ . & . & . & . & . & . \\ . & . & . & . & . & . \\ \bar{n}_{m1} & \bar{n}_{m2} & . & . & . & \bar{n}_{mm} \end{bmatrix}.$$

$$(12.25)$$

where, $K = \left(\bar{k}_{ij}\right)$ and $N = \left(\bar{n}_{ij}\right)$ for $i, j = 1, 2, 3, ..., m$ are the entries of the two fuzzy coefficients matrices $\bar{K}$ & $\bar{N}$, respectively.

For TFN, $\bar{k}_{ij} = \left(k_{ij}^l, k_{ij}^m, k_{ij}^r\right)$ and TrFN $\bar{k}_{ij} = \left(k_{ij}^{l0}, k_{ij}^{l1}, k_{ij}^{r0}, k_{ij}^{r1}\right)$. Similarly, one may define for $\bar{n}_{ij}$ also.

12.4.2 Fuzzy Standard Eigenvalue Problem (FSEP)

Similarly, the FSEP for a fuzzy square matrix $\bar{P}$ may be stated as:

$$\bar{P}\bar{x} = \bar{\lambda}\bar{x} \qquad (12.26)$$

where $\bar{\lambda}$ is the fuzzy eigenvalue of the matrix $\bar{P}$, corresponds to eigenvector $\bar{x}$.

Let the fuzzy matrix $\bar{P}$ be

$$\bar{P} = \begin{bmatrix} \bar{p}_{11} & \bar{p}_{12} & . & . & . & \bar{p}_{1m} \\ \bar{p}_{21} & \bar{p}_{22} & . & . & . & \bar{p}_{2m} \\ . & . & . & . & . & . \\ . & . & . & . & . & . \\ . & . & . & . & . & . \\ \bar{p}_{m1} & \bar{p}_{m2} & . & . & . & \bar{p}_{mm} \end{bmatrix} \qquad (12.27)$$

where $\bar{P} = \left(\bar{p}_{ij}\right)$ for $i, j = 1, 2, 3, ..., m$ are the elements of the coefficient matrix $\bar{P}$. For TFN $\bar{p}_{ij} = \left(p_{ij}^l, p_{ij}^m, p_{ij}^r\right)$ and TrFN $\bar{p}_{ij} = \left(p_{ij}^{l0}, p_{ij}^{l1}, p_{ij}^{r0}, p_{ij}^{r1}\right)$.

12.5   THE PROPOSED METHOD FOR SOLVING FUZZY EIGENVALUE PROBLEM

This section contains a fuzzy expectation-based technique and (α, β)-cut approach to deal with FGEP (or FSEP) and evaluate fuzzy eigenvalue solutions.

Method 1: By using the concept of expectation, FGEP and FSEP may be written as:

$$E\left[\bar{K}\right]x^* = \lambda^* E\left[\bar{N}\right]x^* \qquad (12.28)$$

and,

$$E\left[\bar{P}\right]x^* = \lambda^* x^* \qquad (12.29)$$

where λ^* is the expected value of the eigenvalue $\bar{\lambda}$ and x^* is the expected value of the corresponding eigenvector $\bar{x}$.

Now the expectation-based fuzzy matrices $\bar{K}, \bar{N}$ & $\bar{P}$ may be expressed as follows:

$$E\left[\bar{K}\right] = \begin{bmatrix} E\left[\bar{k}_{11}\right] & E\left[\bar{k}_{12}\right] & \cdot & \cdot & \cdot & E\left[\bar{k}_{1n}\right] \\ E\left[\bar{k}_{21}\right] & E\left[\bar{k}_{22}\right] & \cdot & \cdot & \cdot & E\left[\bar{k}_{2n}\right] \\ \cdot & \cdot & \cdot & \cdot & \cdot & \cdot \\ \cdot & \cdot & \cdot & \cdot & \cdot & \cdot \\ \cdot & \cdot & \cdot & \cdot & \cdot & \cdot \\ E\left[\bar{k}_{n1}\right] & E\left[\bar{k}_{n2}\right] & \cdot & \cdot & \cdot & E\left[\bar{k}_{nn}\right] \end{bmatrix} \qquad (12.30)$$

$$E\left[\bar{N}\right] = \begin{bmatrix} E[\bar{n}_{11}] & E[\bar{n}_{12}] & \cdot & \cdot & \cdot & E[\bar{n}_{1n}] \\ E[\bar{n}_{21}] & E[\bar{n}_{22}] & \cdot & \cdot & \cdot & E[\bar{n}_{2n}] \\ \cdot & \cdot & \cdot & \cdot & \cdot & \cdot \\ \cdot & \cdot & \cdot & \cdot & \cdot & \cdot \\ \cdot & \cdot & \cdot & \cdot & \cdot & \cdot \\ E[\bar{n}_{n1}] & E[\bar{n}_{n2}] & \cdot & \cdot & \cdot & E[\bar{n}_{nn}] \end{bmatrix} \qquad (12.31)$$

and,

$$E\left[\bar{P}\right] = \begin{bmatrix} E[\bar{p}_{11}] & E[\bar{p}_{12}] & \cdot & \cdot & \cdot & E[\bar{p}_{1n}] \\ E[\bar{p}_{21}] & E[\bar{p}_{22}] & \cdot & \cdot & \cdot & E[\bar{p}_{2n}] \\ \cdot & \cdot & \cdot & \cdot & \cdot & \cdot \\ \cdot & \cdot & \cdot & \cdot & \cdot & \cdot \\ \cdot & \cdot & \cdot & \cdot & \cdot & \cdot \\ E[\bar{p}_{n1}] & E[\bar{p}_{n2}] & \cdot & \cdot & \cdot & E[\bar{p}_{nn}] \end{bmatrix} \qquad (12.32)$$

Here the expectation of each entry of the aforementioned said matrices may be obtained by using the formula given in Eqns. (12.20) and (12.21).

Now the expectation of fuzzy eigenvalue of the FGEP may be expressed as:

$$E\left[\bar{K}\right]x^* = \lambda^* E\left[\bar{N}\right]x^*$$
$$\Rightarrow \det\left(E\left[\bar{K}\right] - \lambda^* E\left[\bar{N}\right]\right) = 0 \tag{12.33}$$

Similarly, FSEP may be solved by simplifying,

$$\det\left(E\left[\bar{P}\right] - \lambda^* I_n\right) = 0 \tag{12.34}$$

to obtain the required eigenvalues.

Method 2: Further, FGEP and FSEP may be written by using (α,β)-cut of fuzzy numbers (as discussed in sections 12.2.4 and 12.2.5) as:

$$\bar{K}(\alpha,\beta)\bar{x}(\alpha,\beta) = \bar{\lambda}(\alpha,\beta)\bar{N}(\alpha,\beta)\bar{x}(\alpha,\beta) \tag{12.35}$$

and,

$$\bar{P}(\alpha,\beta)\bar{x}(\alpha,\beta) = \bar{\lambda}(\alpha,\beta)\bar{x}(\alpha,\beta) \tag{12.36}$$

where, $\bar{\lambda}(\alpha,\beta)$ is the corresponding eigenvalue concerning the eigenvector $\bar{x}(\alpha,\beta)$.

By using the (α,β)-cut of fuzzy numbers, the fuzzy matrices $\bar{K}, \bar{N}$ and $\bar{P}$ will become,

$$\bar{K}(\alpha,\beta) = \begin{bmatrix} \bar{k}_{11}(\alpha,\beta) & \bar{k}_{12}(\alpha,\beta) & \cdot & \cdot & \cdot & \bar{k}_{1n}(\alpha,\beta) \\ \bar{k}_{21}(\alpha,\beta) & \bar{k}_{22}(\alpha,\beta) & \cdot & \cdot & \cdot & \bar{k}_{2n}(\alpha,\beta) \\ \cdot & \cdot & \cdot & \cdot & \cdot & \cdot \\ \cdot & \cdot & \cdot & \cdot & \cdot & \cdot \\ \cdot & \cdot & \cdot & \cdot & \cdot & \cdot \\ \bar{k}_{n1}(\alpha,\beta) & \bar{k}_{n2}(\alpha,\beta) & \cdot & \cdot & \cdot & \bar{k}_{nn}(\alpha,\beta) \end{bmatrix} \tag{12.37}$$

$$\bar{N}(\alpha,\beta) = \begin{bmatrix} \bar{n}_{11}(\alpha,\beta) & \bar{n}_{12}(\alpha,\beta) & \cdot & \cdot & \cdot & \bar{n}_{1n}(\alpha,\beta) \\ \bar{n}_{21} & \bar{n}_{22}(\alpha,\beta) & \cdot & \cdot & \cdot & \bar{n}_{2n}(\alpha,\beta) \\ \cdot & \cdot & \cdot & \cdot & \cdot & \cdot \\ \cdot & \cdot & \cdot & \cdot & \cdot & \cdot \\ \cdot & \cdot & \cdot & \cdot & \cdot & \cdot \\ \bar{n}_{n1}(\alpha,\beta) & \bar{n}_{n2}(\alpha,\beta) & \cdot & \cdot & \cdot & \bar{n}_{nn}(\alpha,\beta) \end{bmatrix} \tag{12.38}$$

and,

$$\bar{P}(\alpha,\beta) = \begin{bmatrix} \bar{n}_{11}(\alpha,\beta) & \bar{n}_{12}(\alpha,\beta) & . & . & . & \bar{n}_{1n}(\alpha,\beta) \\ \bar{n}_{21}(\alpha,\beta) & \bar{n}_{22}(\alpha,\beta) & . & . & . & \bar{n}_{2n}(\alpha,\beta) \\ . & . & . & . & & . \\ . & . & . & . & . & . \\ . & . & . & . & . & . \\ \bar{n}_{n1}(\alpha,\beta) & \bar{n}_{n2}(\alpha,\beta) & . & . & . & \bar{n}_{nn}(\alpha,\beta) \end{bmatrix} \qquad (12.39)$$

Hence, by using (α,β)-cut, the FGEP may be obtained and solved by simplifying the following to get fuzzy eigenvalues

$$\bar{K}(\alpha,\beta)\bar{x}(\alpha,\beta) = \bar{\lambda}(\alpha,\beta)\bar{N}(\alpha,\beta)\bar{x}(\alpha,\beta)$$
$$\Rightarrow \det\left\{\bar{K}(\alpha,\beta) - \bar{\lambda}(\alpha,\beta)\bar{N}(\alpha,\beta)\right\} = 0 \qquad (12.40)$$

Similarly, FSEP may also be handled by simplifying,

$$\det\left(\bar{P}(\alpha,\beta) - \bar{\lambda}(\alpha,\beta)I_n\right) = 0 \qquad (12.41)$$

12.6 NUMERICAL EXAMPLES

Two numerical examples of FSEP and FGEP, as well as an uncertain dynamic structural application problem, are solved here.

Example 12.1:

Let us consider a simple 2×2 FSEP,

$$\bar{M}\bar{x} = \bar{\lambda}\bar{x} \qquad (12.42)$$

where

$$\bar{M} = \begin{bmatrix} (1,2,3) & (2,3,4) \\ (3,4,5) & (4,5,6) \end{bmatrix} \qquad (12.43)$$

SOLUTION

Let us start solving the given FSEP by method 1, i.e., by using the expectation of each entry of the matrix.
Equation (12.43) implies

$$E\left[\bar{M}\right] = \begin{bmatrix} 2 & 3 \\ 4 & 5 \end{bmatrix} \qquad (12.44)$$

From Eqn. (12.44), the eigenvalue λ^* may be obtained as:

$$E[\lambda_1] = \lambda_1^* = 7.275$$

and,

$$E[\lambda_2] = \lambda_2^* = -0.275$$

Further, we may solve this FSEP by using (α,β)-cut approach (method 2) to get eigenvalues which are in TFN form,

By using (α,β)-cut approach $(\alpha,\beta \in [0,1])$, each entry of the matrix $\bar{M}$ may be written as:

$$\bar{M}(\alpha,\beta) = \begin{bmatrix} \bar{m}_{11}(\alpha,\beta) & \bar{m}_{12}(\alpha,\beta) \\ \bar{m}_{21}(\alpha,\beta) & \bar{m}_{22}(\alpha,\beta) \end{bmatrix} \qquad (12.45)$$

with

$$\begin{aligned} \bar{m}_{11}(\alpha,\beta) &= \beta(2-2\alpha)+(\alpha+1) \\ \bar{m}_{12}(\alpha,\beta) &= \beta(2-2\alpha)+(\alpha+2) \\ \bar{m}_{21}(\alpha,\beta) &= \beta(2-2\alpha)+(\alpha+3) \\ \bar{m}_{22}(\alpha,\beta) &= \beta(2-2\alpha)+(\alpha+4) \end{aligned} \qquad (12.46)$$

Now, the given FSEP may be written as:

$$\begin{aligned} \bar{P}(\alpha,\beta)\bar{x}(\alpha,\beta) &= \bar{\lambda}(\alpha,\beta)\bar{x}(\alpha,\beta) \\ \Rightarrow \det\left(\bar{P}(\alpha,\beta)-\bar{\lambda}(\alpha,\beta)I_n\right) &= 0 \end{aligned} \qquad (12.47)$$

By simplifying Eqn. (12.47), a quadratic equation will be obtained in $\bar{\lambda}(\alpha,\beta)$, which may be solved to get

$$\bar{\lambda}_1 = (5.3723, 7.2749, 9.2170)$$

$$\bar{\lambda}_2 = (-0.3723, -0.2749, -0.2170)$$

We may now find the expectation of each of the eigenvalues using the concept of expectation in the aforementioned computations.

$$E[\bar{\lambda}_1] = 7.284775, \quad E[\bar{\lambda}_2] = -0.284775$$

Table 12.1 shows the obtained eigenvalues by the present methods, where a comparison is made between the eigenvalues obtained in expectation form (method 1), i.e., expected fuzzy eigenvalues, and the expectation of obtained fuzzy eigenvalues (method 2).

TABLE 12.1

Comparison of Expectation of Calculated Fuzzy Eigenvalue and Expected Fuzzy Eigenvalue (Example 12.1)

Eigenvalue	Fuzzy Eigenvalue (Proposed Method 2)	The Expectation of Obtained Fuzzy Eigenvalue	Expected Fuzzy Eigenvalue (Proposed Method 1)
$\bar{\lambda}_1$	$(5.3723, 7.2749, 9.2170)$	7.284775	7.725
$\bar{\lambda}_1$	$(-0.3723, -0.2749, -0.2170)$	-0.284775	-0.275

Example 12.2 [28]:

Let us take the following eigenvalue problem $\bar{K}\bar{x} = \lambda\bar{N}\bar{x}$, where,

$$\bar{K} = \begin{bmatrix} (17000, 18000, 18200) & (-7300, -7200, -7100) \\ (-7300, -7200, -7100) & (7100, 7200, 7300) \end{bmatrix} \quad (12.48)$$

and,

$$\bar{N} = \begin{bmatrix} 3600 & 0 \\ 0 & 3600 \end{bmatrix} \quad (12.49)$$

SOLUTION

By using the expectation of each entry of two matrices, Eqns. (12.48) and (12.49) may be written as:

$$E[\bar{K}] = \begin{bmatrix} 18000 & -7200 \\ -7200 & 7200 \end{bmatrix}$$

and, (12.50)

$$E[\bar{N}] = \begin{bmatrix} 3600 & 0 \\ 0 & 3600 \end{bmatrix}$$

From Eqn. (12.50), the eigenvalue λ^* may be obtained by using Eqn. (12.31) as:

$$E[\lambda_1] = \lambda_1^* = 1$$

$$E[\lambda_2] = \lambda_2^* = 6$$

Now, by using (α, β)-cut approach $(\alpha, \beta \in [0,1])$, each entry of the matrix $\bar{K}$ & $\bar{N}$

$$\bar{K}(\alpha,\beta) = \begin{bmatrix} \bar{k}_{11}(\alpha,\beta) & \bar{k}_{12}(\alpha,\beta) \\ \bar{k}_{21}(\alpha,\beta) & \bar{k}_{22}(\alpha,\beta) \end{bmatrix} \text{ and } \bar{N}(\alpha,\beta) = \begin{bmatrix} \bar{n}_{11}(\alpha,\beta) & \bar{n}_{12}(\alpha,\beta) \\ \bar{n}_{21}(\alpha,\beta) & \bar{n}_{22}(\alpha,\beta) \end{bmatrix}$$

with

$$\bar{k}_{11}(\alpha,\beta) = 200\alpha + 400\beta - 400\alpha\beta + 17800$$

$$\bar{k}_{12}(\alpha,\beta) = 100\alpha + 200\beta - 200\alpha\beta - 7300$$

$$\bar{k}_{21}(\alpha,\beta) = 100\alpha + 200\beta - 200\alpha\beta - 7300$$

$$\bar{k}_{22}(\alpha,\beta) = 100\alpha + 200\beta - 200\alpha\beta + 7100$$

and,

$$\bar{n}_{11}(\alpha,\beta) = 3600 = \bar{n}_{22}(\alpha,\beta)$$

$$\bar{n}_{12}(\alpha,\beta) = 0 = \bar{n}_{21}(\alpha,\beta)$$

Now, solving the FGEP, i.e., $\bar{K}(\alpha,\beta)\bar{x}(\alpha,\beta) = \bar{\lambda}(\alpha,\beta)\bar{N}(\alpha,\beta)\bar{x}(\alpha,\beta)$

$$\Rightarrow \det\left(\bar{K}(\alpha,\beta) - \bar{\lambda}(\alpha,\beta)\bar{N}(\alpha,\beta)\right) = 0 \qquad (12.51)$$

Solving Eqn. (12.51) for eigenvalue $\bar{\lambda}$, one may get:

$$\bar{\lambda}_1 = (0.9443, 1, 1.0554)$$

$$\bar{\lambda}_2 = (5.9724, 6, 6.0279)$$

Further, by using the concept of expectation in the aforementioned calculations, we may find the expectation of each of the eigenvalues,

$$E\left[\bar{\lambda}_1\right] = 0.999925, \quad E\left[\bar{\lambda}_2\right] = 6.000075$$

In Table 12.2, the obtained eigenvalue is given by the proposed method and a comparison is made between the expectation of obtained fuzzy eigenvalues (method 2) and the eigenvalues obtained in expectation form (method 1), i.e., expected fuzzy eigenvalues. It may be observed that the resulting eigenvalues in expectation form are quite near to the expectation of obtained fuzzy eigenvalues. The graphical representation of obtained fuzzy eigenvalues are plotted in Figures 12.3 and 12.4.

TABLE 12.2

Comparison of Expectation of Calculated Fuzzy Eigenvalue and Expected Fuzzy Eigenvalue (Example 12.2)

Eigenvalue	Fuzzy Eigenvalue (Proposed Method 2)	The Expectation of Obtained Fuzzy Eigenvalue	Expected Fuzzy Eigenvalue (Proposed Method 1)
$\bar{\lambda}_1$	$(0.9443, 1, 1.0554)$	0.999925	1
$\bar{\lambda}_1$	$(5.9724, 6, 6.0279)$	6.000075	6

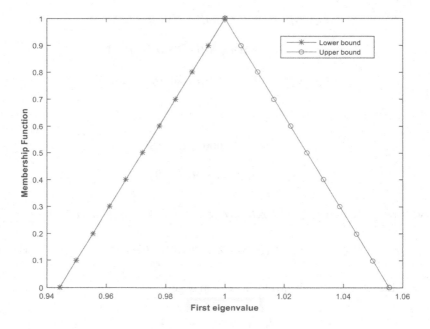

FIGURE 12.3   First fuzzy eigenvalue plot for Example 12.2.

Example 12.3 [28]:

Let us consider an application problem, namely a five-degrees-of-freedom spring-mass structural system as shown in Figure 12.5. Suppose all the structural parameters are considered as TrFN.

Here the fuzzy stiffness and mass parameters in the form of TrFN are given as:

$$\bar{s}_1 = (2000, 2020, 2080, 2100)\,(N/m);$$
$$\bar{s}_2 = (1800, 1815, 1835, 1850)\,(N/m);$$
$$\bar{s}_3 = (1600, 1612, 1618, 1630)\,(N/m);$$
$$\bar{s}_4 = (1400, 1408, 1412, 1420)\,(N/m);$$
$$\bar{s}_5 = (1200, 1203, 1207, 1210)\,(N/m);$$
$$\bar{s}_6 = (1000, 1001, 1007, 1008)\,(N/m);$$

(12.52)

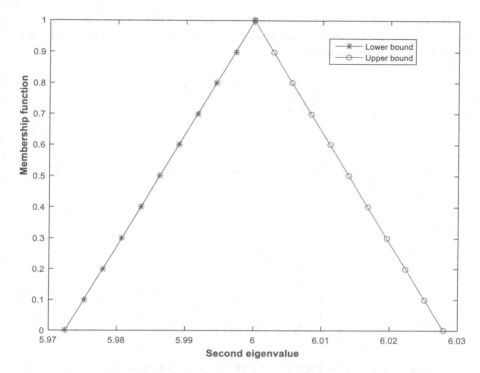

FIGURE 12.4   Second fuzzy eigenvalue plot for Example 12.2.

and,

$$\bar{m}_1 = (10,10.5,11.5,12)(kg);$$
$$\bar{m}_2 = (12,12.4,13.6,14)(kg);$$
$$\bar{m}_3 = (14,14.2,15.8,16)(kg); \qquad (12.53)$$
$$\bar{m}_4 = (16,16.8,17.2,18)(kg);$$
$$\bar{m}_5 = (18,18.6,19.4,20)(kg);$$

SOLUTION

Fuzzy eigenvalues of the FGEP are derived here from the dynamic analysis of the spring-mass structural system.

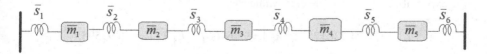

FIGURE 12.5   Spring-mass system of five-degrees-of-freedom.

The dynamic analysis of a five-degree-of-freedom spring-mass structural system with fuzzy parameters yields a FGEP, i.e., $\bar{S}\bar{x} = \lambda \bar{M}\bar{x}$, where the fuzzy stiffness and mass matrices are expressed as:

$$\bar{S} = \begin{bmatrix} \bar{s}_1 + \bar{s}_2 & -\bar{s}_2 & 0 & 0 & 0 \\ -\bar{s}_2 & \bar{s}_2 + \bar{s}_3 & -\bar{s}_3 & 0 & 0 \\ 0 & -\bar{s}_3 & \bar{s}_3 + \bar{s}_4 & -\bar{s}_4 & 0 \\ 0 & 0 & -\bar{s}_4 & \bar{s}_4 + \bar{s}_5 & -\bar{s}_5 \\ 0 & 0 & 0 & -\bar{s}_5 & \bar{s}_5 + \bar{s}_6 \end{bmatrix}. \tag{12.54}$$

and,

$$\bar{M} = \begin{bmatrix} \bar{m}_1 & 0 & 0 & 0 & 0 \\ 0 & \bar{m}_2 & 0 & 0 & 0 \\ 0 & 0 & \bar{m}_3 & 0 & 0 \\ 0 & 0 & 0 & \bar{m}_4 & 0 \\ 0 & 0 & 0 & 0 & \bar{m}_5 \end{bmatrix} \tag{12.55}$$

Let us employ the proposed method to find the required eigenvalues.

Now by implementing parameters α, β in the given data, and then solving for eigenvalues, we may have the following:

$$\bar{\lambda}_1 = (23.4915, 24.7574, 24.8145, 25.8573)$$
$$\bar{\lambda}_2 = (90.5916, 91.5672, 95.5970, 96.8309)$$
$$\bar{\lambda}_3 = (176.6024, 181.7021, 189.7060, 195.8933) \tag{12.56}$$
$$\bar{\lambda}_4 = (279.7002, 286.8253, 308.3415, 316.9578)$$
$$\bar{\lambda}_5 = (452.6226, 467.5860, 509.7944, 529.1678)$$

The graphical representation of the aforementioned obtained eigenvalues are shown in Figures 12.6–12.10.

It may be noted that the values of x for membership degree 1 are very close, (i.e., for $x = 24.7574$ to 24.8145) which may not be clearly seen in Figure 12.6, but the values of x can be seen in Table 12.3.

Now by using the concept of expectation in the above calculations, we may find the expectation of each of the eigenvalues as:

$$E\left[\bar{\lambda}_1\right] = 24.730175, \; E\left[\bar{\lambda}_2\right] = 93.646675, \; E\left[\bar{\lambda}_3\right] = 185.97595,$$
$$E\left[\bar{\lambda}_4\right] = 297.9562, \; E\left[\bar{\lambda}_5\right] = 489.7927$$

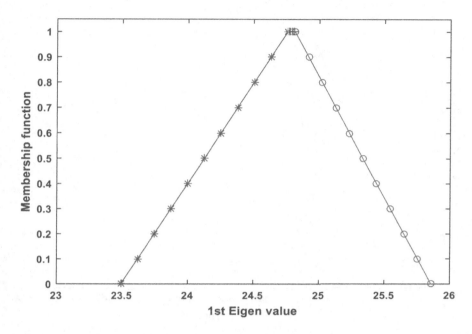

FIGURE 12.6   First fuzzy eigenvalue plot for Example 12.3.

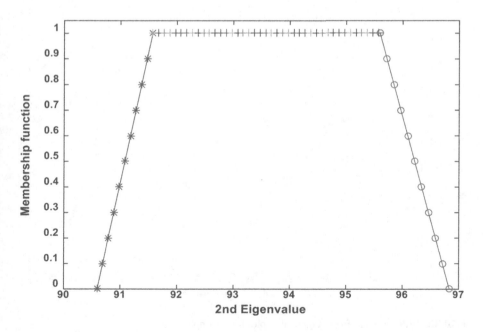

FIGURE 12.7   Second fuzzy eigenvalue plot for Example 12.3.

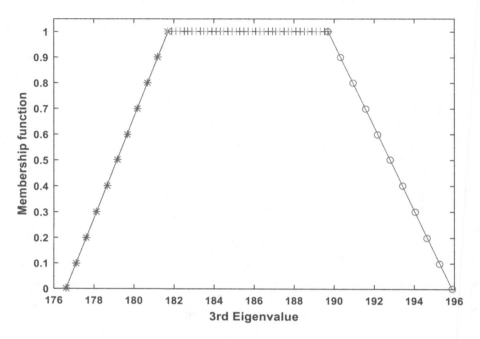

FIGURE 12.8   Third fuzzy eigenvalue plot for Example 12.3.

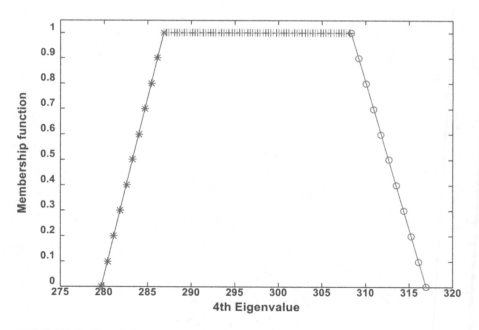

FIGURE 12.9   Fourth fuzzy eigenvalue plot for Example 12.3.

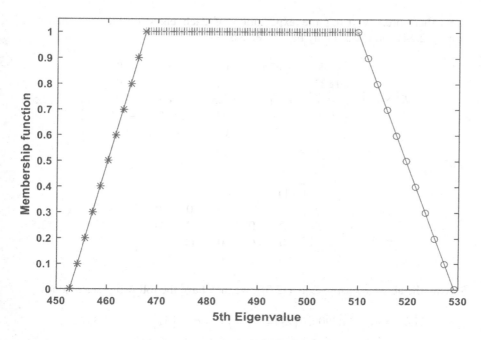

FIGURE 12.10   Fifth fuzzy eigenvalue plot for Example 12.3.

TABLE 12.3

Comparison of Expectation of Calculated Fuzzy Eigenvalue and Expected Fuzzy Eigenvalue (Example 12.3)

Eigenvalues	Fuzzy Eigenvalue (Proposed Method 2)	The Expectation of Obtained Fuzzy Eigenvalue (Proposed)	Expected Fuzzy Eigenvalue (Proposed Method 1)	Fuzzy Eigenvalues [28]	The Expectation of Fuzzy Eigenvalues [28]
$\bar{\lambda}_1$	$\left(\begin{array}{l}23.4915, 24.7574, \\ 24.8145, 25.8573\end{array}\right)$	24.730175	24.7903	$\left(\begin{array}{l}23.5660, 24.1774, \\ 25.4399, 26.1796\end{array}\right)$	24.840725
$\bar{\lambda}_2$	$\left(\begin{array}{l}90.5916, 91.5672, \\ 95.5970, 96.8309\end{array}\right)$	93.646675	93.5129	$\left(\begin{array}{l}88.6697, 90.9004, \\ 96.3431, 99.0332\end{array}\right)$	93.7366
$\bar{\lambda}_3$	$\left(\begin{array}{l}176.6024, 181.7021, \\ 189.7060, 195.8933\end{array}\right)$	185.57595	185.6226	$\left(\begin{array}{l}175.7996, 181.3439, \\ 190.0671, 196.8291\end{array}\right)$	186.00992
$\bar{\lambda}_4$	$\left(\begin{array}{l}279.7002, 286.8253, \\ 308.3415, 316.9578\end{array}\right)$	297.9562	297.0207	$\left(\begin{array}{l}280.4391, 286.8788, \\ 308.3185, 316.1529\end{array}\right)$	297.94732
$\bar{\lambda}_5$	$\left(\begin{array}{l}452.6226, 467.5860, \\ 509.7944, 529.1678\end{array}\right)$	489.7927	487.6949	$\left(\begin{array}{l}456.8997, 469.1946, \\ 508.0278, 524.1465\end{array}\right)$	489.56715

Furthermore, by using the expectation of each entry of the two matrices, Eqns. (12.54) and (12.55) may be written as:

$$E\left[\bar{S}\right] = \begin{bmatrix} 3875 & -1825 & 0 & 0 & 0 \\ -1825 & 3440 & -1615 & 0 & 0 \\ 0 & -1615 & 3025 & -1410 & 0 \\ 0 & 0 & -1410 & 2615 & -1205 \\ 0 & 0 & 0 & -1205 & 2209 \end{bmatrix}. \quad (12.57)$$

and,

$$E\left[\bar{M}\right] = \begin{bmatrix} 11 & 0 & 0 & 0 & 0 \\ 0 & 13 & 0 & 0 & 0 \\ 0 & 0 & 15 & 0 & 0 \\ 0 & 0 & 0 & 17 & 0 \\ 0 & 0 & 0 & 0 & 19 \end{bmatrix} \quad (12.58)$$

Solving the above by using Eqn. (12.31), we may have the following eigenvalues:

$$E\left[\lambda_1\right] = \lambda_1^* = 24.7903, E\left[\lambda_2\right] = \lambda_2^* = 93.5129, E\left[\lambda_3\right] = \lambda_3^* = 185.6226,$$
$$E\left[\lambda_4\right] = \lambda_2^* = 297.0207, E\left[\lambda_5\right] = \lambda_5^* = 487.6949,$$

Table 12.3 shows a comparison among expectation of fuzzy eigenvalues of spring-mass structure (method 2), expected eigenvalue (method 1), and in fuzzy numbers (TrFNs) [28] by obtaining their expectations. The fuzzy solutions obtained by the present method are given in the second column of Table 12.3. In the third column, the expectation of obtained fuzzy eigenvalues are given and the eigenvalues in the expected form obtained by the proposed method are given in column four. The fuzzy solutions of TrFN by [28] and their expectation are mentioned in columns fifth and sixth, respectively. It may be observed that obtained eigenvalues by the present method are close to the results obtained in [28]. Also, it may be noted that the expectations of obtained fuzzy eigenvalues (method 2) are close to the eigenvalues in the expected form (method 1).

12.7 CONCLUSION

In this chapter, two strategies for solving FGEPs and FSEPs using the double parametric form and the idea of expectation are discussed. TFNs and TrFNs are employed to handle fuzzy uncertainty in this case. To support the suggested method, two numerical examples and an application problem of spring-mass systems have been provided.

REFERENCES

[1] Bhat, R. B., & Chakraverty, S. (2004). *Numerical Analysis in Engineering*. Oxford: Alpha Science Int'1 Ltd.
[2] Gerald, C. F. (2004). *Applied Numerical Analysis*. Delhi: Pearson Education India.

[3] Seshu, P. (2003). *Textbook of Finite Element Analysis*. New Delhi: PHI Learning Pvt Ltd.

[4] Humar, J. (2012). *Dynamics of Structures*. Boca Raton, FL: CRC Press.

[5] Hladik, M., David, D., & Elias, T. (2011). A filtering method for the interval eigenvalue problem. *Applied Mathematics and Computer*, 217 (12): 5236–5242.

[6] Hladik, M., Daney, D., & Tsigaridas, E. (2010). Bounds on real eigenvalues and singular values of interval matrices. *The SIAM Journal on Matrix Analysis and Applications*, 31 (4): 2116–2129.

[7] Qiu, Z., Müller, P. C., & Frommer, A. (2001). An approximate method for the standard interval eigenvalue problem of real non-symmetric interval matrices. *Communications in numerical methods in engineering*, 17(4), 239–251.

[8] Rohn, J. (2005). *A Handbook of Results on Interval Linear Problems*. Czech Republic, European Union: Czech Academy of Sciences Prague.

[9] Moore, R. E., Kearfott, R. B., & Cloud, M. J. (2009). *Introduction to Interval Analysis*. Philadelphia: Cambridge University Press.

[10] Alefeld, G., & Herzberger, J. (2012). *Introduction to Interval Computation*. INC, New York: Academic Press.

[11] Leng, H., & He, Z. (2007). Computing eigenvalue bounds of structures with uncertain but non-random parameters by a method based on perturbation theory. *Communications in Numerical Methods in Engineering*, 23 (11): 973–982. DOI: 10.1002/cnm.936.

[12] Leng, H., & He, Z. (2010). Computation of bounds for eigenvalues of structures with interval parameters. *Applied Mathematics and Computation*, 216 (9): 2734–2739. DOI: 10.1016/j.amc.2010.03.121.

[13] Xia, Y., & Friswell, M. (2014). Efficient solution of the fuzzy eigenvalue problem in structural dynamics. *Engineering Computations*, 31 (5): 864–878. DOI: 10.1108/ec-02-2013-0052.

[14] Chakraverty, S., & Behera, D. (2014). Parameter identification of multistorey frame structure from uncertain dynamic data. *Strojniški Vestnik-Journal of Mechanical Engineering*, 60 (5): 331–338. DOI: 10.5545/sv-jme.2014.1832.

[15] Mahato, N. R., & Chakraverty, S. (2016). A filtering algorithm for real eigenvalue bounds of interval and fuzzy generalized eigenvalue problems. *ASCE-ASME Journal of Risk and Uncertainty in Engineering Systems, Part B: Mechanical Engineering*, 2 (4): 044502. DOI: 10.1115/1.4032958.

[16] Chakraverty, S., & Behera, D. (2017). Uncertain static and dynamic analysis of imprecisely defined structural systems. In *Fuzzy Systems: Concepts, Methodologies, Tools, and Applications*, (pp. 1–30). Hershey, PA: IGI Global.

[17] Jeswal, S. K., & Chakraverty, S. (2021). Fuzzy eigenvalue problems of structural dynamics using ANN. In *New Paradigms in Computational Modeling and Its Applications* (pp. 145–161). Cambridge, MA: Academic Press.

[18] Mohapatra, D., & Chakraverty, S. (2022). Type-2 fuzzy linear eigenvalue problems with application in dynamic structures. In *Soft Computing in Interdisciplinary Sciences* (pp. 93–108). Singapore: Springer.

[19] Chakraverty, S., Sahoo, D. M., & Mahato, N. R. (2019). Fuzzy numbers. In: *Concepts of Soft Computing*. Singapore: Springer. DOI: 10.1007/978-981-13-7430-2_3.

[20] Behera, D., & Chakraverty, S. (2015). New approach to solve fully fuzzy system of linear equations using single and double parametric form of fuzzy numbers. *Sadhana*, 40 (1): 35–49.

[21] Zadeh, L. A. (1978). Fuzzy sets as a basis for a theory of possibility. *Fuzzy Sets and Systems*, 1: 3–28.

[22] Zadeh, L. A. (1979). A theory of approximate reasoning. *Machine Intelligence*, 9: 149–194.

[23] Kaufmann, A. (1975). *Introduction to the Theory of Fuzzy Subsets I*. Cambridge, MA: Academic Press.

[24] Nahmias, S. (1978). Fuzzy variables. *Fuzzy Sets and Systems*, 1 (2): 97–111. DOI: 10.1016/0165-0114(78)90011-8.

[25] Wang, W. (1982). Fuzzy contact ability and fuzzy variables. *Fuzzy Sets and Systems*, 8 (1): 81–92.

[26] Liu, B., & Liu, Y. K. (2002). Expected value of a fuzzy variable and fuzzy expected value models. *IEEE Transactions on Fuzzy Systems*, 10 (4): 445–450.

[27] Liu, Y. K., & Liu, B. (2003). Fuzzy random variable: A scalar expected value operator. *Fuzzy Optimization and Decision Making*, 2 (2): 143–160.

[28] Chakraverty, S., & Rout, S. (2020). Affine arithmetic-based solution of uncertain static and dynamic problems. *Synthesis Lectures on Mathematics and Statistics*, 12 (1): 1–170.

13 Dynamical Approach to Forecast Decentralized Currency Exchange Value with Respect to Indian National Rupees

Bhubaneswari Mishra, S. Chakraverty, and Rohtas Kumar

CONTENTS

13.1 INTRODUCTION

An artificial neural network is a data-processing system inspired by the human brain that operates on the same principles as the biological nervous system. They are capable of deriving meaning from complicated and sophisticated data by identifying trends and patterns. Numerous credit risk assessment methodologies have been developed, ranging from statistical models to artificial intelligence technologies. Several techniques including classifier ensembles have been effectively applied to credit scoring issues in recent years, indicating that they are more accurate than single prediction models. However, it remains unclear which basic classifiers should be used in each ensemble to optimize efficiency.

Cryptocurrencies are electronic, decentralized equivalents to government-issued fiat currency. Bitcoin is the first and most well-known cryptocurrency. Cryptocurrencies are used to conduct anonymous and secure online transactions. The decentralization of cryptocurrencies has significantly diminished central authority over them, affecting international commerce and ties. The high volatility of cryptocurrency values underscores the critical need for an accurate model to forecast its value. Cryptocurrency price prediction is a hot topic of discussion among academics. In this area of research, statistical and machine learning methods including Bayesian regression, logistic regression, linear regression, support vector machines, artificial neural networks, deep learning, and reinforcement learning are used. Because Bitcoin has no seasonal impacts, it is difficult to forecast using a statistical technique. While traditional statistical approaches are straightforward to execute and interpret, they rely on a large number of statistical assumptions that may be impractical, leaving machine learning as the most advanced technology in this sector, capable of forecasting pricing based on experience.

Patel et al. [1] pioneered the application of an LSTM and GRU-based hybrid cryptocurrency prediction technique that focused on just two cryptocurrencies, Litecoin and Monero. The outcomes demonstrated that the suggested system correctly forecasts prices with a high degree of precision, indicating that the scheme may be used to forecast the values of a variety of cryptocurrencies. Chowdhury et al. [2] attempted to estimate and predict cryptocurrency values using machine learning methods on the index and components. The objective was to anticipate and forecast the closure (closing) price of the cryptocurrency index 30 and nine of its component cryptocurrencies using machine learning algorithms and models in order to make trading these currencies simpler for individuals. Numerous machine learning approaches and algorithms are applied, and the models are compared against one another to see which produces the best results. The findings indicate that the optimal strategy produces more accurate and competitive results (particularly when ensemble learning is used) than the best studies in the literature, therefore improving the state of the art.

Chen et al. [3] demonstrated that statistical approaches like logistic regression and linear discriminant analysis were capable of predicting Bitcoin daily prices using high-dimensional characteristics with a 66% accuracy, surpassing more difficult machine learning algorithms. After comparing the statistical methods and machine learning algorithms to the benchmark results for daily price prediction, it was ascertained that the statistical methods and machine learning algorithms had the highest accuracies of 66 and 65.3%, respectively, and the best performance rates. The results demonstrate that machine learning models such as random forest, XGBoost, quadratic discriminant analysis, support vector machine, and long short-term memory outperform statistical approaches for Bitcoin 5-minute interval price prediction, with an accuracy of 67.2%.

Between January 1, 2012, and January 8, 2018, Phaladisailoed and Numnonda [4] utilized 1-minute interval trade data from the Bitcoin exchange website Bitstamp. Several alternative regression models were explored with using the scikitlearn and Keras libraries, with the best results indicating that the Mean Squared Error (MSE) was as low as 0.00002 and the R-Square (R2) as high as 99.2%. The prediction of the 12 most liquid cryptocurrencies was examined by Akyildirim et al. [5] using different machine learning classification algorithms, such as support vector machines, logistic regression, artificial neural networks, and random forest algorithms along with historical price data and technical indicators as model features. For all cryptocurrencies, the average categorization accuracy of four algorithms was consistently higher than 50%. On an average, machine learning classification algorithms achieved a predictive accuracy of about 55–65% at daily- or minute-level frequencies, with support vector machines achieving the highest and most consistent predictive accuracy when compared to logistic regression, artificial neural networks, and random forest classification algorithms. Excellent literature on ANNs and soft computing are available [6–7].

In this chapter, the main objective is to use a variety of machine learning approaches to forecast the dynamic exchange rate of Bitcoin in relation to Indian rupee. The results indicate that, when compared to other machine learning techniques such as linear regression and polynomial regression, the support vector regression model performs the best at predicting the dynamicity of bitcoin's exchange rate with respect to Indian national rupees (BTC-INR), with an accuracy of 88.15%.

13.2   METHODOLOGIES

In the upcoming subsections, the undertaken methods and concepts have been discussed briefly.

13.2.1   MACHINE LEARNING [8–9]

Machine learning (ML) is a subfield of computer science that examines algorithms and methodologies for automating the solution of complicated problems that are difficult to solve using traditional programming techniques. Artificial intelligence (AI) is a considerably larger area of research than machine learning (ML), which is primarily concerned with making computers intelligent via a variety of ways, while ML is primarily concerned with a single approach creating robots that can learn to accomplish assignments.

The term "machine learning," or ML for short, was created in 1959 by Arthur Samuel in reference to a machine's ability to solve game of checkers. The phrase refers to a computer programming that is capable of learning to generate behaviour that had not been expressly designed by the programme's inventor. Rather than that, it is capable of revealing behaviour about which the author is fully ignorant.

13.2.2 LINEAR REGRESSION [10]

The main objective of regression analysis is to describe the expectation and dependence of a quantity Y on quantities X_1, X_2..., and so on. A one-directional dependence is assumed. This dependence can be expressed as a general regression function of the following form [10]:

$$E(Y/X) = f(X_1, X_2, X_3, ...) \tag{13.1}$$

The symbol $E(Y/X)$ indicates that the regression function of observed values X_1, X_2, X_3,.., does not correspond to an observed value Y, but rather, it is the average value of Y given the X_i's, which lies on the regression function. The random variables X_1, X_2, X_3,..etc. are referred to as regressors or independent variables. The random variable Y is referred to as regress or dependent variable.

A simple linear regression function has the following form [10]:

$$E(Y_i/X_i) = b_0 + b_1 X_i, \text{ where } i = 1, 2, 3, ...n \tag{13.2}$$

In this equation, X_i represents the observed values of a random variable X(fixed) and b_0 and b_1 are unknown regression parameters. The actual observed values $Y_i (i = 1, 2, ...n)$ can be obtained by summing residual u_i and $E(Y_i/X_i)$.

13.2.3 REGRESSION PARAMETERS [10]

The following definitions apply to the parameters of a simple linear regression function:

b_0 – intercept term (constant)

It denotes the point at where the associated regression line and the y-axis intersect, and it has the same value as variable Y at this point.

b_1 – linear slope coefficient (also a constant)

It provides information on the slope of the relevant regression line. It indicates the magnitude of the change in the anticipated value of random variable Y when the value of variable X is increased by one unit.

13.2.4 POLYNOMIAL REGRESSION [6]

A linear relationship between a dependent (response) and an independent (input) variable indicates a constant rate of change, which may not effectively depict the underlying relationship. Numerous economic time series, such as the inflation index and gross domestic product, demonstrate nonlinear tendencies over time. While the time required to bake a cake may reduce as the oven temperature increases, the decline may not be linear. The rate of change in the mean of the dependent variable (Y) is not constant with respect to the independent variable (X) in any of these cases. The second-order polynomial (quadratic) model is the simplest extension of the straight-line model with one independent variable [6],

$$\varepsilon(Y) = \beta_0 + \beta_1 X + \beta_2 X^2 \tag{13.3}$$

The quadratic model includes the term X^2 in addition to X. Note that this model is a special case of the multiple regression model where $X_1 = X$ and $X_2 = X^2$. Hence, the estimation methods considered are appropriate. Higher-order polynomials of the form [6]

$$\varepsilon(Y) = \beta_0 + \beta_1 X + \beta_2 X^2 + \beta_3 X^3 + \ldots \ldots \beta_p X^p \tag{13.4}$$

allow increasing flexibility of the response relationship and are also special cases of the multiple regression models where $X_i = X^i, i = 1, \ldots .p$

13.2.5 Support Vector Regression [11–12]

The regression problem is an extension of the classification problem in which the model returns a continuous-valued output rather than a finite-valued result. A regression model, in other words, estimates a continuous-valued multivariate function. SVMs address binary classification problems by transforming them into convex optimization problems. SVM is a learning approach that was originally created to fit a linear boundary between the samples of a binary problem, guaranteeing maximal robustness in terms of isotropic uncertainty tolerance. It's worth noting that the border is the furthest away from the nearest point in both classes. Any other dividing boundary will have a class point closer to it than this one. The definition also depicts the classes' closest points to the boundary. These are known as support vectors. In reality, the boundary is determined only by those points. The boundary remains intact even if we delete any other point from the dataset. However, in general, removing any of these special points will cause the boundary to shift.

A hyperplane in R^d is defined as an affine combination of the variables: $\pi \equiv a^T x + b = 0$. A hyperplane splits the space into two half-spaces. Any element belonging to one of the half-spaces has a positive value when the hyperplane equation is evaluated on it. All entries in the opposite half-space have a negative value. The distance of a point $x \in R^d$ to the hyperplane π is defined as [12]

$$d(x, \pi) = \frac{|a^T x + b|}{\|a_2\|} \tag{13.5}$$

During supervised learning, support vectors are the most influential instances that influence the tube shape, and both the training and test variables are assumed to be independent and identical distributed (iid), drawn from the same fixed but unknown probability distribution function [13].

13.2.6   Root Mean Squared Error (RMSE)

Mean Square Error (MSE) indicates the variation between the original and predicted values calculated by squaring the average difference among the datasets.

$$RMSE = \sqrt{MSE} = \sqrt{\frac{1}{N}\sum_{i=1}^{N}(y_i - \hat{y}_i)^2} \tag{13.6}$$

where, $\hat{y}_i$ = predicted value of y_i

13.2.7   R Squared (R²)

It represents the degree to which the values fit the original values. The value between 0 and 1 describes the percentage fit with the initial values. The higher the value, the better the model fits.

$$R^2 = 1 - \frac{\Sigma(y_i - \hat{y}_i)^2}{\Sigma(y_i - \bar{y}_i)^2} \tag{13.7}$$

where, $\hat{y}_i$ = predicted value of y_i and $\bar{y}_i$ = mean value of y_i

13.2.8   Mean Absolute Error (MAE)

It shows the difference between the original and anticipated values as calculated by averaging the absolute difference throughout the dataset.

$$MAE = \frac{1}{N}\sum_{i=1}^{N}\left|(y_i - \hat{y}_i)\right| \tag{13.8}$$

13.3   DATA EXPLANATION

Our dataset includes 30-months (January 2018–June 2020) data of Bitcoin exchange value with respect to Indian national rupees (BTC-INR) on a daily basis. We have obtained the data from [14]. Table 13.1 represents the names of predictive models that are used for prediction. We have considered six attributes in Table 13.2 and divided the dataset into two subsets – testing and training. By creating different

---

TABLE 13.1

Names of the Constituents and Prediction Models under Consideration

Name of Constituent Used	Bitcoin
Predictive models and learners	Linear Regression, Polynomial Regression, Support Vector Regression

TABLE 13.2
Attributes of the Dataset Under Consideration [2]

Sl No	Attributes of Dataset	Remarks
1	Date	A date, often known as a trading date, is a day on which an order to buy, sell, or otherwise acquire a currency is completed in the market.
2	Open Price	The open (opening) price of a currency is the price at which it is initially traded on a specific trading day.
3	High Price	The highest price at which a currency is traded on a given trading day is called the high price.
4	Low Price	The term "low" refers to the lowest price at which a currency is traded in a particular trading day.
5	Close Price	The close price of a currency is the price at which it was last traded on a specific trading day.
6	Volume	The term "volume" or "volume of trade" refers to the total number of contracts exchanged for a certain currency.

models in R programming, we have predicted the price of BTC-INR for the month of May and June 2020, based on historical data. Bitcoin to Indian Rupee Exchange Rates [14] are given in Table 13.3.

TABLE 13.3
BTC-INR Dataset

Sl. No.	Date	Open	High	Low	Close	Volume
1	01-01-2018	900781.8	900781.8	837625.6	869622.3	6.55E+11
2	02-01-2018	867571.9	979959.9	838620	950614.3	1.07E+12
3	03-01-2018	950366.8	989690.4	942885.5	965263.5	1.07E+12
4	04-01-2018	969689.4	997818.3	920925.3	988911.3	1.38E+12
5	05-01-2018	981177.1	1121270	963781.5	1103810	1.51E+12
6	06-01-2018	1105875	1121726	1061702	1109985	1.16E+12
7	07-01-2018	1110004	1113316	1018834	1043526	1E+12
8	08-01-2018	1043438	1047345	901936.6	962846.3	1.17E+12
9	09-01-2018	959901.3	983471.4	918303.9	929216.2	1.06E+12
10	10-01-2018	928776.9	952601.3	871924.1	952601.3	1.18E+12
11	11-01-2018	952276.9	955496.1	835730.5	853681.3	1.05E+12
12	12-01-2018	856744.4	905235.1	837907.8	889166.1	7.67E+11
13	13-01-2018	887372.7	932344.2	887372.7	913308.8	8.12E+11
14	14-01-2018	913982.9	922950.4	843844.8	875899.2	7.05E+11
15	15-01-2018	875600.3	917433.7	867612.1	877695.5	8.1E+11
16	16-01-2018	878730.7	879175.3	652575.6	735506.9	1.21E+12

(Continued)

TABLE 13.3 *(Continued)*
BTC-INR Dataset

Sl. No.	Date	Open	High	Low	Close	Volume
17	17-01-2018	731704.7	745990.6	600618.3	714727.8	1.2E+12
18	18-01-2018	715379.3	773172.2	698350.4	732787.1	9.59E+11
19	19-01-2018	729907	765620.3	713450.3	740784.3	6.85E+11
20	20-01-2018	743898.7	836233.4	743898.7	823226.9	7.53E+11
21	21-01-2018	822588.8	823016.4	720412.9	740318.4	6.34E+11
22	22-01-2018	742424.4	763396.5	654144	698297.9	6.73E+11
23	23-01-2018	699134.7	725435.8	645971	692860.5	6.16E+11
24	24-01-2018	695091.8	732524.2	677356.3	723025.8	6.33E+11
25	25-01-2018	726991.2	748274.1	702366.1	715084.5	5.64E+11
26	26-01-2018	714868.6	740317	665125.8	710165.9	6.2E+11
27	27-01-2018	710388.4	738359.2	698583.4	727285.3	4.82E+11
28	28-01-2018	729484.8	765401.9	729484.8	749255.1	5.31E+11
29	29-01-2018	747297.1	754931.9	710689.7	718338.1	4.52E+11
30	30-01-2018	718999.4	719024.9	639105.3	643569.2	5.5E+11
31	31-01-2018	643690.2	660036.1	622430.6	649448.7	5.11E+11
32	01-02-2018	650478	654830.7	560879.6	586639.4	6.37E+11
33	02-02-2018	584831.7	584831.7	499404.2	566227.7	8.16E+11
34	03-02-2018	567597.9	604699.7	529094.5	588295.3	4.66E+11
35	04-02-2018	588345.9	598551.9	514961.8	530721.9	4.54E+11
36	05-02-2018	530307	536060.8	432832.9	446041.5	5.95E+11
37	06-02-2018	452228.7	504172	388721.7	497806.8	8.99E+11
38	07-02-2018	497902.5	546625.3	464077.3	489820.9	5.89E+11
39	08-02-2018	490885.3	549943.8	490885.3	531394.8	6.01E+11
.						
.						
.						
903	20-06-2020	708456.6	716387.8	705133.5	711612	1.31E+12
904	21-06-2020	711504.1	716855.6	709179.7	709423.6	1.17E+12
905	22-06-2020	709216.7	731543.7	708956.7	731276.8	1.6E+12
906	23-06-2020	730925	730299.8	723541.9	728123	1.29E+12
907	24-06-2020	728311.4	732440.8	702283.5	705452.9	1.44E+12
908	25-06-2020	705492	705446.9	689129.9	699862.6	1.41E+12
909	26-06-2020	699574.2	705131.9	688660.8	692973.1	1.39E+12
910	27-06-2020	693344.2	696368.3	680517.1	684084.8	1.31E+12
911	28-06-2020	684317	695592.1	678801.1	691438.2	1.1E+12
912	29-06-2020	691169.5	697373.9	682942.5	694011.1	1.24E+12
913	30-06-2020	693612.9	695808.3	686641.1	690443.9	1.19E+12

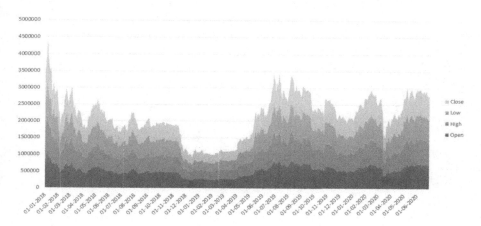

FIGURE 13.1   Area chart plotting of all parameters with respect to time.

13.4   EXPERIMENTAL DESIGN

The following paragraphs discuss the experimental approach and simulation results for the algorithms that have been proposed:

Step 1: Every BTC transaction conducted before June 2020 has been collected from the data accessible on [14]. Following that, the information is stored to a JSON file.

Step 2: For the purpose of data processing, the JSON file has been transformed into a CSV file. Close, date, open, high, low, and volume are all properties or parameters in this dataset. Our objective is to forecast the BTC-INR cryptocurrency's closing price as a function of time or date.

Step 3: The data are plotted in accordance with the attributes of the dataset. The area chart-plotting technique is demonstrated in order to ascertain how values evolve over time. Following that, Figure 13.1 is plotted using the close, low, high, open, and date attributes.

Step 4: Modelling

Multiple predictive regression models have been used to model the dataset. For modelling purposes, linear regression, polynomial regression, and support vector regression models have been employed. Data modelling is the process of organizing data items and standardizing their relationships to one another and to the attributes of real-world entities.

13.5   RESULT AND DISCUSSION

13.5.1   LINEAR REGRESSION

Figure 13.2 shows the actual and predicted value using linear regression model. The blue dots represent the actual data point and the red line indicates the line of best fit.

FIGURE 13.2   Actual and predicted value using linear regression model.

13.5.2   POLYNOMIAL REGRESSION

Figure 13.3 indicates the comparison of the actual and predicted values using polynomial regression model. In the curve shown, the black dots are data points and the blue line represents the best fit line of the curve. This is where polynomial regression comes into play since it predicts the best fit line that follows the pattern (curve).

13.5.3   SUPPORT VECTOR REGRESSION

Figure 13.4 indicates the fitting of actual and predicted value using support vector regression model. The blue dots represent the actual data points and the red curve indicates hyperplane.

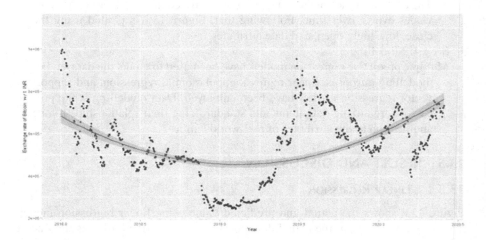

FIGURE 13.3   Actual and predicted value using polynomial regression model.

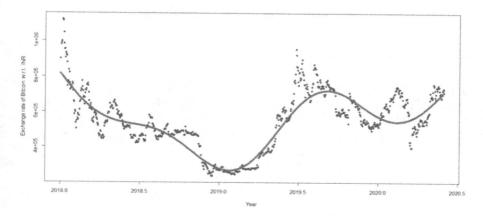

FIGURE 13.4   Actual and predicted value using support vector regression model.

Figure 13.5 describes the gradual performance of support vector regression model with different combination of values of epsilon and cost parameters. Here we have obtained the best SVR model by changing the value of Epsilon from 0 to 1 and value of cost from 0 to 100. We got the best SVR model at epsilon value 0.3 and cost value 100.

Figure 13.6 represents the comparison of Original Data with predicted data obtained from support vector regression and best support vector regression models. The blue line represents support vector regression model data points and the red line represents best support vector regression model data points and black scattered points represents the actual data.

The value of Root Mean Square Error (RMSE) in best support vector regression model is 68971.69. Using this we have obtained better prediction of exchange rate of Bitcoin in comparison to other regression models.

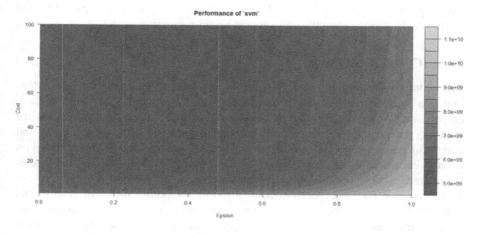

FIGURE 13.5   Performance of support vector regression model.

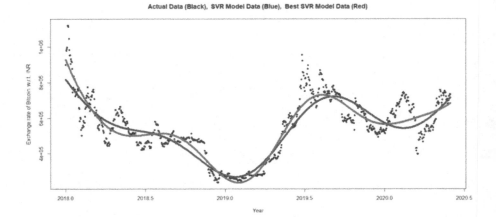

FIGURE 13.6   Best support vector regression model.

As the objective of the model is to estimate the future, the most important problem here is the accuracy of the projections. The predicted exchange value of the aforementioned models has been compared to the original. In this regard, Tables 13.4 and 13.5 illustrate the comparison of BTC-INR exchange rate of May and June 2020, using different models.

A series of performance error calculations have been conducted for linear regression, polynomial regression and support vector regression modelling and the accuracy percentage has been obtained. Table 13.6 shows the comparison between the error values and accuracy of above-discussed regression models.

We have compared the predicted value of BTC-INR for the month of May and June 2020 with original exchange rates. The accuracy percentages of 71.41%, 75.43%, 86.60%, and 88.15% have been obtained for linear regression, polynomial regression, support vector regression, and best support vector regression, respectively. We got the best accuracy from the best support vector regression model.

13.6 CONCLUSION

The underlying technology of cryptocurrency markets has garnered investor attention as a part of the alternative investing market. With investors' growing interest, cryptocurrency markets have become a significant asset class within the alternative investing industry. In this chapter, the predictability of one of the most popular and rapidly increasing decentralized currencies, namely Bitcoin, has been examined on a daily time scale using machine learning regression techniques. Numerical studies with four distinct regression techniques, namely linear, polynomial, and support vector regressions, have been performed to illustrate the predictability of upward and downward price movements. The most accurate and resilient model for forecasting the next day's return is the best support vector regression model, which has an accuracy of 88.15%. In light of the dramatic changes in financial markets over the last

TABLE 13.4
Bitcoin Current and Predicted Exchange Rate for May 2020

Sl No.	Date	Original	Linear	Polynomial	SVR	Best SVR
1	01-05-2020	684844.9	586582.9	728220.8	639530.0	642321.0
2	02-05-2020	683024.1	586683.6	729449.2	641802.0	643353.3
3	03-05-2020	691337.8	586784.3	730680.8	644092.9	644401.3
4	04-05-2020	677503.1	586885.0	731914.7	646400.8	645465.0
5	05-05-2020	684506.8	586985.7	733151.3	648726.0	646545.4
6	06-05-2020	717596.7	587086.4	734390.7	651068.1	647643.1
7	07-05-2020	756565.6	587187.2	735633.3	653427.6	648759.2
8	08-05-2020	757394.3	587287.9	736878.1	655802.3	649893.4
9	09-05-2020	748843.6	587388.6	738125.8	658192.7	651047.0
10	10-05-2020	724504.8	587489.3	739376.1	660598.5	652220.5
11	11-05-2020	685381.2	587590.0	740629.2	663019.0	653414.7
12	12-05-2020	672719.1	587690.7	741885.6	665455.0	654630.6
13	13-05-2020	702900.6	587791.4	743144.2	667904.0	655868.1
14	14-05-2020	739127.4	587892.1	744405.6	670366.7	657128.4
15	15-05-2020	736516.1	587992.9	745669.7	672842.6	658412.2
16	16-05-2020	725703.2	588093.6	746937.0	675332.2	659720.6
17	17-05-2020	745339.9	588194.3	748206.6	677833.2	661053.4
18	18-05-2020	751174.4	588295.0	749479.0	680346.1	662411.9
19	19-05-2020	740495.9	588395.7	750754.1	682870.6	663796.8
20	20-05-2020	742369.9	588496.4	752032.4	685407.0	665209.3
21	21-05-2020	721803.0	588597.1	753313.0	687953.2	666649.2
22	22-05-2020	701472.3	588697.8	754596.4	690509.6	668117.8
23	23-05-2020	706944.5	588798.6	755882.5	693075.8	669615.8
24	24-05-2020	705873.1	588899.3	757171.8	695652.3	671144.5
25	25-05-2020	679157.6	589000.0	758463.4	698236.8	672703.6
26	26-05-2020	680399.8	589100.7	759757.7	700829.9	674294.5
27	27-05-2020	698497.9	589201.4	761054.8	703431.0	675917.8
28	28-05-2020	722656.3	589302.1	762355.1	706040.7	677575.1
29	29-05-2020	724728.5	589402.8	763657.7	708656.8	679265.8
30	30-05-2020	732879.4	589503.5	764963.1	711279.6	680991.4
31	31-05-2020	732601.0	589604.3	766271.2	713908.8	682752.6

decade, which have been facilitated by new technologies in automated trading and financial infrastructure, present findings are particularly important for algorithmic trading, which determines when to purchase and sell cryptocurrencies during the day can be successfully automated using machine learning algorithms.

ACKNOWLEDGEMENT

The first author would like to acknowledge CRISIL Ltd., Powai, Mumbai-400076, India, for funding and support to carry out the present research work.

TABLE 13.5
Bitcoin Future Exchange Rate Prediction for June 2020

Sl No.	Date	Linear	Polynomial	SVR	Best SVR
1	01-06-2020	589705.0	767582.5	716544.9	684550.9
2	02-06-2020	589805.7	768896.1	719185.6	686385.8
3	03-06-2020	589906.4	770212.4	721831.4	688258.7
4	04-06-2020	590007.1	771531.5	724481.8	690170.3
5	05-06-2020	590107.8	772853.8	727137.5	692122.3
6	06-06-2020	590208.5	774178.4	729796.1	694114
7	07-06-2020	590309.2	775505.7	732458	696146.8
8	08-06-2020	590410.0	776835.8	735123	698221.6
9	09-06-2020	590510.7	778168.6	737790.5	700339.1
10	10-06-2020	590611.4	779504.7	740461.1	702500.9
11	11-06-2020	590712.1	780843	743132.5	704706.2
12	12-06-2020	590812.8	782184.1	745805.2	706956.6
13	13-06-2020	590913.5	783527.9	748478.8	709252.8
14	14-06-2020	591014.2	784875	751153.9	711596.5
15	15-06-2020	591114.9	786224.3	753828.1	713986.8
16	16-06-2020	591215.7	787576.4	756502	716425.1
17	17-06-2020	591316.4	788931.2	759175.1	718912.4
18	18-06-2020	591417.1	790289.3	761848.1	721450.2
19	19-06-2020	591517.8	791649.6	764518.6	724037.5
20	20-06-2020	591618.5	793012.7	767187.1	726675.9
21	21-06-2020	591719.2	794378.5	769853.3	729366.2
22	22-06-2020	591819.9	795747.5	772517.7	732110.1
23	23-06-2020	591920.6	797118.8	775178	734906.2
24	24-06-2020	592021.4	798492.9	777834.8	737756.4
25	25-06-2020	592122.1	799869.7	780487.8	740661.2
26	26-06-2020	592222.8	801249.8	783137.3	743622.6
27	27-06-2020	592323.5	802632.1	785781.3	746638.9
28	28-06-2020	592424.2	804017.1	788420.1	749712
29	29-06-2020	592524.9	805404.9	791053.6	752842.5
30	30-06-2020	592625.6	806796	793682.2	756032.2

TABLE 13.6
Error Comparison of Different Regression Models

Sl No.	Model	RMSE	R²	MAE	Accuracy (%)
1	Linear Regression	154816.8	0.04514006	120970.9	71.41
2	Polynomial Regression	133063.8	0.2940922	103688.9	75.43
3	Support Vector Regression	72561.49	0.7932923	55440.2	86.60
4	Best Support Vector Regression	64185.91	0.8383304	50399.95	88.15

REFERENCES

[1] Patel, M.M., Tanwar, S., Gupta, R. and Kumar, N. 2020. A deep learning-based cryptocurrency price prediction scheme for financial institutions. *Journal of Information Security and Applications*, 55, p. 102583.

[2] Chowdhury, R., Rahman, M.A., Rahman, M.S. and Mahdy, M.R.C. 2020. An approach to predict and forecast the price of constituents and index of cryptocurrency using machine learning. *Physica A: Statistical Mechanics and Its Applications*, 551, p. 124569.

[3] Chen, Z., Li, C. and Sun, W. 2020. Bitcoin price prediction using machine learning: An approach to sample dimension engineering. *Journal of Computational and Applied Mathematics*, 365, p. 112395.

[4] Phaladisailoed, T. and Numnonda, T. 2018, July. Machine learning models comparison for bitcoin price prediction. In 2018 10th International Conference on Information Technology and Electrical Engineering (ICITEE) (pp. 506–511). IEEE.

[5] Akyildirim, E., Goncu, A. and Sensoy, A. 2021. Prediction of cryptocurrency returns using machine learning. *Annals of Operations Research*, 297(1), pp. 3–36.

[6] Rawlings, J.O., Pantula, S.G. and Dickey, D.A. 2001. *Applied regression analysis: A research tool*. New York, NY: Springer Science & Business Media.

[7] Chakraverty, S., Sahoo, D.M. and Mahato, N.R. 2019. *Concepts of soft computing: Fuzzy and ANN with programming*. Singapore: Springer.

[8] Rebala, G., Ravi, A. and Churiwala, S. (Eds.). 2019. Machine learning definition and basics. In *An introduction to machine learning* (pp. 1–17). Cham: Springer.

[9] Joshi, A.V. 2020. *Machine learning and artificial intelligence*. Switzerland, AG: Springer Nature.

[10] Härdle, W.K., Klinke, S. and Rönz, B. 2015. *Introduction to statistics: Using interactive MM* Stat elements*. Switzerland, AG: Springer Nature.

[11] Vapnik, V. 1998. The support vector method of function estimation. In Johan A. K. Suykens and Joos Vandewalle, *Nonlinear modeling* (pp. 55–85). Boston, MA: Springer.

[12] Laura, I. and Santi, S. 2017. Introduction to data science: A Python Approach to Concepts, Techniques and Applications.

[13] Awad, M. and Khanna, R. 2015. *Efficient learning machines: Theories, concepts, and applications for engineers and system designers* (p. 268). Berkeley, CA: Springer Nature.

[14] https://www.kaggle.com/

14 Curriculum Learning-Based Approach to Design an Unsupervised Neural Model for Solving Emden–Fowler Type Equations of Third Order

Arup Kumar Sahoo and S. Chakraverty

CONTENTS

14.1 INTRODUCTION

Artificial Neural Network (ANN) has become the focus of numerous researchers owing to its growing applications in diverse fields of science and technology. In recent years, there has been a flurry of research and rapid improvement in machine learning methods for dealing models with singularity. There are few existing analytic and numerical schemes to handle such singular non-linear models. All of these methods have their own intrinsic worth, applicability, and limitations. However, these methods are sometimes problem-dependent. In most of the numerical techniques, the solution is discrete in nature and/or a solution of finite differentiability,

DOI: 10.1201/9781003328032-14

whereas the neural network model-based solution is continuous and can be used as a black box.

In 1943, Warren S. McCulloch and Walter Pitts proposed neuron activities that merged the studies of neurophysiology and mathematical logic. In their historical paper [1], they developed the first elementary model of ANN, in which neurons were escorted by the "all-or-none" process. In 1990, Lee and Kang first proposed a Hopfield neural model for solving differential equation (DE) [2]. Then ANN gets attention from the research community for the development of ODE and PDE solvers. Lagaris et al. suggested neural network approaches for tackling boundary value problems with irregular boundaries [3, 4]. For finding the solution of DE, Parisi et al. [5] developed a neural model using an unsupervised feed-forward neural network. Habiba and Pearlmutter proposed two new ODE-based RNN models (GRU-ODE and LSTM-ODE), which can compute the hidden state and cell state at any point of time using an ODE solver [6]. Gaeta et al. introduced functional networks which can deal with fuzziness and developed fuzzy functional networks to handle some non-linear regression problems [7]. Okereke et al. proposed a novel method using feed-forward Multi-layer Perceptron Neural Network (MLPNN) for solving DE [8]. Many essential works on neural network model to solve DE has been done in recent years, and a good number of research papers have been composed by various authors [9–13].

The Emden–Fowler (E-F) equation, which was developed by Fowler [14], may be used to express many scientific and engineering applications in the literature of quantum mechanics [15] and mathematical physics [16]. It may be written as

$$\psi''(t) + \frac{\delta}{t}\psi'(t) + f(t)g(\psi) = h(t), t \in (0,1], \delta > 0, \qquad (14.1)$$

subject to initial conditions $\psi(0) = a, \psi'(0) = 0$, where $f(t)$ and $h(t)$ are functions of t, $g(\psi)$ is function of ψ, and α is called the shape factor.

For solving these classical singular models, various numerical techniques have been developed. Few reported schemes in this regard are Adomian decomposition by Shawagfeh [17], and Wazwaz [18, 19], analytic scheme by Liao [20], variational iteration method by Dehghan et al. [21], homotopy perturbation method by Yildirim et al. [22]. Simultaneously, some ANN models have also been developed to counter these singular problems. Some potential models are the spherical gas cloud model by Ahmad et al. [23], ChNN model by Mall and Chakraverty [24, 25], multi-layer perceptron neural network methods by Verma and Kumar [26]. With the increasing global requirements, scientists have a strong need to develop new and efficient ANN models for solving higher-order singular value problems. In this regard, Zulqurnain et al. [27] introduced an integrated intelligent computing paradigm model for solving third order E-F equations. Motivated by the above considerations, it is challenging to propose new and efficient ANN algorithms to understand the dynamic behaviour of systems with singularity.

14.2 CURRICULUM LEARNING

Learning becomes easy for a student when a teacher starts from basic examples and gradually increases the difficulty of the task. Like the human brain, the neural

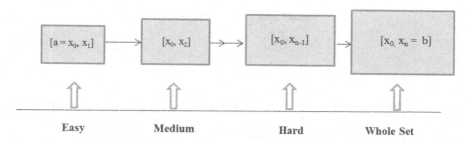

FIGURE 14.1 Schematic diagram of curriculum training of neural network in a given domain $[a,b]$ where $a,b \in R$.

network predicts more accurately when we start training from small data set and steadily increase the complexity level [28]. In the context of machine learning, this is known as "curriculum learning" [29].

In 1993, Elman [28] proposed that curriculum learning may accelerate a neural network's training for solving more difficult tasks. Due to the increase in complexity of problems, curriculum learning has recently become popular in the field of machine learning. In recent work, Sahoo and Chakraverty [30] have used curriculum learning models with neural networks for solving differential equations. In this chapter, we have discussed curriculum learning techniques used in an unsupervised neural model for solving third-order Emden–Fowler equations.

In this experiment, we have used curriculum learning techniques for solving titled DE in the given domain $[a,b]$. With this aim, we have divided the given domain $[a,b]$ into n different groups, viz. $[a,x_1],[a,x_2],[a,x_3],....,[a,x_{n-1}],[a,b]$,

where, $a = x_0 < x_1 < x_2 .... < x_{n-1} < b$, and $a,b,x_i \in R$.

First, we have trained the network in $[a,x_1]$, then trained in $[a,x_2]$, and gradually expanded the domain and continued the training process as demonstrated in Figure 14.1.

14.3 PRELIMINARIES OF ARTIFICIAL NEURAL NETWORK

ANN is an exciting form of Artificial Intelligence (AI) that mimics the human brain's training process to predict patterns from the given historical data. Neural networks are processing devices of mathematical algorithms that can be implemented by computer programming [30].

ANN is a machine learning approach that acquires knowledge from different parameters and learning algorithms [31–34]. The neural network is made up of layers and neurons as shown in Figure 14.2a, b. Perceptron or single-layer neural network (Figure 14.2a) is the most basic and the oldest model of neuron. It simply takes set of inputs, sums them up, then passes through an activation function and sends the results to the output layer.

In feed-forward neural network (Figure 14.2c), all nodes are fully connected by weights; there exists one or more hidden layer between input and output layer. The input layer receives the signals, multiplies them by the weights of interneuron connections, then adds them together and sends them to one or more hidden layers.

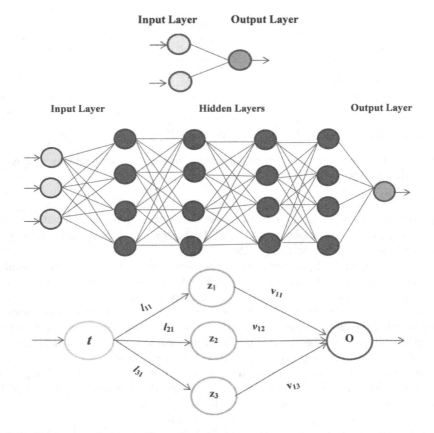

FIGURE 14.2 (a) Single-layer neural network. (b) Multi-layer neural network. (c) Schematic diagram of multi-layer feed-forward neural network with single input, single output, and single hidden layer with three nodes.

The summed value passes through mathematical equations known as activation function that shapes the output of a neural network. The activation functions have influenced the expressivity of the network. In this regard, some potential activation functions are Sigmoid, Tanh, arctan, relu, and Swish [35] which are defined next:

a. **Sigmoid:**

Sigmoid activation function is bounded, monotonic, and differentiable function. It is defined as $\nabla(z) = \dfrac{1}{1+\exp(-z)}$, and range is $(0,1)$.

b. **Tanh:**

Hyperbolic tangent function (Tanh) is bounded, monotonic, and infinitely differentiable function. Tanh activation function may be expressed as:

$$\nabla(z) = \frac{\exp(z) - \exp(-z)}{\exp(z) + \exp(-z)}, \ \text{Range} = (-1,1).$$

c. **Arctan:**

Simply known as $\tan^{-1}[z]$. $\nabla_{\text{arctan}} : R \rightarrow \left(-\dfrac{\pi}{2}, \dfrac{\pi}{2} \right)$.

Arctan is infinitely differentiable in its domain.

d. **Relu:**

Relu (Rectifier Linear Unit) is also known as positive-part function and may be written as $\nabla(z) = \max\{0, z\}$. This activation function is unbounded, monotonic, and differentiable. Range of Relu is $[0, \infty[$.

e. **Swish:**

Swish is state-of-art activation function developed by Google AI lab. It is defined as $\nabla(z) = \dfrac{z}{1 + \exp(-az)}, a \in [0, \infty]$. Range of Swish activation function is $(-\infty, \infty)$. It is infinitely differentiable function on its domain.

ANN handles data in the same way as the human brain does. It reads the training data set's input values and updates the values of parameters to minimize the error of neural-forecasted value. One complete cycle of passing training algorithm and updating parameters is called an epoch. The error is minimized using an optimal number of epochs. But when a network trains for a long time, it loses its capacity to generalize. The accuracy of a neural network is improved through optimizing hyper-parameter tweaking.

14.4 EMDEN–FOWLER EQUATION

The general form of E-F equation can be written [36] as:

$$t^{-k} \frac{d^m}{dt^m} \left(t^k \frac{d^n}{dt^n} \right) t + f(t)g(\psi) = h(t) \tag{14.2}$$

where $f(t)$ and $h(t)$ are real-valued given functions of t, $g(\psi)$ is given function of ψ, and k is called the shape factor.

In order to find E-F equation of third-order set $m + n = 3, m, n \in N, m, n \geq 1$ in Eqn. (14.2). It leads to generate the following two types of third-order E-F equation:

14.4.1 CASE I

Substitute $m = 1, n = 2$ in Eqn. (14.2) to obtain third-order E-F equation and may be written as [36]:

$$\psi'''(t) + \frac{\alpha}{t} \psi''(t) + f(t)g(\psi) = h(t), t \in D \subseteq R, \alpha > 0, \tag{14.3}$$

subject to initial conditions $\psi(0) = a, \psi'(0) = \psi''(0) = 0$, where $f(t), h(t), g(\psi)$ are given functions, α is called the shape factor, $a \in R$, and $D \subseteq R$ is the given domain.

Here $t = 0$ is a singular point of Eqn. (14.3) with shape factors α.

The corresponding system of first-order equation may be obtained by taking

$$\begin{cases} \psi_1(t) = \psi(t) \\ \psi_2(t) = \psi'(t) \\ \psi_3(t) = \psi''(t) \end{cases} \tag{14.4}$$

Then Eqn. (14.3) becomes

$$\begin{cases} \psi_1'(t) = \psi_2(t) \\ \psi_2'(t) = \psi_3(t) \\ \psi_3'(t) = -\dfrac{\alpha}{t}\psi_3(t) - f(t)g(\psi_1) + h(t) \end{cases}, \tag{14.5}$$

subject to the new initial conditions

$$\begin{cases} \psi_1(0) = a \\ \psi_2(0) = 0 \\ \psi_3(0) = 0 \end{cases}. \tag{14.6}$$

The general solutions of system for Eqn. (14.5) may be obtained in the following form:

$$\begin{cases} \psi_1(t) = \chi_1(t) \\ \psi_2(t) = \chi_2(t) \\ \psi_3(t) = \chi_3(t) \end{cases}. \tag{14.7}$$

14.4.2 CASE II

Similarly, set $m = 2, n = 1$ in Eqn. (14.2), so we can write another type third-order E-F equation as:

$$\psi'''(t) + \frac{2\alpha}{t}\psi''(t) + \frac{\alpha(\alpha-1)}{t^2}\psi'(t) + f(t)g(\psi) = h(t), t \in D \subseteq R, \alpha > 0, \tag{14.8}$$

subject to the initial conditions $\psi(0) = a, \psi'(0) = \psi''(0) = 0$, where $f(t), h(t), g(\psi)$ are given functions, α is called the shape factor, $a \in R$, $D \subseteq R$ is the given domain.

From Eqn. (14.8), it can be noticed that $t = 0$ is the singular point appeared twice with shape factors 2α and $\alpha(\alpha - 1)$.

In order to transform Eqn. (14.8) involving higher-order derivatives into a system of first-order equations, let us consider

$$\begin{cases} \psi_1(t) = \psi(t) \\ \psi_2(t) = \psi'(t) \\ \psi_3(t) = \psi''(t) \end{cases}. \tag{14.9}$$

Then we have a corresponding system of first-order equations as follows:

$$\begin{cases} \psi_1'(t) = \psi_2(t) \\ \psi_2'(t) = \psi_3(t) \\ \psi_3'(t) = -\dfrac{2\alpha}{t}\psi_3(t) - \dfrac{\alpha(\alpha-1)}{t^2}\psi_2(t) - f(t)g(\psi_1) + h(t) \end{cases} \tag{14.10}$$

with the initial conditions

$$\begin{cases} \psi_1(0) = a \\ \psi_2(0) = 0 \\ \psi_3(0) = 0 \end{cases} \tag{14.11}$$

Proceeding similarly as for Eqn. (14.5), the general solutions of system for Eqn. (14.10) may be written as

$$\begin{cases} \psi_1(t) = \chi_1(t) \\ \psi_2(t) = \chi_2(t) \\ \psi_3(t) = \chi_3(t) \end{cases} \tag{14.12}$$

These E-F equations are used to model various phenomena in fluid mechanics, thermal behaviour of a spherical cloud of gas, thermionic currents, and many other scientific problems.

14.5 FORMULATION FOR SYSTEM OF FIRST-ORDER DIFFERENTIAL EQUATIONS

Let us consider a first-order system of ODE [37] as:

$$\begin{cases} \dot{\upsilon}_1 = \varphi_1\left(t, \upsilon_1(t), \upsilon_2(t), \ldots \ldots \upsilon_k(t)\right), \, m \in N, \, t \in [a,b] \\ \dot{\upsilon}_2 = \varphi_2\left(t, \upsilon_1(t), \upsilon_2(t), \ldots \ldots \upsilon_k(t)\right), \, m \in N, \, t \in [a,b] \\ \cdot \\ \cdot \\ \cdot \\ \dot{\upsilon}_m = \varphi_m\left(t, \upsilon_1(t), \upsilon_2(t), \ldots \ldots \upsilon_k(t)\right), \, m \in N, \, t \in [a,b] \end{cases} \tag{14.13}$$

subject to the initial conditions $\upsilon_k(a) = \hat{\alpha}_r, k = 1, 2, \ldots, m$.

Here, $\dot{v}_k, k = 1,2,...,m$ are first-order differential operators, $\varphi_k, k = 1,2,...,m$ is a function that defines the structure of a differential equation, and $t = (t_1, t_2, ... t_l) \in R^l$.

ANN trail solution $v_\tau(t, p)$ for first-order DE, with initial or boundary conditions can be written as the sum of two terms

$$v_\tau(t, p) = \alpha + F(t, \Omega(t, p)), \quad (14.14)$$

where, α satisfies initial or boundary conditions of given DE. $\Omega(t, p)$ is an output of feed-forward neural network with parameters, and the second term $F(t, \Omega(t, p))$ does not contribute to the initial and boundary conditions.

In this case, ANN approximate trial solution can be written as the sum of two terms such that $v_\tau(t, p)$ satisfies the initial condition of the given first-order system of ODE. It may be defined as [38]

$$\begin{cases} v_{\tau_1}(t, p_1) = \hat{\alpha}_1 + (1 - e^{-(t-a)})\Omega_1(t, p_1), \\ v_{\tau_2}(t, p_2) = \hat{\alpha}_2 + (1 - e^{-(t-a)})\Omega_2(t, p_2), \\ \cdot \\ \cdot \\ \cdot \\ v_{\tau_m}(t, p_m) = \hat{\alpha}_m + (1 - e^{-(t-a)})\Omega_m(t, p_m), \end{cases} \quad (14.15)$$

where, $\Omega_k(t, p_k), k = 1,2,...,m$ is the neural output with one input data t with parameters $p_k, k = 1,2,...,m$ and may be defined as

$$\Omega_k(t, p_k) = \sum_{j=1}^{r} v_j \nabla(z_j), \quad (14.16)$$

such that

$$z_j = \sum_{i=1}^{n} \ell_{ji} t_i + \tilde{b}_j, \quad (14.17)$$

where, ℓ_{ji} and v_j represent the weights from the input unit i to the hidden unit j and the hidden unit j to the output unit, respectively, $\tilde{b}_j$ is the biases and $\nabla(z_j)$ stands for an activation function. The schematic diagram of feed-forward neural model is depicted in Figure 14.2c.

The corresponding error function can be written as:

$$E(p) = \sum_{i=1}^{l} \sum_{k=1}^{m} \left(\frac{dv_{\tau_m}(t_i, p_k)}{dt} - f_k(t_i, v_{\tau_1}(t_i, p_1), ..., v_{\tau_m}(t_i, p_k)) \right)^2 \quad (14.18)$$

14.6 SIMULATION RESULTS

In this section, we have considered the following two problems of third-order Emden–Fowler-type equations. These problems have been solved by the curriculum learning neural network technique. A multi-layer network has been constructed and trained for the prediction of solutions at testing points. Sigmoid activation function has been used during the training of the model. However, after selecting the basic framework [13], the optimal number of hidden layers and the number of nodes in each hidden layer are selected after several experiments with a different number of hidden layers and nodes. To evaluate the performance indices of the model, MAE, MSE, and RMSE have been calculated with respect to the exact solution as follows:

$$\text{MAE} = \frac{1}{N} \sum_{i=1}^{N} \left| Exact\,Solution(\psi_i(t)) - ANN(\psi_i(t)) \right|$$

$$\text{MSE} = \frac{1}{N} \sum_{i=1}^{N} \left(Exact\,Solution(\psi_i(t)) - ANN(\psi_i(t)) \right)^2$$

$$\text{RMSE} = \sqrt{\frac{1}{N} \sum_{i=1}^{N} \left(Exact\,Solution(\psi_i(t)) - ANN(\psi_i(t)) \right)^2}$$

14.6.1 PROBLEM 1

To demonstrate Case I, a third-order E-F equation [36] has been considered here by substituting $\alpha = 4, a = 1$ and by fixing $f(t)g(\psi) = -(10 + 10t^3 + t^6)\psi, h(t) = 0$ in Eqn. (14.3)

$$\psi'''(t) + \frac{4}{t}\psi''(t) - \left(10 + 10t^3 + t^6\right)\psi = 0, t \in (0,1], \qquad (14.19)$$

subject to the initial conditions $\psi(0) = 1, \psi'(0) = \psi''(0) = 0$.

The analytic solution of Eqn. (14.19) is $\psi(t) = \exp\left(\frac{t^3}{3}\right)$ [36].

Now the corresponding first-order system of equation for Eqn. (14.19) may be obtained as by substituting

$$\begin{cases} \psi_1(t) = \psi(t) \\ \psi_2(t) = \psi'(t) \\ \psi_3(t) = \psi''(t) \end{cases} \qquad (14.20)$$

Accordingly, we have the system of DE as follows:

$$\begin{cases} \psi_1'(t) = \psi_2(t) \\ \psi_2'(t) = \psi_3(t) \\ \psi_3'(t) = -\dfrac{4}{t}\psi_3(t) + (10 + 10t^3 + t^6)\psi_1 \end{cases}, \qquad (14.21)$$

with the initial conditions

$$\begin{cases} \psi_1(0) = 1 \\ \psi_2(0) = 0 \\ \psi_3(0) = 0 \end{cases}. \qquad (14.22)$$

In order to find the neural solution of the aforementioned second kind third-order E-F equation ($\psi_1(t)$), let us consider the ANN trail solution that satisfies the given initial conditions as:

$$\begin{cases} \upsilon_1(t, p_1) = 1 + (1 - e^{-(t)})\Omega_1(t, p_1) \\ \upsilon_2(t, p_2) = (1 - e^{-(t)})\Omega_2(t, p_2) \\ \upsilon_3(t, p_3) = (1 - e^{-(t)})\Omega_3(t, p_3) \end{cases}. \qquad (14.23)$$

To counter this system of equations, a neural model has been constructed such as single input, a single output layer, and two hidden layers, each having 512 nodes. First, we have predicted the neural results with traditional training. For traditional training, the network has been trained for 100 equidistant points from t = 0 s to t = 1 s. Table 14.1 shows the comparison between the exact solution and the neural solution by traditional training. The system of solutions which are predicted by neural model has been depicted in Figure 14.3.

For curriculum training, we have trained the network again for 100 equispaced points from t = 0 s to t = 0.5 s and gradually the domain is increased as shown in Figure 14.1. Comparison between the exact solution and the neural results of system of equations at few testing points by curriculum training has been presented in Table 14.2. These results are compared graphically in Figure 14.4.

Finally, the resemblance of ANN solution using curriculum training with the exact solution of given second kind third-order E-F equation (Eqn. (14.19)) has shown in Figure 14.5. The absolute error at a few testing points is found, and those are plotted in Figure 14.6. Table 14.3 and Figure 14.7 indicate the performance measures between both training models. Figure 14.8 demonstrates the curves of training and validation loss over 5000 epochs for the given model. It may be perceived that the loss function is minimized over the increasing training period. From these figures and tables, it may be observed that neural results by curriculum training are showing good convergence in comparison to the traditional training.

TABLE 14.1

Comparison between Exact and Neural Solution (Using Traditional Training) (Eqn. 14.21) (Problem 1)

Sl. No.	t(s)	Exact Solution $\psi_1(t)$	ANN Solution $\psi_1(t)$	Exact Solution $\psi_2(t)$	ANN Solution $\psi_2(t)$	Exact Solution $\psi_3(t)$	ANN Solution $\psi_3(t)$	Absolute Error $\psi_1(t)$
0	0.00	1.000000	1.000000	0.000000	0.000000	0.000000	0.000000	0.000000
1	0.05	1.000042	0.999062	0.002500	0.002200	0.100010	0.100054	0.000979
2	0.10	1.000333	0.998488	0.010003	0.009378	0.200166	0.199619	0.001845
3	0.15	1.001126	0.998472	0.022525	0.021563	0.300844	0.299468	0.002653
4	0.20	1.002670	0.999257	0.040106	0.038816	0.402672	0.400489	0.003414
5	0.25	1.005222	1.001123	0.062826	0.061230	0.506537	0.503708	0.004099
6	0.30	1.009041	1.004378	0.090813	0.088933	0.613597	0.610317	0.004663
7	0.35	1.014394	1.009347	0.124263	0.122101	0.725298	0.721715	0.005047
8	0.40	1.021563	1.016352	0.16345	0.160981	0.843402	0.839560	0.005211
9	0.45	1.030841	1.025691	0.208745	0.205927	0.970027	0.965835	0.005150
10	0.50	1.042547	1.037607	0.260636	0.257432	1.107706	1.102933	0.004940
11	0.55	1.057025	1.052255	0.319750	0.316147	1.259451	1.253766	0.004770
12	0.60	1.074655	1.069730	0.386875	0.382886	1.428861	1.421892	0.004925
13	0.65	1.095862	1.090229	0.463001	0.458612	1.620239	1.611656	0.005633
14	0.70	1.121126	1.114332	0.549351	0.544473	1.838758	1.828324	0.006794
15	0.75	1.150993	1.143142	0.647433	0.641894	2.090670	2.078196	0.007851
16	0.80	1.186095	1.177955	0.759101	0.752727	2.383577	2.368738	0.008140
17	0.85	1.227167	1.219655	0.886628	0.879368	2.726772	2.708969	0.007512
18	0.90	1.275069	1.268398	1.032805	1.024776	3.131696	3.110647	0.006670
19	0.95	1.330815	1.323710	1.201060	1.192357	3.612506	3.589910	0.007105
20	1.00	1.395612	1.384720	1.395612	1.385709	4.186837	4.160760	0.010892

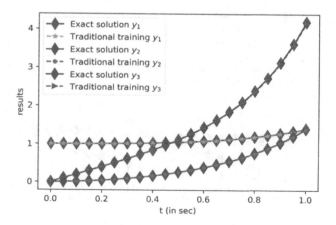

FIGURE 14.3 Plot of exact solution and neural solution of system of equations (Eqn. 14.21) by traditional training (Problem 1).

TABLE 14.2

Comparison between Exact and Neural Solution (Using Curriculum Training) (Eqn. 14.21) (Problem 1)

Sl. No.	t(s)	Exact Solution $\psi_1(t)$	ANN Solution $\psi_1(t)$	Exact Solution $\psi_2(t)$	ANN Solution $\psi_2(t)$	Exact Solution $\psi_3(t)$	ANN Solution $\psi_3(t)$	Abs. Error $\psi_1(t)$
0	0.00	1.000000	1.000000	0.000000	0.000000	0.000000	0.000000	0.000000
1	0.05	1.000042	1.000738	0.002500	0.002388	0.100010	0.100046	0.000696
2	0.10	1.000333	1.001781	0.010003	0.009641	0.200166	0.200191	0.001447
3	0.15	1.001126	1.003224	0.022525	0.021866	0.300844	0.301172	0.002098
4	0.20	1.002670	1.005188	0.040106	0.039168	0.402672	0.403582	0.002518
5	0.25	1.005222	1.007829	0.062826	0.061658	0.506537	0.508055	0.002607
6	0.30	1.009041	1.011341	0.090813	0.089467	0.613597	0.615487	0.002301
7	0.35	1.014394	1.015968	0.124263	0.122770	0.725298	0.727157	0.001574
8	0.40	1.021563	1.022008	0.16345	0.161816	0.843402	0.844768	0.000446
9	0.45	1.030841	1.029825	0.208745	0.206958	0.970027	0.970454	0.001016
10	0.50	1.042547	1.039853	0.260636	0.258686	1.107706	1.106787	0.002694
11	0.55	1.057025	1.052598	0.319750	0.317647	1.259451	1.256833	0.004427
12	0.60	1.074655	1.068637	0.386875	0.384650	1.428861	1.424247	0.006018
13	0.65	1.095862	1.088610	0.463001	0.460669	1.620239	1.613407	0.007252
14	0.70	1.121126	1.113192	0.549351	0.546856	1.838758	1.829587	0.007933
15	0.75	1.150993	1.143063	0.647433	0.644602	2.090670	2.079155	0.007930
16	0.80	1.186095	1.178854	0.759101	0.755669	2.383577	2.369788	0.007242
17	0.85	1.227167	1.221083	0.886628	0.882367	2.726772	2.710812	0.006084
18	0.90	1.275069	1.270087	1.032805	1.027690	3.131696	3.113949	0.004982
19	0.95	1.330815	1.325950	1.201060	1.195256	3.612506	3.594445	0.004866
20	1.00	1.395612	1.388451	1.395612	1.388847	4.186837	4.167174	0.007162

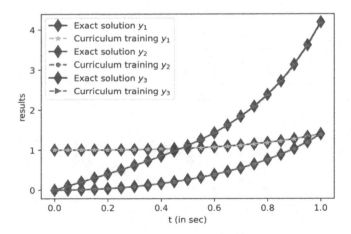

FIGURE 14.4   Plot of exact solution and neural solution of system of equations (Eqn. 14.21) by curriculum training (Problem 1).

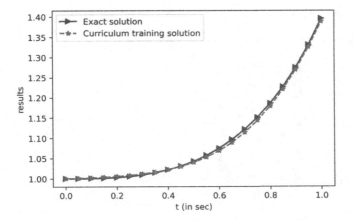

FIGURE 14.5 Comparison plot between ANN solution using curriculum training and exact solution of given third-order E-F equation (Problem 1).

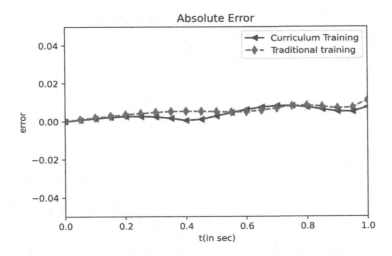

FIGURE 14.6 Error plot between ANN solution (using traditional training and curriculum training) and exact solution of given third-order E-F equation (Eqn. 14.19) (Problem 1).

TABLE 14.3

Performance Indices of Curriculum Training and Traditional Training (Problem 1)

Error	Traditional Training	Curriculum Training
MAE	5.156886190476199 E-03	3.8710438095238184 E-03
MSE	3.276220093450956 E-05	2.1897651112304993 E-05
RMSE	5.723827472461898 E-03	4.6794926126990520 E-03

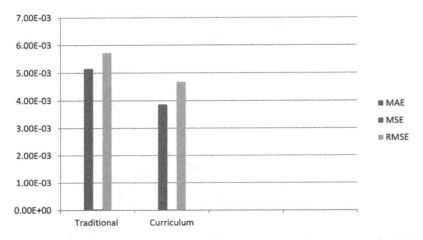

FIGURE 14.7 Plot of performance indices of curriculum training and traditional training (Problem 1).

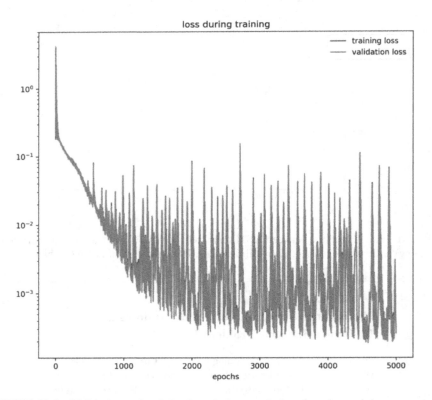

FIGURE 14.8 Validation and training loss during curriculum learning training process of neural network (Problem 1).

14.6.2   PROBLEM 2

We conclude this section by considering a third-order E-F equation [36] as described in Case II by substituting $\alpha = 1$, $a = 1$ and by fixing $f(t)g(\psi) = -\dfrac{9}{8}(8 + t^6)\psi^{-5}$, $h(t) = 0$ in Eqn. (14.8)

$$\psi'''(t) + \frac{2}{t}\psi'(t) - \frac{9}{8}(8 + t^6)\psi^{-5} = 0, \tag{14.24}$$

subject to initial conditions $\psi(0) = 1, \psi'(0) = \psi''(0) = 0.$

The exact solution [36] for Eqn. (14.24) is $\psi(t) = \sqrt{1 + t^3}$.

The corresponding system of first-order equation of Eqn. (14.24) can be written as by replacing

$$\begin{cases} \psi_1(t) = \psi(t) \\ \psi_2(t) = \psi'(t) \\ \psi_3(t) = \psi''(t) \end{cases} \tag{14.25}$$

Then Eqn. (14.24) may be expressed as:

$$\begin{cases} \psi_1'(t) = \psi_2(t) \\ \psi_2'(t) = \psi_3(t) \\ \psi_3'(t) = -\dfrac{2}{t}\psi_2(t) + \dfrac{9}{8}(8 + t^6)\psi_1^{-5} \end{cases}, \tag{14.26}$$

with the initial conditions

$$\begin{cases} \psi_1(0) = 1 \\ \psi_2(0) = 0 \\ \psi_3(0) = 0 \end{cases} \tag{14.27}$$

As discussed in Section 14.5, we may write the ANN trail solution as:

$$\begin{cases} v_1(t, p_1) = 1 + (1 - e^{-(t)})\Omega_1(t, p_1) \\ v_2(t, p_2) = (1 - e^{-(t)})\Omega_2(t, p_2) \\ v_3(t, p_3) = (1 - e^{-(t)})\Omega_3(t, p_3) \end{cases} \tag{14.28}$$

To handle this singular problem, we have been constructed a neural model such as a single input, a single output layer, and three hidden layers, each having 512 nodes. Sigmoid activation function has been used during the training of the network. The network has been trained for 128 equispaced points from t = 0 s to t = 0.3 s and

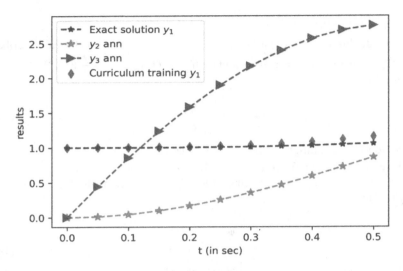

FIGURE 14.9   Plot of exact solution (Eqn. 14.24) and neural solution of system of equations (Eqn. 14.26) by curriculum training (Problem 2).

gradually the domain is increased. Figure 14.9 and Table 14.4 show the resemblance between the exact solution and the neural solution by curriculum training at few testing points. Proceeding similarly as Problem 1, one may find exact solutions of system of equations. The absolute errors have been found at testing points, and those are depicted in Figure 14.10. One may very well observe from the performance indices noted in Table 14.5 that curriculum learning technique during the training of

TABLE 14.4
Comparison between Exact and Neural Solution (Using Curriculum Training) (Eqn. 14.26) (Problem 2)

Sl. No.	t(s)	Exact Solution $\psi_1(t)$	ANN Solution $\psi_1(t)$	ANN Solution $\psi_2(t)$	ANN Solution $\psi_3(t)$	Abs Error $\psi_1(t)$
0	0.00	1.000000	1.000000	0.000000	0.000000	0.000000
1	0.05	1.000063	1.000828	0.010377	0.441919	0.000766
2	0.10	1.000500	1.003668	0.04358	0.854426	0.003168
3	0.15	1.001686	1.008613	0.096741	1.236356	0.006926
4	0.20	1.003992	1.015933	0.167601	1.585635	0.011941
5	0.25	1.007782	1.026208	0.254509	1.899000	0.018426
6	0.30	1.013410	1.040464	0.356015	2.171819	0.027054
7	0.35	1.021213	1.060188	0.470360	2.398401	0.038976
8	0.40	1.031504	1.086873	0.595141	2.573336	0.055369
9	0.45	1.044569	1.120842	0.727327	2.693751	0.076273
10	0.50	1.060660	1.160108	0.863654	2.760755	0.099448

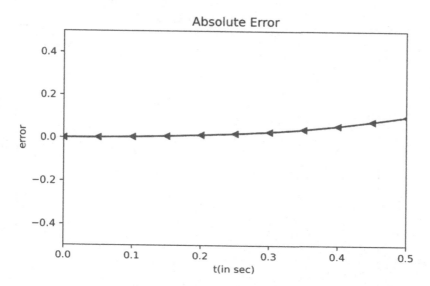

FIGURE 14.10 Error plot between neural solution (using curriculum training) and exact solution (Eqn. 14.24) (Problem 2).

TABLE 14.5
Performance Indices of Curriculum Training (Problem 2)

MAE	0.030758860077397810
MSE	0.001960449120089887
RMSE	0.044276959246202610

---

network makes the model efficient. The training and validation loss graphs (Figure 14.11) expressed a consistent and significant decrease over the period of time, which shows the accuracy of the model.

14.7 CONCLUSION

The present experiment shows the effect of an artificial neural network for solving the third-order E-F equation. The neural model successfully overcomes the singular behaviour of the problem. The advantages of this learning method are examined by solving two different kinds titled DE, which describe various phenomena in mathematical physics. The excellent agreement between the neural results and different numerical methods shows the robustness of our model. Lastly, it is worth mentioning that the use of curriculum learning techniques during the training of neural network is easy to implement and makes the model convergent. This machine

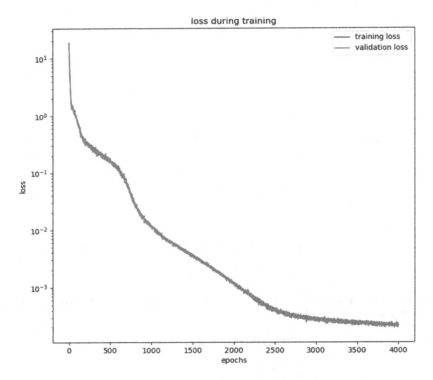

FIGURE 14.11　Validation and training loss during curriculum learning training process of neural network (Problem 2).

intelligence-based method can be used to solve relevant singular problems emerging in any science and engineering applications.

Nomenclatures

$\tilde{b}$	Bias
v, ℓ	Weight
∇	Activation function
α	Shape factor

Abbreviation

E-F	Emden–Fowler
MSE	Mean squared error
ANN	Artificial neural network
MAE	Mean absolute error
ODE	Ordinary differential eq.
RMSE	Root mean squared error
PDE	Partial differential eq.
FFNN	Feed-forward neural network

ACKNOWLEDGEMENT

The first author would like to acknowledge the Council of Scientific and Industrial Research (CSIR), New Delhi, India (File no.: 09/983(0042)/2019-EMR-I), for the support and funding to pursue the present research work.

REFERENCES

[1] W.S. McCulloch and W. Pitts. A logical calculus of the ideas immanent in nervous activity. *Bulletin of Mathematical Biophysics*: 5(4), 115–133, 1943.

[2] H. Lee, I.S. Kang. Neural algorithms for solving differential equations. *Journal of Computational Physics*: 91(1), 110–131, 1990.

[3] I.E. Lagaris, A. Likas, D.I. Fotiadis. Artificial neural networks for solving ordinary and partial differential equations. *IEEE Transactions on Neural Networks*: 9(5), 987–1000, 1998.

[4] I.E. Lagaris, A.C. Likas, and D.G. Papageorgiou. Neural network methods for boundary value problems with irregular boundaries. *IEEE Transactions on Neural Networks*: 11(5), 1041–1049, 2000.

[5] D.R. Parisi, M.C. Mariani, M.A. Laborde. Solving differential equations with unsupervised neural networks. *Chemical Engineering and Processing*: 42 (8–9), 715–721, 2003.

[6] M. Habiba and B.A. Pearlmutter. Neural ordinary differential equation based recurrent neural network model. *Proceedings of 31st Irish Signals and Systems Conference (ISSC)*: 1–6, 2020.

[7] M. Gaeta, V. Loia, and S. Tomasiello. A fuzzy functional network for nonlinear regression problems. *International Journal of Knowledge Engineering and Soft Data Paradigms*: 4(4), 351, 2014.

[8] R. Okereke, O. Maliki, B. Oruh. A novel method for solving ordinary differential equations with artificial neural networks. *Applied Mathematics*: 12, 900–918, 2021. doi: 10.4236/am.2021.1210059

[9] B. Choi and J.H. Lee. Comparison of generalization ability on solving differential equations using back-propagation and reformulated radial basis function networks. *Neurocomputing*: 73(1–3), 115–118, 2009.

[10] S. Mall and S. Chakraverty. Comparison of artificial neural network architecture in solving ordinary differential equations. *Advances in Artificial Neural Systems*: 2013, 1–24, 2013.

[11] S. Mall and S. Chakraverty. Regression-based neural network training for the solution of ordinary differential equations. *International Journal of Mathematical Modelling and Numerical Optimization*: 4(2), 136–149, 2013.

[12] S. Chakraverty and S. Mall. Regression based weight generation algorithm in neural network for solution of initial and boundary value problems. *Neural Computing and Applications*: 25(3), 585–594, 2014.

[13] F. Chen et al. NeuroDiffEq: A Python package for solving differential equations with neural networks. *Journal of Open Source Software*: 5(46), 1931, 2020. doi: 10.21105/joss.01931

[14] R.H. Fowler. Further studies of Emdens and similar differential equations. *Quarterly Journal of Mathematics*: 2(1), 259–288, 1931.

[15] J.I. Ramos, Linearization methods in classical and quantum mechanics. *Computer Physics Communications*: 153(2), 199–208, 2003.

[16] A.H. Bhrawy, A.S. Alofi, and R.A. Van Gorder. An efficient collocation method for a class of boundary value problems arising in mathematical physics and geometry. *Abstract and Applied Analysis*: 2014, 425648, 2014.

[17] N.T. Shawagfeh. Non perturbative approximate solution for Lane–Emden equation. *Journal of Mathematical Physics*: 34(9), 4364–4369, 1993.

[18] A.M. Wazwaz. A new algorithm for solving differential equations of Lane–Emden type. *Applied Mathematics and Computation*: 118(2), 287–310, 2001.

[19] A.M. Wazwaz. Adomian decomposition method for a reliable treatment of the Emden–Fowler equation. *Applied Mathematics and Computation*: 161(2), 543–560, 2005.

[20] S. Liao. A new analytic algorithm of Lane–Emden type equations. *Applied Mathematics and Computation*: 142(1), 1–16, 2003.

[21] M. Dehghan and F. Shakeri. Approximate solution of a differential equation arising in astrophysics using the variational iteration method. *New Astronomy*: 13, 53–59, 2008.

[22] A. Yildirim and T. Ozis. Solutions of singular IVPs of Lane–Emden type by homotopy perturbation method. *Physics Letters A*: 369, 70–76, 2007.

[23] I. Ahmad, M.A.Z. Raja, M. Bilal, and F. Ashraf. Neural network methods to solve the Lane–Emden type equations arising in thermodynamic studies of the spherical gas cloud model. Neural Computing and Applications: 28(1), 929–944, 2017.

[24] S. Mall and S. Chakraverty. Chebyshev neural network based model for solving Lane–Emden type equations. *Applied Mathematics and Computation*: 247, 100–114, 2014.

[25] S. Mall and S. Chakraverty. Numerical solution of non-linear singular initial value problems of Emden–Fowler type using Chebyshev neural network method. *Neurocomputing*: 149, 975–982, 2015.

[26] A. Verma and M. Kumar. Numerical solution of third-order Emden–Fowler type equations using artificial neural network technique. *European Physical Journal – Plus*: 135, 751, 2020. doi: 10.1140/epjp/s13360-020-00780-3

[27] Z. Sabir, M. Umar, J.L. Guirao, M. Shoaib, and M.A.Z. Raja. Integrated intelligent computing paradigm for non-linear multi-singular third-order Emden–Fowler equation. *Neural Computing and Applications*: 1–20, 2020.

[28] J.L. Elman. Learning and development in neural network: The importance of starting small. *Cognition*: 48, 781–799, 1993.

[29] Y. Bengio, J. Louradour, R. Collobert, and J. Weston. Curriculum learning. In proceeding of the 26th International Conference on Machine Learning. ACM: 41–48, 2009.

[30] A.K. Sahoo and S. Chakraverty. Curriculum learning-based artificial neural network model for solving differential equations. In: S. Chakraverty (eds.), *Soft Computing in Interdisciplinary Sciences. Studies in Computational Intelligence*: vol. 988. Singapore: Springer, 2022. doi: 10.1007/978-981-16-4713-0_6

[31] J.M. Zurada. *Introduction to Artificial Neural Systems*. St. Paul, MN: West Publishing Co., 1992.

[32] S. Chakraverty, D.M. Sahoo, and N.R. Mahato. *Concepts of Soft Computing: Fuzzy and ANN with Programming*. Singapore: Springer, 2019.

[33] S. Chakraverty and S. Mall. *Artificial neural networks for engineers and scientists: Solving ordinary differential equations*. Boca Raton: Taylor and Francis, CRC Press, 2017

[34] S. Chakraverty and S.K. Jeswal. Applied artificial neural network methods for engineers and scientists solving algebraic equations. *World Scientific*: Singapore: World Scientific Publishing Company, 2021. doi: 10.1142/12097

[35] J. Lederer. Activation functions in artificial neural networks: A systematic overview. *arXiv preprint arXiv*: 2101, 09957, 2021.

[36] A.M. Wazwaz. Solving two Emden–Fowler type equations of third order by the variational iteration method. *Applied Mathematics and Information Sciences*: 9(5), 2429, 2015.

[37] A. Malek and R. Shekari Beidokhti. Numerical solution for high order differential equations using a hybrid neural network-Optimization method. *Applied Mathematics and Computation*: 183(1), 260–271, 2006.

[38] M. Mattheakis, P. Protopapas, D. Sondak, M. Di Giovanni, and E. Kaxiras. Physical symmetries embedded in neural networks. *Preprint at arXiv*: 2019. https://arxiv.org/abs/1904.08991

Index

Printed in the United States
by Baker & Taylor Publisher Services